A Mathematical Foundation

for Elementary Teachers

Class Test Edition

Patricia Jones

Kathleen D. Lopez

Lee Ellen Price

University of Southwestern Louisiana

An imprint of Addison Wesley Longman, Inc.

Reading, Massachusetts • Menlo Park, California • New York • Harlow, England
Don Mills, Ontario • Sydney • Mexico City • Madrid • Amsterdam

Reprinted with corrections, March 2000

This project was supported in part by a grant from LaCEPT (sponsored by NSF and Louisiana's Board of Regents, Board of Elementary and Secondary Education, and the State Department of Education).

Reproduced by Addison Wesley Longman from camera-ready copy supplied by the authors.

ISBN: 0-201-34716-4

5 6 7 8 9 10 VG 00

Preface for Instructors

Our goal in writing this text was to provide a mathematical foundation for K-8 teachers which is faithful to the recommendations of A Call for Change and the vision of the NCTM Standards.

We are convinced that college mathematics departments must provide courses which are designed to meet the special needs of elementary teachers — not watered-down algebra and trigonometry, or sixth grade arithmetic couched in reform language — but important, challenging mathematics which is relevant to a Standards-based K-8 curriculum.

These materials are rooted in our long-term involvement in, and commitment to, mathematics education reform. They reflect what we have learned in over twenty years of developing and teaching mathematics courses for elementary teachers, as well as participation and leadership in national, state, and regional reform efforts.

Five years ago, we set out to write materials for a three-semester sequence of mathematics content courses which, together with appropriate methods courses, would prepare prospective elementary teachers to implement the Standards in their own classrooms. We began with a list of what we believe are the most important things for teachers to accomplish in these courses:

- acquire an understanding of mathematics which is considerably broader and deeper than what they will be expected to teach
- develop number sense, spatial sense, and mathematical intuition
- be able to explain mathematical concepts, relationships and processes, clearly and correctly
- see elementary school mathematics as a Big Picture: important, interconnected ideas developed and applied from K through 8
- learn, from experience, that mathematical understanding begins with concrete models, then progresses through verbal, numerical and, finally, abstract stages
- become confident, successful problem-solvers
- develop appreciation for the power and the beauty of mathematics, and enthusiasm for teaching it

These goals have led us far from the beaten path of traditional "Mathematics for Elementary Teachers", and they have motivated approaches which are significantly different from other post-Standards texts. Please consider the features described below; we believe they are particularly important and innovative.

Homework Sets

We consider homework to be an integral part of these courses — not simply practice exercises intended to reinforce what was "covered in class". Each question is different in some significant way from the others in that Set, and each has a specific purpose. Every Set includes questions which : (a) lead to understanding of the concepts in that section; (b) review and connect material learned in previous sections; (c) require students to use number sense, explain or describe mathematical ideas and relationships and solve real problems; and (d) include explorations and/or activities which motivate subsequent material.

We strongly recommend that the entire Homework Set be assigned for each section, and that adequate class time be allowed for whatever discussion, extension, and connections are motivated by the Homework questions. You will notice that some sections include just one or two new ideas; this is to allow time for discussion of particularly "rich" Homework Sets.

Roles of Language

The importance of language in the learning and teaching of mathematics has been clearly recognized in reform literature. However, attention has been focused on only two of the four crucial roles of language in mathematics education: understanding and use of mathematical terms and symbols, and "writing to learn". We recognize the importance of these two aspects of learning mathematics, and we give them proper consideration. We want to call special attention, however, to the two aspects of language which are relevant to teaching mathematics, and which we feel have been badly neglected, both in reform literature and in "Mathematics for Elementary Teachers" textbooks.

First, it seems obvious that a mathematics teacher must be able to explain mathematical ideas, relationships, and processes to her students. This is quite different from "writing to learn", which can achieve its purpose even when the communication is imprecise, incomplete, and not entirely correct. Communication from teacher to students, however, must be clear and correct.

It has been our experience that preservice teachers need considerable practice and guidance in this regard. Consequently, a significant number of discussion questions have been included throughout the Homework Sets, Reviews, and Sample Tests. To provide even more opportunities for verbal expression, Journal questions are suggested in each section of Chapters I and II.

We are convinced that clear and correct teacher-to-student communication is an essential aspect of reform. We urge you to give it prominence in your courses, and to hold prospective teachers to a high standard of performance.

The second role of language in teaching is the verbal component of concept development. Words provide a bridge for learners between things and symbols but, for many years this bridge has been missing from elementary school mathematics classrooms. So, most students in our preservice courses are not aware of its importance. You will notice that, throughout these materials, we have been zealous in our emphasis on words in the development and connection of mathematical concepts.

Pedagogy

Several pedagogical issues seem relevant for this Introduction. First, we believe you will agree that these materials model Standards-based pedagogy. Each concept is developed from a concrete and verbal foundation, to numerical relationships and applications, and finally to generalizations. Students are expected to be active learners — guided by carefully selected questions, problems, and activities. Whole class discussions, small group activities and explorations, group and individual problem-solving, as well as instructor explanation, are all included whenever they are appropriate. Use of calculators is encouraged throughout, and particular attention is given to selecting the most convenient method of calculation in a given situation (mental, paper and pencil, or calculator).

Individual problem-solving and language skills receive more emphasis in Chapters I and II, whereas group activities are more prominent in subsequent chapters. There are two reasons for this: (i) in order to contribute to, and benefit from, group learning activities, individual students must have developed some basic competence in critical thinking and communication of ideas; and (ii) measurement, geometry, and data collection are naturally suited to collaborative work.

In addition to modeling reform methodology, everything in the text is related to a Standards-based K-8 curriculum. Model curricular goals and guidelines, as well as pertinent excerpts from 1997 TIMSS materials are included in the Appendices. Teaching Notes call prospective teachers' attention to significant mathematical transitions and connections or suggest helpful strategies.

We want to stress, however, that the primary purpose of this text is to help preservice teachers understand, appreciate, use, and communicate important mathematics; it is not intended to provide extensive examples of elementary school materials and strategies. These should be the focus of the mathematics methods classes which must accompany and compliment content courses. Our commitment to reform, particularly to the recommendations of the MAA's A Call for Change and the vision of the NCTM's Standards, is reflected in everything we have written; but we have been careful to provide a college-level text which is appropriate for adult learners.

Faculty Development

Many instructors in preservice mathematics courses are just beginning to learn about, and respond to, calls for reform in the preparation of elementary teachers. Few of them have the experience or the time to design new courses and write materials. They have no choice but to depend on a textbook. This was certainly the case at our University and at others we are familiar with. So our primary motivation for this project was to provide our faculty with comprehensive course materials which were really faithful to the spirit of reform.

The response from our Department, as well as from colleagues at other universities who have piloted draft versions of the manuscript, has been overwhelmingly positive and gratifying. Without exception, these instructors have told us that the materials helped them understand what standards-based teaching is all about. They are enthusiastic about teaching the courses, and they have begun to read reform literature and participate in conferences and workshops related to mathematics education.

Our deepest hope is that this book will make a real difference in the mathematical preparation of elementary teachers — for the prospective teachers themselves and for their university instructors.

Introduction for Students

This text was developed to meet the special needs of prospective elementary school teachers. You may find it very different from your previous experiences with mathematics textbooks and courses. We hope you will be open to seeing and doing mathematics in new ways. In particular, you will be asked to focus on understanding, thinking, explaining, and connecting mathematical concepts to each other and to the real world. This is dramatically different from viewing mathematics as a collection of isolated rules, formulas, and definitions; it is also more challenging, more interesting, and more relevant to our lives.

Our goals are that these materials will help you to:

- understand important mathematical concepts and relationships
- express mathematical ideas clearly and correctly
- develop number sense and spatial sense
- become a successful, confident problem-solver
- understand how mathematical strands are developed and connected within the K-8 curriculum
- develop appreciation for the power and the beauty of mathematics, and enthusiasm for teaching it.

Table of Contents

Chapter III: Attributes, Units, and Measurement

Chapter V: Area and Volume

CHAPTER I

Section 1: Numeration

A whole number answers the question: How many things are in this set? (As you know, the whole numbers are: 0, 1, 2, 3, ...) The process of counting, which is an ordering of the whole numbers, emphasizes that the set is discrete — for each whole number, there is a next larger whole number.

The concept of number, and the means of communicating mathematical ideas, develop in children (as they did for mankind) by this process:

(i) awareness that there is something the same about four people, four forks, four knives, etc.
(ii) using fingers to indicate "how many"
(iii) using numbers words
(iv) counting
(v) using written symbols to represent whole numbers

Just as different languages and alphabets evolved in various cultures, so did different symbols for representing numbers. We will look at a few examples of ancient numeral systems in order to compare and contrast their features with our familiar Hindu-Arabic numerals.

Original Roman Numerals (no subtractive feature)

Symbols:	I,	X,	C,	M,	etc.
	one	ten	hundred	thousand	

Some things to notice:

(i) There is a different symbol for each power of ten. (So in order to represent the U.S. Gross National Product, the Romans would have needed 19 different symbols!!)

(ii) The symbols had a fixed value — C meant one hundred regardless of where it appeared in a numeral.

(iii) There is no Roman symbol for zero.

Example 1:

(i) MMCCCCCCXXXXXIII represents two thousand six hundred fifty-three things.

(ii) Romans didn't write numerals as shown at the right. But if someone would have written this, it would have been understood as representing nine hundred seventy-six things.

C I C X X I C
C I C X C I X
X C C I X X I C

It's obvious that writing and calculating with these numerals was very tedious.

Original Babylonian Numerals

Symbols: V and <
one ten

Some things to notice:

(i) Since there are only two symbols, increasingly larger numbers were represented by giving the symbols different meanings depending on their position in the numeral. (This type of system is said to have place value.)

The first fifty-nine counting numbers were represented by simply writing the necessary tens and ones. For example:

V V V V V V V	< V V < V	<< V V V V << V V V V
seven	twenty-three	forty-eight

For numbers greater than fifty-nine, the symbols were put into groups having different values based on powers of sixty. Symbols in the first position on the right represented their face value; symbols in the next position represented sixty times their face value; symbols in the third position represented thirty-six hundred (60^2) times their face value, etc. For example:

V V	< V V V V	< V V < V
two thirty-six hundreds	fourteen sixties	twenty-three ones

This numeral represents eight thousand sixty-three things: 23 + 14(60) + 2(3600).

(ii) There could be many symbols in each position.

(iii) The system had no zero. Therefore, it was often impossible to tell, just by looking at a numeral, what value the groups of symbols had. For example:

V V < V These symbols could represent
two sixties and eleven ones
or
two thirty-six hundreds and eleven sixties
or
two thirty-six hundreds and eleven ones

Example 2:

< V V	<<V V < V V	<<V V <<V
thirty-six-hundreds	sixties	ones

The symbols in the thirty-six-hundreds place represent twelve — a ten and two ones. So that position represents 12 x 3600 **things**. In the sixties place, the symbols designate thirty-four — three tens and four ones. So that position indicates 34 x 60 **things**. Finally, in the ones place, the symbols represent forty-three **things**. Hence the numeral stands for 43,200 + 2040 + 43 **things**.

Example 3:

Write the Babylonian numeral that represents three thousand four hundred ninety- seven things.

Since this number is less than thirty-six hundred, the biggest place in the numeral will have value sixty. So all we need to do is fill-in symbols in these two places:

sixties ones

The first question is, How many bundles of sixty can be made from three thousand four hundred ninety-seven things? This is a division question, and the answer is fifty-eight bundles, with seventeen things left over. So the numeral is:

```
<<VVV    <VVV
<<VVV     VVV
 <VV      V
sixties   ones
```

Mayan Numerals

Symbols: • ___ Θ

one five zero

Some things to notice:

(i) The Mayan numeral system probably had place value based on a combination of twenty and eighteen, with places representing groups of one, twenty, eighteen times twenty, eighteen times twenty squared, eighteen times twenty cubed, etc. This has not been definitely determined, however. Therefore, for simplicity, we will suppose that the Mayan numerals had place value based on twenty, with groups of symbols from right to left representing ones, twenties, four hundreds, eight thousands, etc.

(ii) There was a symbol for zero, so the value of each group was clear.

(iii) There were many symbols in each position.

(iv) Numerals were written vertically.

Example 4:

In this numeral, the symbols in the eight-thousands place mean two; so this place represents 2 x 8000 **things**. The zero symbol in the second position signifies that there are no four-hundreds to be counted. The symbols in the twenties place represent fourteen — two fives and four ones. So their value is 14 x 20 **things**. Finally, in the ones place, the symbols signify 18 **things** — three fives and three ones.

Hence, this numeral represents 2 x 8000 plus 14 x 20 plus 18 — which is sixteen thousand two hundred ninety-eight **things**.

Example 5:

Write a Mayan numeral to represent two thousand four hundred seventeen things.

First we must decide what places are needed. Since the number we want to represent is less than eight thousand, the biggest position will have value four hundred. Thus we must fill-in these blanks with Mayan symbols:

four-hundreds twenties ones

The beginning question is: How many bundles of four-hundred can be made from two thousand four hundred seventeen things? Division gives a quotient of six and remainder seventeen; so we have six bundles of four hundred with seventeen things left. Now, since no bundles of twenty can be made from seventeen things, we can complete the numeral:

Note: A calculator is not always efficient for the arithmetic required in these examples because we need to know the **quotient** and the **remainder** for each division process. The process for finding these numbers with a calculator sometimes requires more time than paper and pencil division.

Hindu-Arabic Numerals (our system)

Symbols: 0, 1, 2, 3, 4, 5, 6, 7, 8, 9

Some things to notice:

(i) There is a symbol for zero.

(ii) Place value is based on ten. (Symbols, from right to left, represent ones, tens, hundreds, thousands, etc.)

(iii) There is only one symbol (digit) in each position. This is the feature which distinguishes our system from all the others and makes it much simpler and more efficient. (This is undoubtedly the reason why Hindu-Arabic numerals have survived and are used world-wide.)

Note: Our number **words** name the value of each digit from largest to smallest place. 8375 is read: eight **thousand**, three **hundred** seven**ty** five. ("ty" is a corruption of "ten".) If zero is in a place, that position isn't read. 308 is simply read three hundred eight — no mention of tens.

Homework: Section 1

1. How many things does each numeral represent?

   a) M M C C C C X X X I I I I

   b) V V <V V << V
V V V V <<<

   c) V V V << V V << V V V V
< V << V V V V V

2. Write a Babylonian numeral to represent each of these numbers:
   a) one hundred seventy-three
   b) seven thousand two hundred fifty-three
   c) ten thousand three hundred eighty

3. How many things does each numeral represent?

a)	••	Θ	• — —	••• — — —	
b)	•	•	•• — — —	•••• — —	•

4. Write a Mayan numeral to represent each number:
   a) three thousand two hundred seventeen
   b) nine thousand

5. Use near-by, easy numbers to approximate:

   a) $593 + 209 + 348$ b) $79 \div 4$ c) $140{,}000 - 1968$

   d) 211×95 e) 50% of 793 f) $\frac{11}{16}$ of 80

   g) $2\frac{7}{9} \div 5\frac{1}{32}$ h) $19 + 32 + 28 + 58 + 11$ i) $\$20 - (\$5.95 + \$8.79)$

6. **Estimate:**

   a) $3857 + 2136 + 992 - 819$ b) $1\frac{3}{8}$ of 17

   c) $9\frac{1}{2}$% of 860 d) $17\frac{5}{6} \div 3$

7. Find exact answers **mentally:**

   a) $1\frac{2}{3}$ of 21 b) $10 - 6.95$ c) $3\frac{1}{5} \div 4$

   d) 40% of 35 e) $28 + 57 + 92$

8. If a numeral system had place value based on fifteen, how many symbols would it have to have so that there would be exactly one symbol in each place? Explain.

9. If people had six fingers on each hand, what do you think would be different about our numeral system? How would we represent two hundred thirty-eight things in such a system?

Answer in Journal:

1. Explain what it means to say that a numeral system has place value.
2. What is a digit?
3. If a machine caps 600 bottles of Coke in one hour and fifteen minutes, at what **rate** does the machine cap bottles?
4. **Explain** why a calculator may not be efficient in answering this question: Farmer Brown has 105 eggs. They will be packed in cartons that hold two dozen each. How many cartons will he get, and how many eggs will be left over?
5. Explain what the word average means. For example, what does it mean in each of these statements:
   a) Suzy's scoring average this season is 12 points per game.
   b) The average annual salary for TopCo employees is $28,000.
   c) In driving from Lafayette to Houston, I averaged 60 mph.
   d) The average American teenager watches TV for 20 hours per week.

Section 2: Bases Other Than Ten

The purpose of this section is to emphasize the meaning of place value, and the significance of having exactly one symbol in each position. For each example we assume that the numeral system has place value based on the number of fingers, and that each position has exactly one symbol.

A. Eight-finger system
   * Symbols: 0, 1, 2, 3, 4, 5, 6, 7
   * Place value based on eight
   * Exactly one symbol in each place

Example 1:

2	1	5	7
five-hundred-twelves	sixty-fours	eights	ones

This numeral, 2157_{eight}, represents 2 five-hundred-twelves, plus 1 sixty-four, plus 5 eights, plus 7. This is one thousand one hundred thirty-five.

Example 2: Write the eight-finger numeral which represents one thousand five-hundred ninety-six **things.**

First we must determine what is the biggest place needed. Since the values of positions in this system are one, eight, sixty-four, five-hundred-twelve, four-thousand-ninety-six, etc., the largest place needed is five-hundred-twelve, so these are the positions to be filled-in:

five-hundred-twelves sixty-fours eights ones

Now we must find out how many bundles of five-hundred-twelve can be made from one thousand five hundred ninety-six things. Division shows that there are 3 bundles with sixty things left over.

3			
five-hundred-twelves	sixty-fours	eights	ones

Next we notice that no bundles of sixty-four can be made from the sixty left-over things, so we have

3	0		
five-hundred-twelves	sixty-fours	eights	ones

Finally, we see that the sixty left-overs make 7 bundles of eight and 4 individual things (**ones**). This completes the numeral:

3074_{eight}

B. Fifteen-finger system

* Symbols: 0, 1, 2, 3, 4, 5, 6, 7, 8, 9, T, E, ☆, ♧, ☺
* Place value based on fifteen
* Exactly one symbol in each position (Notice we have to "invent" new symbols!)

Example 3:

2	1	E	☺
three-thousand-three-hundred-seventy-five	two-hundred-twenty-five	fifteen	ones

This numeral represents **two** thirty-three-hundred-seventy-fives, **one** two-hundred-twenty five, **eleven** fifteens, and **fourteen** ones — which is a total of seven thousand one hundred fifty-four.

Example 4: Write the fifteen-finger numeral which represents one thousand, eight hundred thirteen **things**.

First we notice that the largest place needed is two-hundred-twenty-fives, so these blanks must be filled in:

two-hundred-twenty-five	fifteen	ones

How many bundles of two-hundred-twenty-five can be made? The division process shows there are 8 bundles with thirteen things left over:

8		
two-hundred-twenty-five	fifteen	ones

Now, since no bundles of fifteen can be made from the thirteen left-overs, we can complete the numeral:

80♧ $_{\text{fifteen}}$

Homework: Section 2

1. What are the **digits** in a base 2 numeral system? In a base 7 system? In a base 13 system?

2. How many apples does each of these represent?

   a) 3205_{seven} b) 10211_{three} c) 1689_{twelve}

   d) 23034_{five} e) 1000101_{two} f) $387_{thirteen}$

3. Write **each** of these in base 2, base 5, base 9 and base 15.

   a) forty-seven b) ninety-five c) one hundred sixty-three

4. Find exact answers **mentally:**

   a) \$20 - \$14.56 b) $\frac{5}{8}$ of 32 c) $28.16 \div 4$

   d) $54 + 29 - 43 + 36 + 11 - 17$

5. **Estimate**:

   a) 76 % of 24 b) $19\frac{11}{12} \div 5$ c) 2.03 x 3.972

6. The average weight of four boys is 150 lb and the average weight of six girls is 110 lb. What is the average weight of the ten children?

Answer in Journal

1. Explain what an **equation** is. (Don't use the word "equal" in your explanation.)
2. Explain how you got the answer to #4 in the Homework Exercises.

Section 3: Divisibility of Whole Numbers

Remember that the result of multiplying numbers is called a **product**. If the numbers which are multiplied are whole numbers, they are called **factors** or **divisors**. For example, since 7 times 5 is 35, 7 and 5 are factors (or divisors) of 35.

There isn't any whole number that can be multiplied by 8 to get the product 94; so 8 is not a divisor (or factor) of 94.

Example 1: Consider these questions

(i) If you want to arrange 60 chairs in equal rows, could there be 12 rows?

(ii) A class of 60 people is to be separated into groups. Could there be 12 people in each group?

(iii) Suzy's mother bought 60 cookies for Suzy's birthday party. If there are 12 children at the party, can she give out the cookies so that each child gets the same number?

These questions all ask the same thing: Is 12 a factor (or divisor) of 60? Since 12 x 5 = 60, the answer is yes.

Example 2: If 60 chairs are to be arranged in equal rows, how many chairs could be in each row?

This is asking: What are all the divisors (or factors) of 60? It's important to notice that factors come in pairs — if j is a factor of m, then there is a whole number k that multiplies by j to give the product m. So the easiest way to find all the factors of a number is to list them, in a systematic way, in pairs:

1 x 60
2 x 30
3 x 20
4 x 15
5 x 12
6 x 10

Since 7, 8, and 9 are not divisors of 60, and 10 is already listed, we know we have found them all. The factors of 60 are 1, 2, 3, 4, 5, 6, 10, 12, 15, 20, 30, 60.

Example 3: Find all the factors of 17.
It's obvious that there are only two, 1 and 17.

A whole number which has exactly two factors, itself and 1, is called a **prime number** (or simply a **prime**).
The prime numbers which are less than sixty are: 2, 3, 5, 7, 11, 13, 17, 19, 23, 29, 31, 37, 41, 43, 47, 53, 59.

Example 4: Find all numbers which are factors of both 42 and 54

Factors of 42	Factors of 54
1 x 42	1 x 54
2 x 21	2 x 27
3 x 14	3 x 18
6 x 7	6 x 9
1, 2, 3, 6, 7, 14, 21, 42	1, 2, 3, 6, 8, 18, 27, 54

The numbers which are factors of both 42 and 54 are 1, 2, 3, and 6. They are called **common factors** of 42 and 54.

The greatest common factor of the two numbers is 6.

Notice that all the common factors are themselves factors of the greatest common factor.

The word **divisible** is often used in discussing factors and products. Saying that 15 is divisible by 3, means that 3 is a factor of 15.

Example 5:

a) 21 is divisible by 7
b) 39 is divisible by 13
c) 36 is not divisible by 8
d) 9 is not divisible by 18

In general k is divisible by j if j is a factor of k.

A whole number is **even** if it is divisible by 2; it is **odd** if it's not divisible by 2. Notice that every even number can be written as 2 times some whole number:

0 = 2 x 0
2 = 2 x 1
4 = 2 x 2
6 = 2 x 3, etc.

In general, every even number can be written as $2k$, where k represents some particular whole number.

Every odd number is an even number plus 1:

$$1 = 0 + 1$$
$$3 = 2 + 1$$
$$5 = 4 + 1$$
$$7 = 6 + 1, \qquad \text{etc.}$$

So, every odd number can be written as $2m + 1$, where m represents some particular whole number. The word **multiple** is also frequently used in relation to factors, products, and divisibility. Factors and multiples are inverses of each other; if k is a factor of n, then n is a multiple of k.

Example 6:

a) 12 is a multiple of 3 (because 12 = 3 x 4)
b) 63 is a multiple of 7 (because 63 = 7 x 9)
c) 253 is a multiple of 11 (because 253 = 11 x 23)

Example 7: List all the multiples of 8, of 15, and of 42

Multiples of 8	Multiples of 15	Multiples of 42
0 (8 x 0)	0 (15 x 0)	0 (42 x 0)
8 (8 x 1)	15 (15 x 1)	42 (42 x 1)
16 (8 x 2)	30 (15 x 2)	84 (42 x 2)
24 (8 x 3)	45 (15 x 3)	126 (42 x 3)
32 (8 x 4)	60 (15 x 4)	168 (42 x 4)
etc.	etc.	etc.

Note: All multiples of 3 can be written as $3j$ (3 times some whole number); all multiples of 8 can be written as $8k$ (8 times some whole number); all multiples of 45 can be written as $45m$ (45 times some whole number); etc.

Example 8: Find all the numbers which are multiples of both 12 and 15.

Multiples of 12		Multiples of 15	
0	108	0	135
12	120	15	150
24	132	30	165
36	144	45	180
48	156	60	195
60	168	75	etc.
72	180	90	
84	192	105	
96	etc.	120	

We see that 0, 60, 120, 180, etc. are multiplies of both 12 and 15 — they are called **common multiples** of the two numbers.

Notice that 60 is the smallest positive common multiple of 12 and 15. It is called (for good reason) the least common multiple of the two numbers.

Also notice that **every** common multiple is a multiple of the least common multiple.

Homework: Section 3

1. Write 52 as the product of three whole numbers in four different ways.
2. Suppose p is a prime number and $p = jk$. Could j and k both be whole numbers which are bigger than 1? Explain.
3. Which of these numbers are divisors of 76?

   2, 76, 1, 0, 38, 152
4. Find all the factors of 140.
5. Circle all the multiples of 21:

   3, 21, 0, 1, 84
6. What is the second largest divisor of 296?
7. List all the numbers which are factors of both 72 and 132. What is the greatest of these common factors?
8. What is the smallest positive number which is a multiple of both 18 and 45? Name 5 common multiples of 18 and 45.
9. What number is a factor of every whole number?
10. What number is a multiple of every whole number?
11. Which of these are prime numbers?
    79, 57, 2918, 835, 127
12. Write each of these numbers as a product in which all the factors are primes: 140, 324, 441.
13. Seven hundred people are lined up waiting for a 4th of July parade. A veteran walks down the line and gives every fifth person a small flag. A clown follows behind him and gives every twelfth person a balloon. How many people get both a flag and a balloon?

Answer in Journal

Explain what each statement **means.**

1. j is a divisor of 38.
2. m is odd.
3. y is a multiple of 38.
4. The greatest common factor of t and v is 24.
5. The least common multiple of j and k is 24.

Section 4: Factorization

We have been using the word **factor** as a noun; **factor** is also used as a verb. To factor a whole number means to write it as the product of whole numbers.

Example 1: Factor 24 in as many ways as possible.

First, we will agree on these things regarding factoring:

(i) The only time 1 will be used as a factor is when we write 1 times the original number.

(ii) Changing the **order** of factors does not give a different factorization (3 x 8 is the same factorization as 8 x 3, and 2 x 4 x 3 is the same as 3 x 4 x 2).

Therefore, these are all the different factorizations of 24:
1 x 24, 2 x 12, 3 x 8, 4 x 6, 2 x 3 x 4, 6 x 2 x 2, 2 x 2 x 2 x 3

Example 2: Factor each number into the product of primes: 98, 210, 1035

98	210	1035
2 x 49	21 x 10	5 x 207
2 x 7 x 7	3 x 7 x 2 x 5	5 x 3 x 3 x 23

There are other procedures for factoring these numbers into primes. Each person should do what seems most obvious at each stage of the process.

The important thing to note is that no matter how an individual goes about factoring a number into primes, the result is always the same.

Example 3: Prime factor each of these numbers: 162, 64

162	64
2 x 81	8 x 8
2 x 9 x 9	2 x 4 x 2 x 4
2 x 3 x 3 x 3 x 3	2 x 2 x 2 x 2 x 2 x 2
(or 2×3^4)	(or 2^6)

Teaching Note: Examples such as these motivate the introduction of exponents in about the sixth grade. Exponents simply provide an efficient notation for writing products which contain several equal factors. The words "exponent" and "power" need not be used with fifth and sixth graders; they must simply understand is that 2^6 means "two multiplied six times".

If a number greater than 1 cannot be written as the product of primes, then the number itself is a prime. How can we determine whether a particular whole number is prime?

First, we check to see if the number is divisible by 2, 3, or 5. If not, we proceed by testing each succeeding prime. If we find a prime which is smaller than the number, then the number isn't prime. If we don't find a prime divisor which is smaller than the number, then the number is prime.

Now suppose we are trying to determine whether 2741 is prime. Must we test every prime number less than 2741? The answer is **no**. Once we get a quotient which is less than the prime we are testing, and none of the preceding primes is a factor of the number, then we can be certain that the number is prime.

For 2741, we first notice that it is not divisible by 2, 3, or 5. Next, we find (by dividing) that 7, 11, 13, 17, ..., 47 are not factors of 2741. The next prime is 53. On a calculator, 2741÷ 53 is 51.716981. Since this quotient is not a whole number, we know that 2741 is not divisible by 53. And since this quotient is **less than** 53, we need not continue to test primes. We are certain that 2741 is a prime number.

The reasoning is that if some prime bigger than 53 were a factor of 2741, then there would have to be a whole number smaller than 51.716981 which multiplies by 53 to give the product 2741. But we know this isn't possible because we have already tested all the primes which are less than 51.

Prime factorization provides an efficient method for finding greatest common factors and least common multiples.

Example 4: Find the greatest common factor of 104 and 182.

First, we prime factor the two numbers.

182	104
2 x 91	2 x 52
2 x 7 x 13	2 x 2 x 2 x 13

Since 2 and 13 are the only common prime factors of the two numbers, then 2 x 13 must be their greatest common factor.

Example 5: Find the greatest common factor of 48 and 35.

48	35
6 x 8	5 x 7
2 x 3 x 2 x 4	
2 x 3 x 2 x 2 x 2	

These have no common prime factors. Therefore, their greatest common factor is 1.

Numbers which have no common prime factors are said to be **relatively prime**.

Example 6: Find the least common multiple of 18 and 48.

We begin by prime factoring the two numbers.

18	48
2 x 9	6 x 8
2 x 3 x 3	2 x 3 x 2 x 4
	2 x 3 x 2 x 2 x 2

So, in order for a number to be a multiple of 18, it must have one 2 and two 3s as prime factors; and in order for it to be a multiple of 48, it must have four 2s and one 3 as prime factors. Therefore, the **fewest** prime factors that a number can have, and be a multiple of both 18 and 48, is four 2s and two 3s: $2^4 \cdot 3^2$.

Notice that this number is 18 x 4 (so it's a multiple of 18), and it's 48 x 3 (so it's a multiple of 48).

Teaching Note: Remember, short-cuts in mathematics should ideally be discovered by the students. But if, as in this case, it isn't likely that they will discover them, we should introduce the short-cut only **after** the students **understand the concepts involved.**

Homework: Section 4

1. Which of these are prime numbers: 965, 137, 2001, 667
2. List four common multiples of 35 and 98.
3. What is the biggest number that divides both 84 and 140?
4. Write a base fourteen numeral to represent three thousand nine.
5. How many factors does 36 have?
6. Circle the numbers which are divisible by 11:

   131, 1, 11, 781, 2959, 3621

7. Which pairs of numbers are **relatively prime**?

   8 and 9, 6 and 15, 7 and 31, 46 and 69

8. Write a 7-digit number which is a multiple of 3.
9. How many odd numbers less than 100 are divisible by 3?
10. How many even numbers between 18 and 120 are multiples of 7?
11. Find **mentally:** What must be multiplied by $2^3 \cdot 3^5 \cdot 7^2$ to get the product $2^4 \cdot 3^5 \cdot 5 \cdot 7^3$?
12. Find **mentally:** Circle the numbers which are multiples of 36:

    $2^3 \cdot 3 \cdot 5^2$, $2^2 \cdot 3^4 \cdot 11$, $2 \cdot 3^3 \cdot 17$, $2^5 \cdot 3^5 \cdot 19^2$

13. Find **mentally:** For each of these products, when the number is written as a single whole number, how many zeros are at the end?

    a) $2^5 \cdot 7 \cdot 13$ b) $2 \cdot 5^3 \cdot 11^2$ c) $2^4 \cdot 5^3 \cdot 17^2 \cdot 19$

    (Hint: If a numeral has one zero at the end, the number is divisible by 10. What prime factors must it have?
    If a numeral has two zeros at the end, the number is divisible by 100. What prime factors must it have?

14. The sequence of numbers shown below is formed by beginning with two given numbers. Each new number is twice the sum of the previous two numbers. Find the first and last numbers.

    ? , ____ , ____ , ____ , ____ , 108 , 296 , ? .

15. The average of Joe's eight test scores is 85. The teacher disregards the highest and the lowest scores, which are 54 and 92. What is the average of the remaining set of scores?

Answer in Journal

1. **Explain** what you would **do** to find out whether or not 2617 is prime.
2. Write a sentence in which the word **factor** is used as a noun, and write another sentence in which **factor** is used as a verb.

Section 5: Building Numbers with Primes

Activity

1. All the numbers in the Bag A are ______________________.

2. All the numbers in the Bag B are ______________________.

3. From Bag A take out one each of these: 2, 3, and 5. Find their product. What number did you "build"?

   ____

4. Now find three 2's, and two 5's. What number do these building blocks form? __________

5. Using numbers from either bag, "build" the number 18. How many different ways can this be done?

   ____

6. a) Now build 18 using only colored squares. Did everyone in your group get the same answer?

   b) If you are not allowed to use the "1", must everyone in your group get the same answer?

The two previous exercises illustrate the Fundamental Theorem of Arithmetic: Every natural number greater than one is either a prime number or can be written as a product of prime numbers in a <u>unique</u> way. This product is called the "prime factorization" of the number. (Remember: One is not prime.)

7. What is the prime factorization of 66? ____________

8. List all the factors of 66. ____________________

When we know the prime factorization of a number, we can easily see all of its factors. For example: 1 is a factor of any number. Use each prime factor one at a time and we see that 2, 3, 11 are all factors of 66. Use the prime factors two at a time to get $2 \cdot 3 = 6$, $2 \cdot 11 = 22$, and $3 \cdot 11 = 33$. Using all the prime factors yields $2 \cdot 3 \cdot 11 = 66$, the number itself which is its largest divisor.

9. a) Prime factor 90. ________

   b) List all the factors of 90. ________________________

   c) Can 14 be a factor of 90? _____ How do you know?

   d) Is 18 a divisor of 90? _____ How do you know?

   e) Use your knowledge of the building blocks of 90 and 18 to determine the quotient of 90 and 18 without dividing.

   ____

10. a) Write 250 as a product of primes. ________

    b) Without dividing, determine if 20 a factor of 250? ______

    c) Find all the divisors of 250. ________________________________

11. a) What are the common factors of 90 and 250? __________

    b) What is the largest common factor? _______

    c) What prime factors do 90 and 250 have in common? __________

    d) List all the numbers which can be built using the primes from c).______________________

12. a) Build 312 and 260 with your blocks. ____________________

    b) Find the common factors of 312 and 260 by considering the prime factors that they have in common.

    ____

    c) Do you see that the largest common divisor of 312 and 260 is 52? How do the common prime factors of 312 and 260 help you find 52?

13. a) Pull 52 out of 312 to find the quotient of 312 and 52. _______
    b) Pull 52 out of 260 to find the quotient of 260 and 52. _______
    c) What factors do these quotients have in common? ______

When the greatest common factor of two or more numbers is one, we say that the numbers are "relatively prime".

Knowing the prime factorization of a number can help you find its multiples as well as its factors. For a number to be a multiple of 90, it must have 90 as a factor so the multiple must contain all of the prime factors of 90 as often as they occur in 90.

14. Build 90 again. Now pull any number out of either bag. (Try to get a variety within your group.) Place that square next to your original 90. What is the multiple of 90 that you created? __________

15. A common multiple of 90 and 250 must contain all the necessary prime factors of 90 and of 250. Build some common multiples of 90 and 250:
    a) Use all of the prime factors of both numbers together. You will get their product which is a common multiple.

    ____

    b) Combine any other square from either bag with your answer in a) and you will get another common multiple of 90 and 250.

    ____

Can we find a smaller common multiple than the product of the two? Build a common multiple this way:

Start with $\underline{(2 \cdot 3 \cdot 3 \cdot 5)}$. (This makes 90 a divisor of the number we are building.) Now put in the prime factors of 250 that do not occur often enough in our new number $\underline{(2 \cdot 3 \cdot 3 \cdot 5)\ (5 \cdot 5)}$.

Do you "see" 250 and 90 in this number? Notice that we did not include any unnecessary factors. So this product which is 2250 is the least common multiple of 90 and 250.

c) What is 2250 divided by 90? ________ What is 250 divided into 2250? ________
What is the relationship between these two quotients?

____

d) What is the relationship between 2250 and your answer in b)?

16. a) What is the smallest positive number that is a multiple of 60, 312, 260? ___________
b) What is the quotient when each of these numbers is divided into their least common multiple?
60 ________, 312 ________, 260 ________

How are these three answers related?

c) Name three other common multiples of 60, 312 and 260. __________

17. Recall that $250 = 2 \cdot 5^3$ and 250 has eight factors.

a) What is the prime factorization of 24? __________

b) How many different prime factors does 250 have? _______ 24? _______
How many of each of these primes are there in each number? __________

c) Without listing them, how many factors does 24 have? ________

d) Recall that $90 = 2 \cdot 3^2 \cdot 5$ and 90 has twelve factors. Without listing them, how do you know that this is true?

e) How many factors does $140 = 2^2 \cdot 5 \cdot 7$ have? _________

f) How many factors does $360 = 2^3 \cdot 3^2 \cdot 5$ have? _________

Section 6: Divisibility Tests and Generalizations

Group Exploration

I. Answer the following questions.

1. How can you tell whether a whole number is even just by looking at it?

2. How can you tell whether a whole number is divisible by 5 just by looking at it?

3. Which of these numbers are divisible by 3?

   78, 95, 231, 387, 414, 265, 742, 654

4. For each number in #3(above), find the sum of its digits. What seems to be true about the sum of the digits of numbers which are divisible by 3?

5. Which of these numbers are divisible by 11?

   374, 589, 869, 627, 158

6. For each number in #5, find the sum of the first and last digit, then subtract the middle digit. What pattern do you notice among the numbers which are multiples of 11?

II. Decide whether each statement is true or false.

1. If j and k are both even, then $j + k$ is even and jk is even.

2. If m and v are both odd, then $m + v$ is odd and mv is odd.

3. If c is even and t is odd, then $c + t$ is odd and ct is even.

4. If j is divisible by both 4 and 9, then j has to be divisible by 36.

5. If m is a multiple of both 6 and 9, then m has to be a multiple of 54.

6. Every divisor of 42 is also a divisor of 84.

7. Every multiple of 39 is also a multiple of 13.

8. Every whole number bigger than 1 has at least one prime number as a factor.

Homework: Section 6

1. Write a base twelve numeral to represent seven thousand five hundred six.
2. How many **things** does this base seven numeral represent? 2056_{seven}
3. Which of these numbers are prime?

   7965, 28341, 97736, 181, 259
4. List all the multiples of 14.
5. Find all the common factors of 60 and 75.
6. Name 4 odd factors of 140.
7. How many even numbers are there between 39 and 257?
8. **Estimate:**

   a) 3 x $14.98 b) 15% of $41.07 c) $\frac{3}{8}$ of 55
9. A museum shows two short subject films in different rooms. One lasts 48 minutes while the other runs for 56 minutes. If they both began at 11:00 a.m., what time will it be when they begin simultaneously again?
10. The average of five different positive whole numbers is 12. What is the largest possible value of any of these numbers?

Answer in Journal

1. Suppose 9 and 12 are both divisors of *m*. Can you be sure that 108 is a divisor of *m*? **Explain.**
2. Suppose 9 and 14 are both factors of *t*. Can you be sure that 126 is a factor of *t*? **Explain.**
3. Explain how you found the answer to #10 in the Homework Exercises.

Section 7: Meaning of Fractions

Fractions represent equal portions or parts of things. If something has been separated into eleven **equal** parts and I have 3 of those parts, then the fraction $\frac{3}{11}$ represents the portion that I have. The equal parts into which a thing is separated are called halves, thirds, fourths, fifths, sixths, etc. To read a fraction we simply say how many pieces and the name of the pieces - 2 **thirds,** 1 **seventh**, 38 **sixtieths,** etc.

Example 1: For each figure, what fraction represents the shaded portion?

a) b)

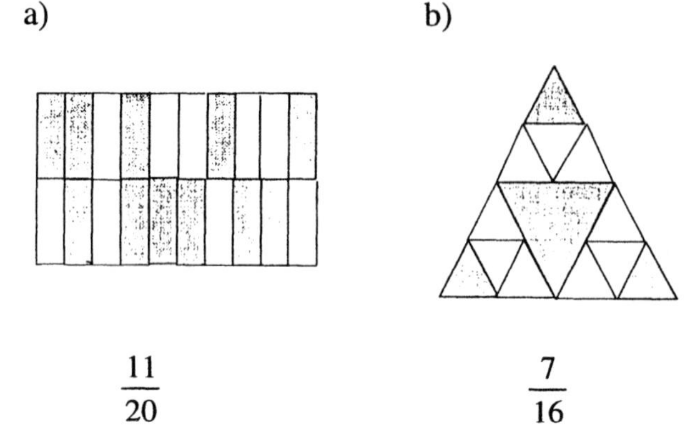

$\frac{11}{20}$ $\frac{7}{16}$

Example 2: There are 13 boys and 15 girls in Mr. Smith's class.
So, of the 28 students, 13 are boys.
Hence, we can say that 13 twenty-eighths of the class are boys.

Example 3: In this figure, each small segment is 1 fourth of an inch.

Hence, the entire segment is 15 fourths or $\frac{15}{4}$.

Example 4: In this figure, all the circles are the same and they have each been separated into three equal pieces. So each piece is a third of a circle. Therefore the 5 shaded pieces can be represented as 5 thirds or $\frac{5}{3}$.

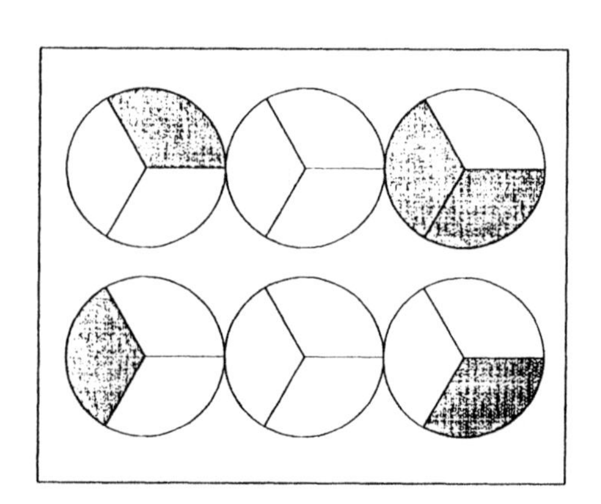

Also, notice that the picture consists of 6 whole circles — or 18 **thirds**. And since 5 **thirds** are shaded, it's reasonable to say that $\frac{5}{18}$ of the picture is shaded.

Homework: Section 7

1. Write a fraction to represent the shaded part of each figure:

a) 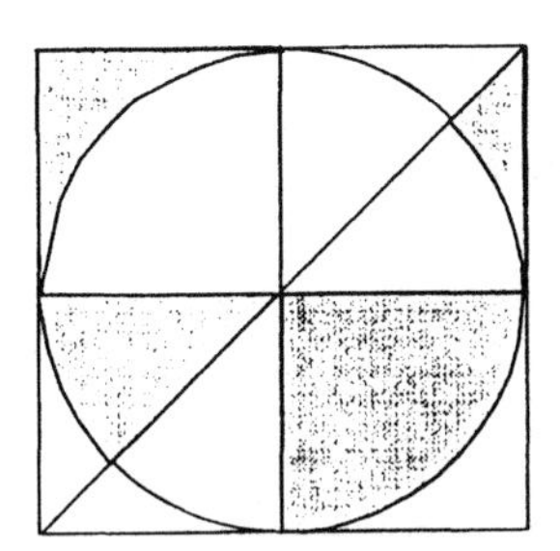

b)  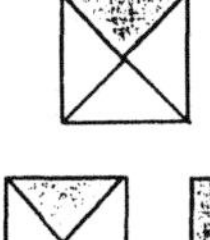 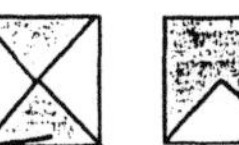 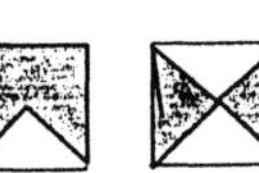

c) 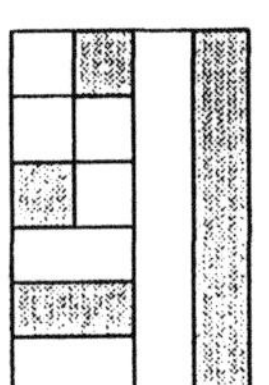

2. Find each of these mentally:

a) $\frac{2}{3}$ of 15 b) $\frac{4}{9}$ of 63 c) $\frac{5}{12}$ of 48

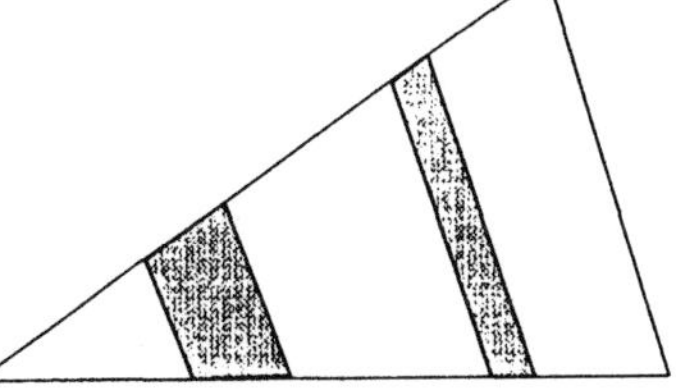

3. The figure at right is separated into five parts, and two parts are shaded. Explain why $\frac{2}{5}$ does not represent the shaded portion of the figure.

4. Mr. Brown separated his LP record collection into 7 equal shares to give his children and grandchildren. Joe got 38 LPs, which was 2 shares. How many LPs were in the whole collection?

5. Nine people each contributed a pie for the Church Fair. The pies were cut into eighths and sold for 65 cents per slice. All but 3 slices were sold.
   a) Write a fraction to represent the **quantity** of pie that was sold.
   b) Write a fraction to represent the **portion** of contributed pie which was sold.

6. Three out of every eight customers at Burger Heaven ask for catsup. If there are 12,000 customers per week, how many requests are there per week for catsup?

7. Find the answer **mentally**: Tom got two-thirds of the answers on a test correct. If he answered 26 correctly, how many questions were on the test?

8. The Louisiana Legislature is currently made up of 62 Democrats, 18 Republicans, and 10 Independents. It is predicted that in the next election 6 Democrats will be defeated by Republicans, an Independent will win the seat in the newly-formed District, and all the other members will be re-elected. True or False: If this prediction is correct, then $\frac{8}{13}$ of the new Legislature will be Democrats.

9. In what column is the number 1 million?

J	K	M
1	4	9
16	25	36
•	•	•
•	•	•
•	•	•

10. Which one of the following bar graphs could represent the data from the circle graph?

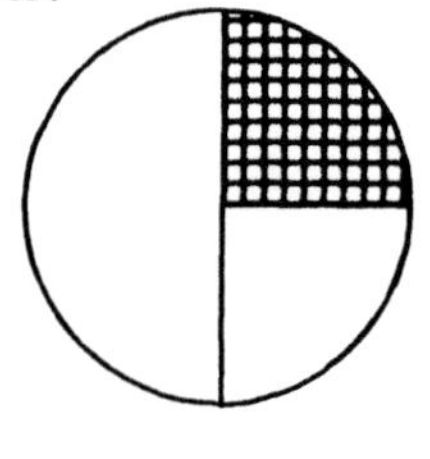

(A)

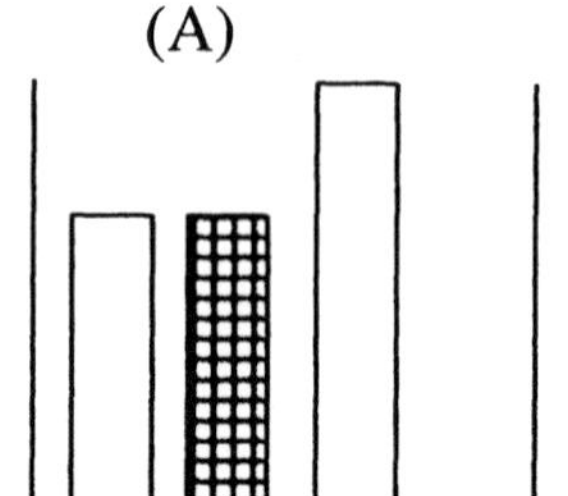

(B)

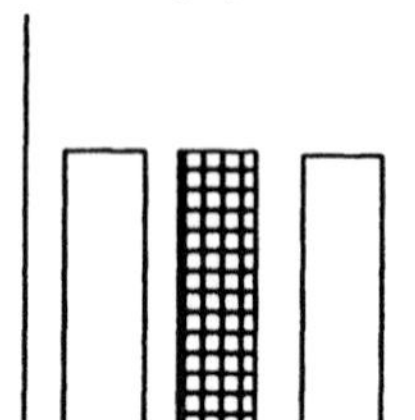

(C)

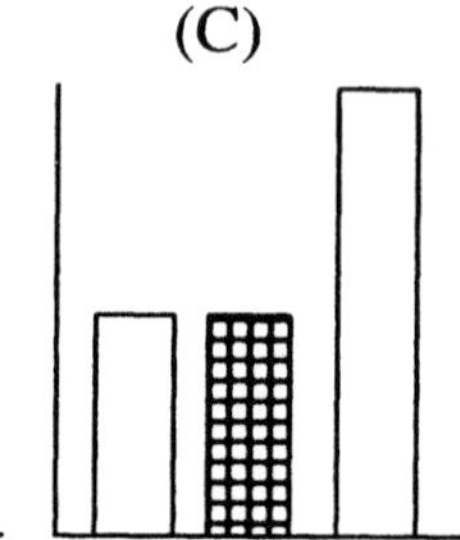

(D)

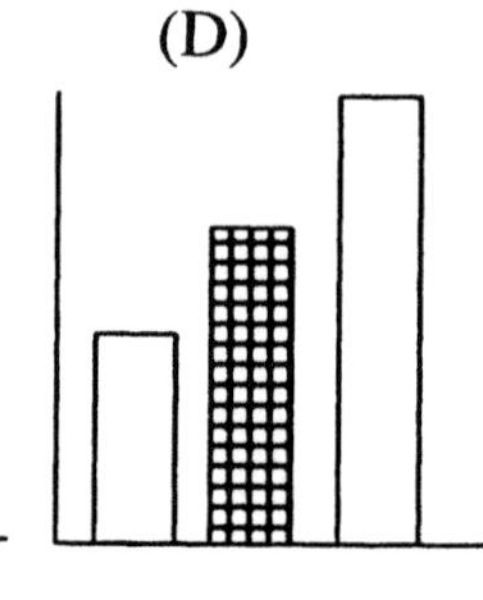

(E)

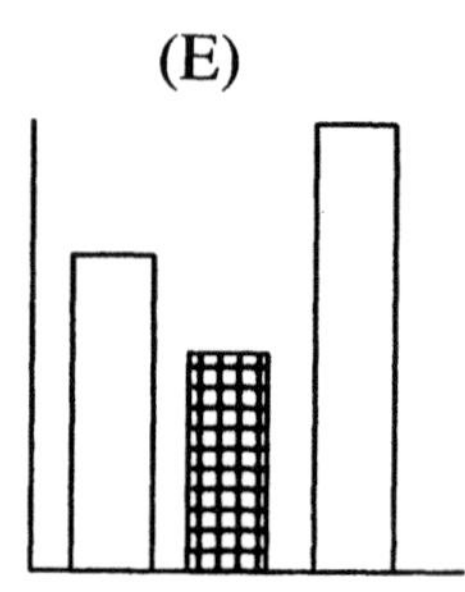

11. Shade $\frac{5}{6}$ of the figure to the right. (You may subdivide parts of the figure.)

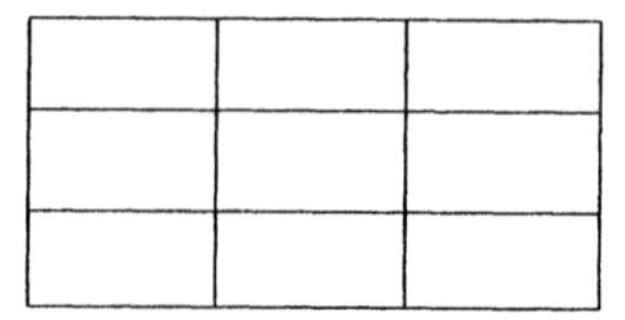

Answer in Journal

1. For #1(b) in the Exercises, what fraction did you write? What other fraction could be associated with this picture? Write a question which has this second fraction as its answer.
2. A city council is formed so that half the members are Asian-American, one-third are African-American, and the rest are Mexican-American. What is the smallest number of people that could constitute this council? **Explain** how you figured out the answer.
3. Explain how you found the answer to #7 in the Homework Exercises.

Section 8: Equivalent Fractions

In this section we explore different ways to represent the same portion of something.

A. Separating into smaller pieces (making the denominator **bigger**)

The rectangle is separated into four equal parts, and 3 of those parts are shaded. So, $\frac{3}{4}$ represents the shaded portion.

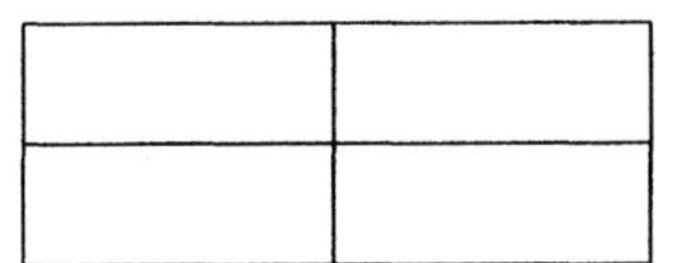

If each part of the rectangle is cut in half, then there are 8 equal pieces, and 6 of them are shaded. Hence $\frac{6}{8}$ also represents the shaded portion.

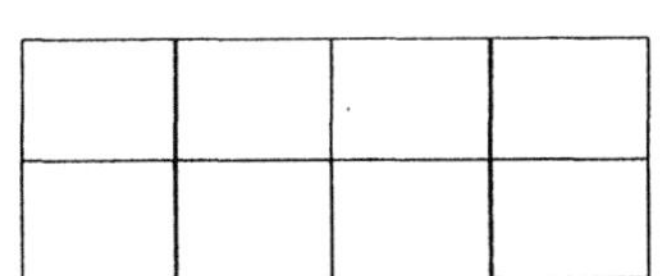

Similarly, if each of the original sections of the rectangle are cut into thirds, then the whole figure is now separated into 12 equal pieces and 9 of them are shaded. So $\frac{9}{12}$ is another representation of the shaded region.

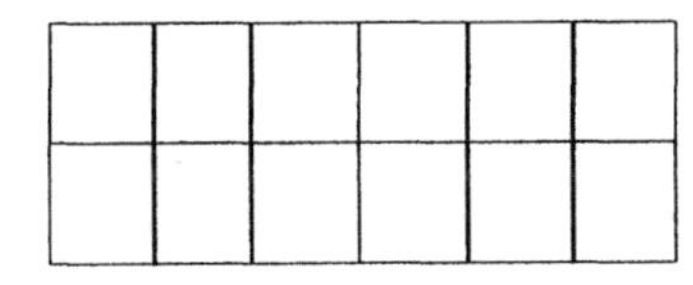

It's easy to see that what we're doing is **multiplying** — making more pieces; and whatever we do to the whole figure (denominator), we obviously also do to the shaded region (numerator). So cutting all the original sections in half has the effect of multiplying both denominator and numerator of the original fraction by 2. This is a concrete explanation of why $\frac{3}{4} = \frac{6}{8}$. And since the original sections can be re-cut into **any** number of equal pieces, there are infinitely many fractions which represent the same number as $\frac{3}{4}$: $\frac{6}{8}, \frac{9}{12}, \frac{12}{16}, \frac{15}{20}$, etc. Does this list contain all the fractions which are equal to $\frac{3}{4}$?

The equation $\frac{3}{4} = \frac{6}{8}$ states that the symbols $\frac{3}{4}$ and $\frac{6}{8}$ **represent the same number.**

This is what an equation is — a statement that two different symbols represent the same number.

B. Grouping pieces together (making the denominator **smaller**). The following activity suggests a concrete basis for the process of "reducing" fractions.

Class Activity

Use Bag (1):

1. How many squares are in the bag?
2. How many are red?
3. What fraction of the squares are red?
4. Arrange the squares into **pairs** - both red or both green. How many pairs are there?
5. How many pairs are red?
6. What fraction of the pairs are red? Does this fraction also represent the portion **of all the squares** which are red?
7. Separate the squares into groups of four so that each group is all the same color. How many groups are there?
8. How many groups are red?
9. What fraction of the groups are red? Does this fraction also represent the portion **of all the squares** which are red?
10. Since $\frac{12}{20}$, $\frac{6}{10}$, and $\frac{3}{5}$ all represent the same thing (the fraction of all the squares which are red), how are these numbers related?
11. Are there other ways to separate the squares into equal groups so that each group is all the same color? How do you know?

Use Bag (2):

1. How many squares are in the bag?
2. How many are red?
3. What fraction of the squares are red?
4. Find all the ways that the squares can be arranged in equal groups so that each group is all the same color.
5. For each arrangement, do you notice any relationship between the number of squares in each group and the original fraction $\frac{12}{18}$?
6. For each arrangement, what fraction represents the portion of the groups which are red?
7. Since the fractions in #6 all represent the same thing (the portion of all the squares which are red), what can you say about how these numbers compare to each other?
8. For each of these fractions, do you notice any relationship between the fraction and the number of groups in the arrangement it corresponds to?

Use Bag (3)

1. How many squares are there?
2. How many are red?
3. What fraction of the squares are red?
4. Can the squares be arranged in equal groups so that each group is all the same color?

Recap of Activity

1. Suppose you have 6 white and 10 blue marbles. Explain how you could use them to show a child that the fractions $\frac{10}{16}$ and $\frac{5}{8}$ represent the same number.
2. Explain how you can tell whether a particular fraction is equal to a fraction with smaller numerator and denominator.
3. Which of these fractions would be easiest to "deal with:" $\frac{2}{3}$, $\frac{34}{51}$, $\frac{70}{105}$, or $\frac{186}{279}$? Why?
4. What is a fraction **in simplest form**?
5. If a fraction is not in simplest form, how can it be "reduced?"

(Notice that we use the word "reduce" to mean: write an equivalent fraction which has a smaller numerator and denominator.)

Beyond Concrete Models

1. A fraction is written as $\frac{j}{k}$ where j and k are whole numbers and is k not 0. (This, of course, is extended to include other kinds of numbers.)
2. There are infinitely many fractions which represent the same number as $\frac{j}{k}$.
3. $\frac{2j}{2k}$, $\frac{3j}{3k}$, $\frac{4j}{4k}$, etc are all equal to $\frac{j}{k}$
4. If j and k have a **common factor** (bigger than 1), then there is a fraction equal to $\frac{j}{k}$ which has a numerator smaller than j and a denominator smaller than k. This simpler fraction is found by dividing both j and k by their common factor.
5. If j and k have no common factor (bigger than 1), then $\frac{j}{k}$ is said to be in **simplest form**.
6. **If $\frac{m}{t}$ is in simplest form**, then $\frac{2m}{2t}$, $\frac{3m}{3t}$, $\frac{4m}{4t}$, etc are all the fractions which are equal to $\frac{m}{t}$.

   So in order to represent $\frac{m}{t}$ as a fraction with denominator d, d **must be** a multiple of t.

Homework: Section 8

1. Write each number as a fraction with denominator 15. $\frac{4}{10}, \frac{2}{3}, \frac{8}{30}, \frac{34}{51}, \frac{13}{39}$
2. Which of these numbers can be written as a fraction with denominator 42?
   $\frac{4}{7}, \frac{3}{84}, \frac{55}{66}, \frac{14}{16}, \frac{1}{3}$
3. On a trip to Memphis, Suzy drove 493 miles in 7 hours and 15 minutes. What was her average speed?
4. For each pair of numbers, write both of them as fractions with the same denominator.

   a) $\frac{2}{3}$ and $\frac{3}{4}$, b) $\frac{5}{8}$ and $\frac{1}{6}$, c) 2 and $\frac{3}{7}$, d) $\frac{8}{18}$ and $\frac{35}{42}$
5. Write a **base seven** numeral which represents five thousand thirty-nine things.
6. How many apples does 1098_{twelve} represent?
7. **Estimate:**
   a) 98 x 73 b) 6921 - 1035 c) 26% of 789
8. To control her thyroid level, Suzy takes one half of a pill every other day. If one supply of medicine contains 60 pills, then the supply will last approximately how many months?
9. When five gallons of gasoline are added to a tank that is half full, the tank is then two-thirds full. How much more gasoline must be added to fill the tank?
10. Three sets of blinking lights are on a Christmas tree. The set of red lights blink 10 times per minute, the set of blue lights blink 12 times per minute, and the set of yellow lights blink 20 times per minute. If all blink at exactly 8 PM, how many seconds will elapse before they next blink together?

Answer in Journal

1. Explain what a fraction **in simplest form** is.
2. Explain why $\frac{14}{63}$ cannot be written as a fraction with denominator 35.

Section 9: Mixed Numbers and Comparison of Fractions

Determining Which of Two Unequal Fractions is Greater

Children should learn a variety of techniques for judging which of two unequal fractions is greater. These are some of the most useful strategies.

i) If two fractions have the same denominator, then it's obvious that the one with larger numerator is the larger number. The symbol > is read "is greater than" or "is larger than." So we can write:

$$17 > 12, \ \frac{5}{7} > \frac{4}{7}, \ \frac{11}{30} > \frac{8}{30}, \text{ etc.}$$

Also, the symbol < is read "is less than" or "is smaller than." So we can write

$$12 < 17, \ \frac{4}{7} < \frac{5}{7}, \ \frac{8}{30} < \frac{11}{30}, \text{ etc.}$$

ii) To compare $\frac{5}{7}$ and $\frac{5}{8}$, the important thing to notice is that **sevenths** are bigger than **eighths** (for the same object). So 5 sevenths is more than 5 eighths; and we can write $\frac{5}{7} > \frac{5}{8}$ or $\frac{5}{8} < \frac{5}{7}$.

Similarly, $\frac{9}{10} > \frac{9}{11}$ and $\frac{197}{205} < \frac{197}{203}$.

iii) We can often determine **mentally** which of two fractions if greater by comparing each of them to some easy number such as $\frac{1}{2}$ or $\frac{1}{3}$.

For example, it's clear that $\frac{3}{5} > \frac{4}{11}$, because $\frac{3}{5}$ is more than one-half and $\frac{4}{11}$ is less than one-half. By the same reasoning, we know that $\frac{2}{9} < \frac{7}{15}$ because $\frac{2}{9}$ is less than one-third and $\frac{7}{15}$ is more than one-third.

iv) Now consider the fractions $\frac{7}{8}$ and $\frac{9}{11}$. They don't have the same numerator or the same denominator, and there isn't any easy fraction to compare them to. How can we determine which is larger? The simplest technique is to represent both numbers with the same denominator, then compare numerators. In this case, 88 is an obvious common denominator; and we can **mentally** calculate that:

$\frac{7}{8}$ is the same as $\frac{77}{88}$ and $\frac{9}{11}$ is the same as $\frac{72}{88}$. So it's clear that $\frac{7}{8} > \frac{9}{11}$.

Mathematical statements that say which of two numbers is greater (such as 4 >1 or $\frac{5}{11} > \frac{5}{12}$) are called **inequalities**.

Relationship of Fractions and Mixed Numbers

As always, we begin with **concrete** models. Suppose that some pies, all the same size, have each been cut into 4 equal pieces. Then each sector is a **fourth** of a pie. The picture below shows 11 of these sectors - or, we could say 11 **fourths**. So the fraction $\frac{11}{4}$ represents the quantity of pie shown in the picture.

Now, if we rearrange the sectors (put them back into pie plates), we get:

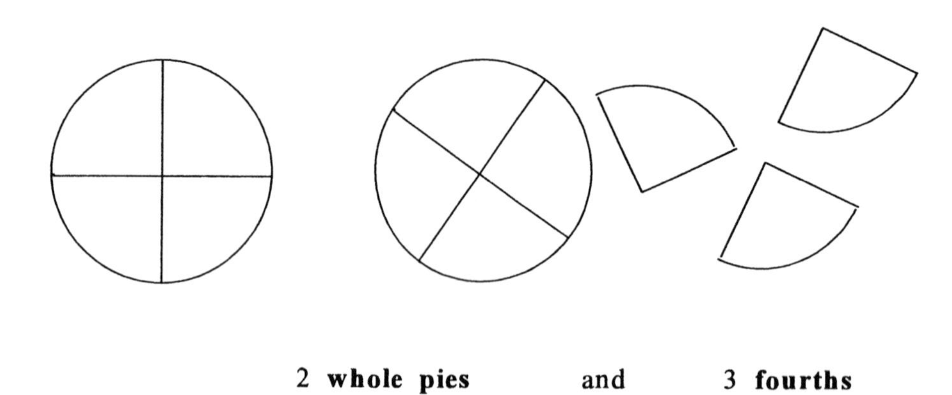

2 whole pies and **3 fourths**

or 2 and $\frac{3}{4}$. In writing this, we shorten it to $2\frac{3}{4}$. But the mixed number is read "two and three fourths."

In this picture, each small segment is 1 **half** inch. Since there are 7 of them, we can write 7 half inches or $\frac{7}{2}$ inches as the length of the entire segment.

But we can also think of the picture as 3 **whole** inches and 1 **half** inch. Therefore the length could be given as 3 and $\frac{1}{2}$, or $3\frac{1}{2}$ inches.

Example 1: Express $\frac{19}{6}$ as a mixed number

The number represent 19 **sixths**. We know that 6 **sixths** is a whole thing, and there are 3 bundles of 6 **sixths** in 19 **sixths**. So 19 **sixths** is 3 whole things and 1 **sixth** left over - or 3 and $\frac{1}{6}$. We shorten this to the mixed number $3\frac{1}{6}$.

Example 2: Express $5\frac{2}{3}$ as a fraction.

This number represents 5 **whole things** and 2 **thirds**. If the 5 **whole things** are each cut into thirds, the result is 15 **thirds**. Now it's easy to see that 15 **thirds** and 2 **thirds** is 17 **thirds**. So the fraction is $\frac{17}{3}$.

Teaching Note: Once children understand what fractions and mixed numbers **mean**, they will discover the "shortcuts" for translating from one to the other by themselves. Our role, as teachers, is to provide activities which will lead them to these discoveries.

Homework: Section 9

1. Find the answer **mentally:** Five out of eight USL students live off-campus. If there are 16,000 students, how many of them live on-campus?

2. For each pair of numbers, determine mentally which is greater.

   a) $\frac{3}{8}$ and $\frac{3}{7}$ b) 8 and $\frac{25}{3}$ c) $\frac{19}{40}$ and $\frac{127}{250}$

   d) $5\frac{1}{6}$ and $\frac{26}{5}$ e) $\frac{973}{16}$ and $\frac{973}{19}$

3. List these numbers in order from smallest to largest: $\frac{17}{7}$, $2\frac{18}{36}$, $2\frac{4}{5}$, $\frac{17}{6}$, $\frac{35}{15}$

4. For each number, write it as a fraction with denominator 24, if possible:
   $1\frac{5}{8}$, $\frac{28}{42}$, $\frac{27}{6}$, $\frac{36}{80}$, $\frac{51}{34}$, 19, $2\frac{15}{72}$

5. Write each number as a fraction with smallest possible denominator:
   $1\frac{5}{8}$, $\frac{28}{42}$, $\frac{27}{6}$, $\frac{36}{80}$, $\frac{51}{34}$, $2\frac{5}{6}$, $1\frac{12}{15}$

6. In the game Friday night, Harry carried the ball 18 times and his average yards-gained-per-carry was $3\frac{1}{2}$. What was Harry's total yards-gained in the game?

7. An advertisement claims that two out of three doctors recommend aspirin for headaches. If this is true, and there are 177,000 doctors in the U.S., how many of them recommend aspirin for headaches?

8. A robot assembles 136 computer circuits in 12 minutes. How many circuits does the robot assemble in an 8-hour work day?

9. Write a thirteen-finger numeral to represent six thousand seven hundred fifty-seven things.

10. Three out of every ten high school seniors in a certain city have used cocaine at least once. If 1644 seniors have used cocaine, how many have never used it?

11. **Estimate:**

    a) $\frac{10}{19}$ of 780 b) 3.17×29.8 c) $5\frac{1}{4}\%$ of 7915

12. A group of students pass around a bag of 49 cookies. Each person takes a cookie and then passes the bag to the next person. If Suzy takes the first and last cookie, how many students were in the group? Give all possible answers.

13. Al, Bob, Carl, Dot, and Ellen were hired by the new Zippo Company. On Opening Day, only Al and Carl worked. After that, the five employees worked according to this schedule: Al works if, on the previous day, Bob worked and Carl didn't. Bob works if, on the previous day, Carl worked and Dot didn't. Carl works if, on the previous day, Dot worked but Ellen didn't. Dot works if, on the previous day, Ellen worked but Al didn't. Ellen works if, on the previous day, Al worked but Bob didn't. Who works on day 105? On day 463?

14. The circle graph to the right shows the distribution of $20,000 in bonus money to five salesmen. Which of the salesmen received an amount closest to $5000? Who received approximately $1500?

DISTRIBUTION OF $20,000
IN BONUS MONEY

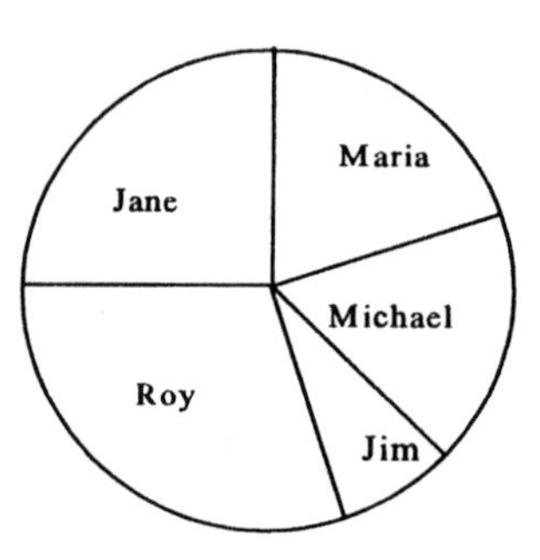

Answer in Journal

Explain, in a way that fifth or sixth graders could understand, how to find answers for #1, 11, and 14 in the Exercises.

Section 10: Fractions Which Represent Probabilities

In the Class Activity on p. 24, Bag 1 contained 12 red squares and 8 green squares. Suppose you reach into Bag 1, without looking and take a square. What are the chances that it will be green? Since there are 20 squares, and it is **equally likely** that you will get any one of them, your chances of drawing a green squares are 8 out of 20, or $\frac{8}{20}$. The probability of drawing a red square is $\frac{12}{20}$.

Now suppose a box contains 15 blue marbles and nothing else. If you reach in and take a marble, what is the probability it will be blue? Clearly, the answer is 15 out of 15, or $\frac{15}{15}$. It is important to notice that if something is a **certainty** (nothing else can possibly happen), then the probability of it happening is 1. If several different outcomes are possible in a certain situation, then their individual probabilities must add up to 1. For example, when you draw a square from Bag 1, you must get green or red. So the probability of getting red plus the probability of getting green must equal 1.

green or red is certain

$$\frac{8}{20} + \frac{12}{20} = 1$$

Considering the box of 15 blue marbles again, what is the probability of reaching in and pulling out a white marble? Common sense tells you there is **no chance** of getting a white marble. Mathematically, we say the probability is 0. (We could also say $\frac{0}{15}$.)

In general, if an event can happen in m equally likely ways, and a particular outcome can happen in j of those ways, then the probability of that particular thing happening is the fraction $\frac{j}{m}$.

Example 1: The circle at the right is separated into 8 equal parts, so if it is spun, each wedge has an equal chance of facing the arrow. Since there is only one green wedge, the chance that the arrow will point to green is 1 out of 8. Therefore, the probability that the arrow points to green is $\frac{1}{8}$. By this same reasoning, the probability that the arrow points to red is $\frac{3}{8}$ and the probability that it points to blue is $\frac{2}{8}$ or $\frac{1}{4}$.

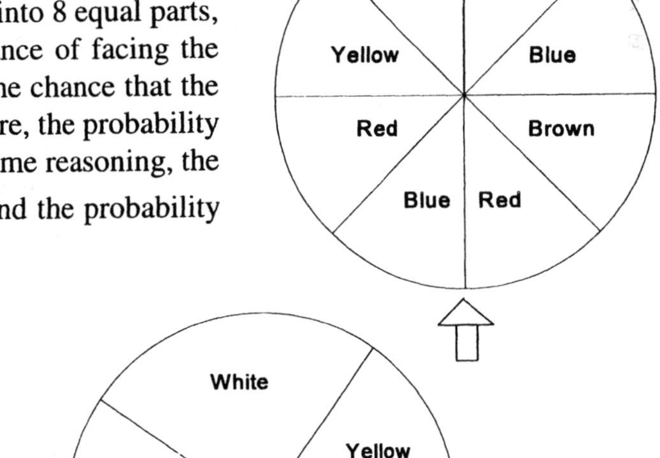

Example 2: For this circle, it's impossible to tell what fraction is painted in various colors. Hence we cannot determine the probability of the arrow pointing to a particular color. We can say, however, that it is more likely to point to white than to green, and the probability that it points to yellow is less than the probability that it points to red.

Example 3: When you throw a die, it is equally likely that you get 1, 2, 3, 4, 5, or 6. So the probability of getting 3 is $\frac{1}{6}$; and the probability of getting an even number is $\frac{3}{6}$, or $\frac{1}{2}$. Now suppose you throw two dice, one red and one blue. What can happen?

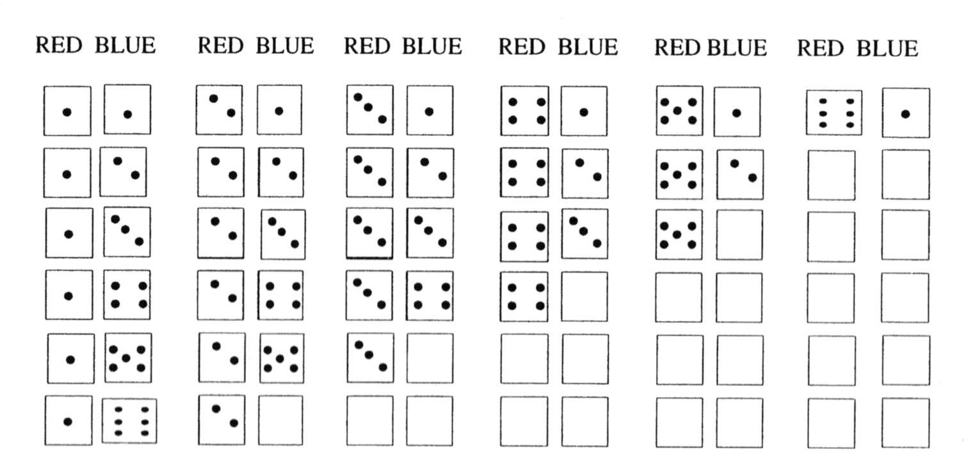

Fill in the rest of the dots to show the 36 things that can happen. All of these results have the same chance of happening — they are **equally likely**.

Of the 36 things that can happen, 4 of them give the sum 5:

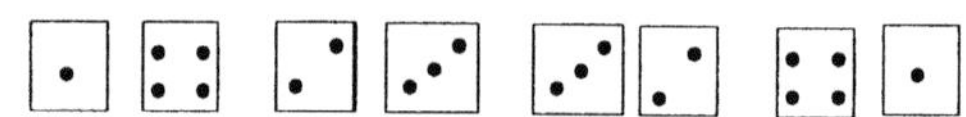

So the probability of getting the sum 5 when you throw a pair of dice is 4 out of 36 or $\frac{4}{36}$ or $\frac{1}{9}$.
What is the probability of getting the sum 8?

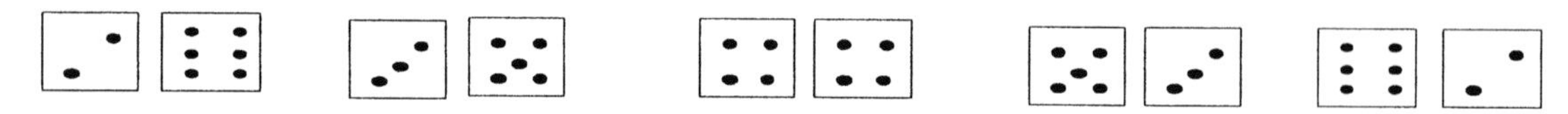

Since five pairs have the sum 8, the probability is $\frac{5}{36}$.

Now, suppose you throw a pair of dice thirty-six times. Will you definitely get a sum of 8 on five of the throws? Of course not — you could get 8 on fifteen throws or you could not get 8 at all.

So what does the probability $\frac{5}{36}$ **mean**? It means that if pair of dice is thrown 36 million times, the sum will be 8 about 5 million times.

Probability can predict what is "more likely" to happen in a particular situation for a very large number of trials. It does not tell us exactly what will happen.

Example 4: The medical histories of ten million heavy smokers (at least a pack a day for ten years or more), showed that seven million of them got lung cancer. From this large number of cases, we can say that the probability that a heavy smoker will get lung cancer is $\frac{7,000,000}{10,000,000}$ or $\frac{7}{10}$. This doesn't mean that if you pick 10 heavy smokers from among your friends and family, 7 of them will definitely get lung cancer. Maybe all of them will get it, and maybe none of them will get it. But "the chances are" that about 7 of them will get it.

Homework: Section 10

1. There are 7 Snickers, 8 Milky Ways, and 5 Almond Joys in a paper bag. If you pick one without looking, what is the probability that you will get a Milky Way?

2. As people come in to a meeting, their names are put into a box. During the meeting one name will be drawn for a "door prize." If 250 men and 170 women attend the meeting, what is the probability that a man will receive the prize?

3. When you throw a pair of dice, what is the probability of getting "doubles" (two 1's, two 2's, two 3's, etc.)?

4. If today is Monday, what is the probability that tomorrow will be Tuesday? What is the probability that tomorrow will be Friday?

5. When a contestant spins the Wheel of Fortune:
   a) What is the probability that she will win $100?
   b) What is the probability that she will win $200?
   c) What is the probability that she will lose all her money?

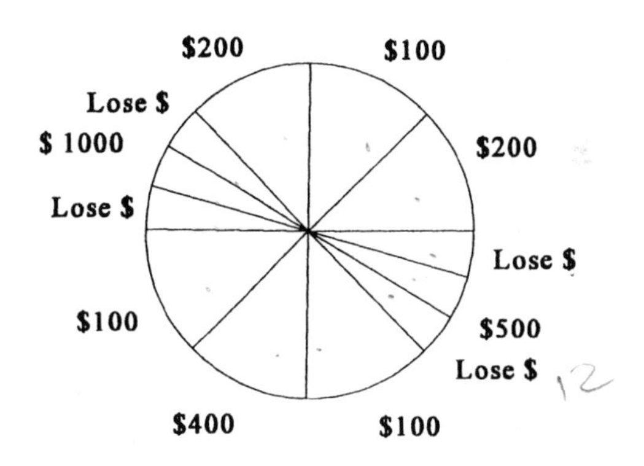

6. If the Wheel of Fortune is spun 48,000 times per year, about how many times will a contestant win $1000?

7. A life insurance company studied the medical records of 2 million men, beginning when each individual was 35 years old. 1,400,000 of the men lived to be 50 years old. Based on this study, what is the probability that a 35 year old man will live to be 50?

8. My recipe for roux is "two-thirds as much oil as flour."
   a) If I use 6 cups of oil, how much flour should I use?
   b) If I use 6 cups of flour, how much oil should I use?
   c) Could I describe my recipe as "one-and-a-half times as much flour as oil?"
   d) What is the ratio of oil to flour in my recipe?

9. A bag contains only blue balls and green balls. There are 6 blue balls. If the probability of drawing a blue ball at random from this bag is ¼, then the number of green balls in the bag is _______ ?

10. What is the probability that a randomly selected three-digit number is divisible by 5? Express your answer as a common fraction.

11. On a trip, a car traveled 80 miles in an hour and a half, then was stopped in traffic for 30 minutes, then traveled 100 miles during the next 2 hours. What was the car's average speed in miles per hour for the 4-hour trip?

12. Using only **mental** arithmetic, list these numbers in order of size, beginning with the largest.

   a) $13579 + \frac{1}{2468}$ b) $13579 - \frac{1}{2468}$ c) $13579 \times \frac{1}{2468}$

   d) $13579 \div \frac{1}{2468}$ e) 13579.2468

Answer in Journal

You are a contestant on a game show, there are three boxes with these contents: Box 1 contains two real \$100 bills and seven fake bills. Box 2 contains three real \$100 bills and ten fakes. Box 3 contains five real \$100 bills and seventeen fakes. You can draw a bill from any box. Which box would you choose? **Explain why.**

Section 11: Fractions Which Represent Ratios

A ratio is a comparison of two quantities. If there are twice as many apples as oranges in a bag, then the ratio of apples to oranges is 2 to 1.

What does it mean to say that the ratio of girls to boys in Mrs. Broussard's class is 3 to 4? There are two concrete ways to think about it:

i) The class can be separated into 7 equal groups such that 3 groups are all girls and 4 groups are all boys.

ii) The class can be separated into groups of 7 pupils, with 3 girls and 4 boys in each group.

(GGGBBBB) (GGGBBBB) (GGGBBBB) (GGGBBBB) ...

From either point of view, it's clear that $\frac{3}{7}$ of the class is girls and $\frac{4}{7}$ is boys. It's important to notice that although these fractions are determined from the ratio of girls to boys, they do not represent that ratio — they relate girls to whole class and boys to whole class. The fraction $\frac{3}{4}$, which directly represents the ratio, means that there are $\frac{3}{4}$ as many girls as **boys** in the class.

When the two quantities being compared in a ratio are measurements (rather than the result of counting), the units of measurement must be considered very carefully.

Example 1: In the triangle at right, the side labeled j is 2 inches long, and the side labeled k is 3 inches long. Since the units are the same, we can simply say that the ratio of j to k is 2 to 3 — or that j is $\frac{2}{3}$ of k.

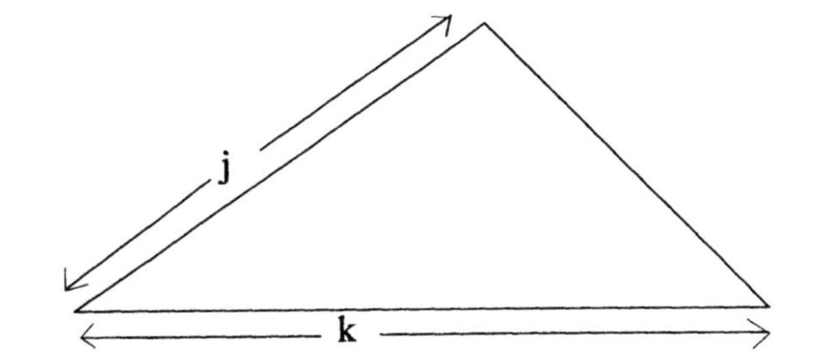

Example 2: A recipe for oatmeal cookies calls for 4 cups of flour and 3 teaspoons of baking powder. If Suzy wants to make extra cookies, and she uses 6 cups of flour, how much baking powder should she use?

The underlying idea in this situation is clearly the ratio of baking powder to flour. But there is a significant difference between this scenario and the one in Example 1 — the units for the flour and baking powder are not the same.

We can say that for every 3 teaspoons of baking powder, there must be 4 cups of flour; and we can say that the ratio of baking powder to flour is 3 teaspoons to 4 cups. But we cannot say that there is $\frac{3}{4}$ as much baking powder as flour!

Fractional notation is used to represent the ratio; but the units must be included - we write $\frac{3 \text{ teas}}{4 \text{ cups}}$.

(It is important for you to understand that this notation extends the concept of fraction beyond its **concrete** origins. One of the most important and difficult aspects of teaching elementary school mathematics is to help children see connections and make transitions from the concrete models to abstract ideas; and in order to do this, a teacher must recognize where those transitions are.)

Now, back to the question. Since 4 cups of flour requires 3 teaspoons of baking powder, then $1\frac{1}{2}$ times as much flour (6 cups) will require $1\frac{1}{2}$ times as much baking powder ($4\frac{1}{2}$ teaspoons).

In fractional notation, we multiply numerator and denominator by the same number — which results in an equivalent fraction. Or, in this case, we can say it results in the same ratio.

$$\frac{3 \text{ teas}}{4 \text{ cups}} = \frac{\left(1\frac{1}{2}\right)(3) \text{ teas}}{\left(1\frac{1}{2}\right)(4) \text{ cups}} = \frac{4\frac{1}{2} \text{ teas}}{6 \text{ cups}}$$

Example 3: The ratio of smokers to non-smokers who come to Harry's Hamburger Heaven is 2 to 7. If the restaurant serves an average of 180 people per night, about how many of them are smokers?

First, we notice that the customers can be put into groups of 9, with 2 smokers and 7 non-smokers in each group. Since there are 180 customers, there are 20 groups. And since there are 2 smokers in each group, we can estimate that about 40 smokers will come in each night.

Example 4: The ratio of American-made to foreign-made cars sold in Lafayette is 5 to 3. If 1200 foreign-made cars were sold last month, how many American-made cars were sold during that time?

All the cars sold during the month can be put into groups of 8, with 5 American-made and 3 foreign-made. Since 1200 foreign cars were sold, there are 400 groups. And since each group has 5 American cars, there were 2000 American cars sold.

Homework: Section 11

1. Five out of every twelve children at Camp Happy Days earned a Jr. Lifesaving Badge. There were 156 children at the Camp. How many badges were given out?

2. Write a base thirteen numeral to represent four thousand four hundred six horses.

3. Find the answers **mentally**: The ratio of cats to dogs brought into Elmwood Veterinary Clinic yesterday was 2 to 3.
   a) If 30 cats were brought in, how many dogs were treated?
   b) If 24 dogs came in, how many cats were treated?
   c) If 18 dogs were brought in, how many total cats and dogs came in?
   d) If 40 cats and dogs (total) came in, how many were cats?

4. Three-sixteenths of the students at Wisdom High School have not received a measles vaccine. What is the ratio of students who have received the vaccine to those who have not?

5. List all the factors of 210.

6. In a recent poll, 120 people said they take Tylenol for headaches; 90 people said they take aspirin, and 150 said they take Advil. Of the people in this poll:
   a) What fraction take aspirin?
   b) What is the ratio of those who take Tylenol to those who take Advil?
   c) If you met one of these people on the street, what is the probability that she takes Tylenol?
   d) Would it be honest for an add to say: "Five out of twelve people prefer Advil to other pain relievers?" **Explain.**

7. In Lafayette Parish, the ratio of adult women who work outside the home to those who don't, is 8 to 5. There are 18,900 women who do not have outside jobs. How many do?

8. How many two digit numbers are multiples of 6?

9. If I draw a card from a full deck, what is the probability that it will be:
   a) red
   b) a spade
   c) a queen
   d) a black jack
   e) a five or a six
   f) a face card (J, Q, K)

10. Write =, >, or < between each pair of numbers:

a) $\frac{5}{16}$ $\frac{5}{17}$ b) $2\frac{7}{15}$ $2\frac{21}{40}$ c) $\frac{3}{4}$ $\frac{57}{76}$

11. There are 85 men and 68 women at a conference
    a) Write each answer as a fraction in simplest form:
    What portion of the people are men?
    What is the ratio of men to women?
    If a name is drawn at random from among all the participants, what is the probability it will be a women's name?
    b) The participants are to be separated into equal-numbered discussion groups. What are the possibilities for these groups (how many groups and how many people in each group)?
    c) If the discussion groups are to consist of all men or all women, how many people must be in each group?

12. Write 84 as a product of three whole numbers, each bigger than 1. In how many different ways can this be done?

13. This graph compares the quantity of fiction and non-fiction books in Jonesville Public Library.
    Which of these statements are true?
    a) The ratio of fiction to non-fiction is 3 to 2.
    b) There are 1 ½ times as many fiction as non-fiction.
    c) If there are 17,892 fictional books, then whole library contains 29,820 volumes.
    d) It would be possible to arrange the whole library into groups of 70 books each, so that every group had exactly 42 fictional books.

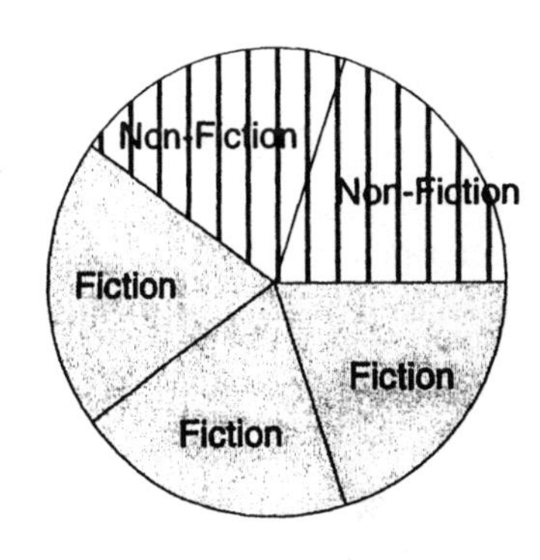

14. What is the ratio of shaded to unshaded area in each picture?

a)

b)

c)

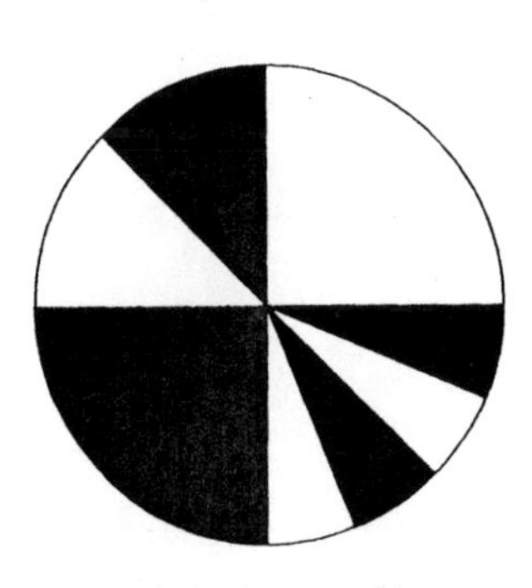

15. You know that the distance from Lafayette to Dallas is 380 miles. You don't know how far it is from Lafayette to Chicago. Using a big wall map, you measure the highway distance from Lafayette to Dallas ($4\frac{3}{4}$ inches) and from Lafayette to Chicago ($16\frac{5}{8}$ inches). How far is Lafayette from Chicago?

Answer in Journal

Explain, in a way that a 6th grader could understand, how to find the answer to Homework exercises #1, #4, and #9(e).

Section 12: Fractions Which Represent Rates

Some ratios, particularly those which involve time, are called rates: rate of travel (miles per hour); heart rate (beats per minute); rate of production (cars per month); salary (dollars per week), etc. There are also many common rates which don't involve time: rate of gas consumption (miles per gallon); gaining ground in football (yards per carry); nutritional value of food (calories per gram); etc.

Sometimes we are interested in what's happening at a particular moment or for an **isolated** measurement: I'm traveling right now at 68 miles per hour or, my pulse rate right now is 82 beats per minute.

Some rates are **constant**: a machine sorts 135 envelopes every 15 minutes; butter provides 100 calories per tablespoon; an A earns 4 quality points per semester hour.

Constant rates have an important graphic significance. For example, the machine that sorts 135 envelopes every 15 minutes will sort 270 envelopes in a half-hour, 540 envelopes in an hour, 1080 envelopes in 2 hours, etc. When this data is plotted on coordinate axes, the result is a **line** and the rate of the machine is the **slope** of the line.

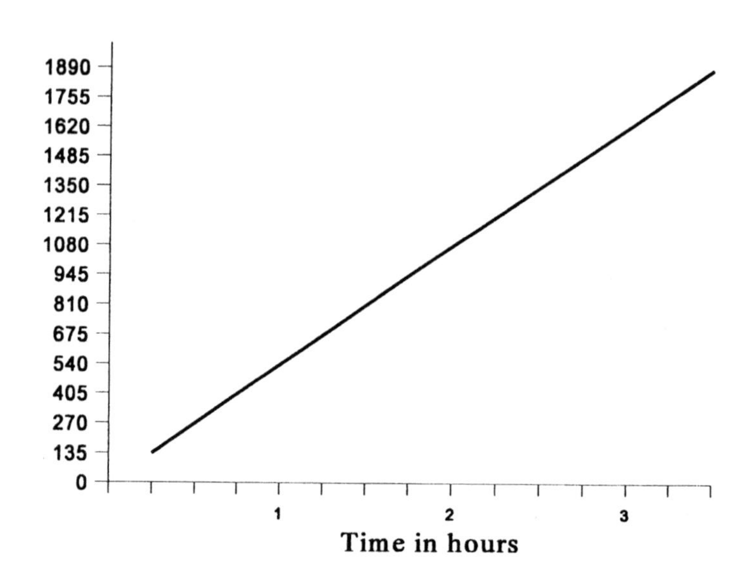

Many familiar rates are neither isolated nor constant — they are **average** rates. For example:

i) If Suzy drives for 4 hours and she travels 260 miles, then her average speed (rate) is 65 miles per hour. That doesn't mean she drove 65 miles per hour for the whole trip.

ii) If Tom scores a total of 140 points in the first ten games of the season, then his scoring rate (average) is 14 points per game - but he may not have scored exactly 14 points in any of the games.

Rates are described in three different ways:

1. a number with a verbal description of rate:

   70 words per minute, $3\frac{1}{2}$ cups per recipe, $4.85 per hour

2. a number with fractional description of rate:

   $$72\ \frac{\text{mi}}{\text{hr}},\quad 8\frac{1}{2}\ \frac{\text{cal}}{\text{gram}},\quad 26.3\ \frac{\text{mi}}{\text{gal}}$$

   [Note: fractional rates are **read** the same as verbal rates: miles per hour, calories per gram, miles per gallon.]

3. fraction with labels in numerator and denominator

   $$\frac{28\ \text{laps}}{43\ \text{min}},\quad \frac{68\ \text{earned runs}}{25\ \text{games}},\quad \frac{2976\ \text{bottles capped}}{12\ \text{machines}}$$

Example 1: If Sally drives 145 miles in $2\frac{1}{2}$ hours, then her average speed (rate) can be written as

$$\frac{145\ \text{mi}}{2\frac{1}{2}\ \text{hr}} \quad \text{or} \quad 58\ \frac{\text{mi}}{\text{hr}} \quad \text{or} \quad 58 \text{ miles per hour.}$$

Example 2: If Joe types 70 words per minute, how many words will he type in 3 hours?

Since 3 hours is 180 minutes, Joe will type 180 x 70 words in 3 hours.

Example 3: The drilling rate for a machine is $\frac{92\ \text{holes}}{15\ \text{minutes}}$. How long will it take the machine to drill 460 holes?

First, we find how many groups of 92 there are in 460. This is a division question, and the answer is 5. Therefore the job will take 5 fifteen-minute intervals — which is 75 minutes, or $1\frac{1}{4}$ hours, or 1 hour and 15 minutes.

Homework: Section 12

1. If a machine assembles 45 pairs of shoes in 18 minutes, what is its rate per minute? per hour? per 8-hour day? per 40-hour week? What kind of rate is this (isolated, constant, or average)?
2. Joe counted 7 pulse beats in 12 seconds. What was his pulse rate per second? per minute? What kind of rate is this?
3. A secretary types 80 words per minute. How long will it take him to type a 1500-word essay?
4. A company produces 300 electric irons in each 8-hour day. Yesterday, the company was shut down for 2 hours because of a power failure. How many irons were produced yesterday?

5. Name the three largest factors of 3522.
6. A basketball team has made 119 out of 140 free throws so far this season. At this rate, how many free throws would the team have to shoot in order to make 17 of them?
7. Bill's car goes 180 miles on 12 gallons of gas. If gas costs 85 cents per gallon, how much will Bill pay for gasoline on a 1620-mile trip?
8. Name three even numbers which are multiples of 29. Are there any odd numbers which are multiples of 42? Explain.
9. The monorail at Disney World travels the 8 miles from Epcot to the Magic Kingdom in 14 minutes. How long does it take to travel the 10 miles from Epcot to Disney Village? (Assume constant rate).
10. A hospital patient is to receive 4 mg of tetrohydrate for each 25 pounds of body weight. If the patient weighs 168 pounds, how much of the drug should he receive? (Round to the nearest mg.)

11. For each graph, answer these questions:
    a) What is the ratio of lions to tigers?
    b) What percent of the large animal population is lions?

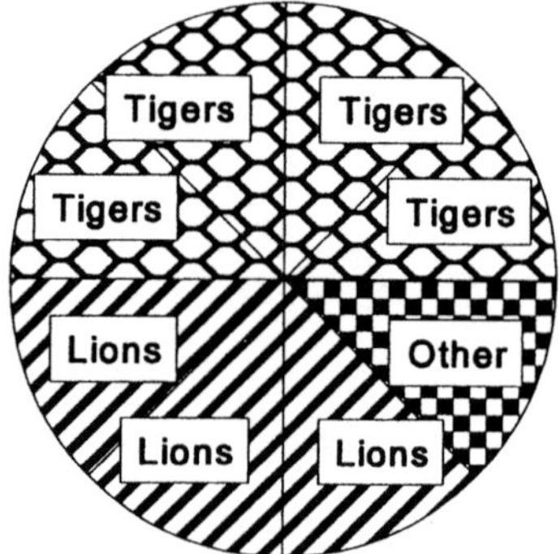

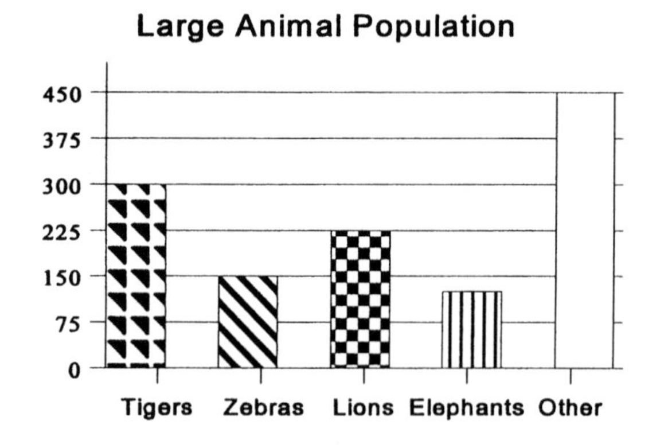

12. The graph shows the rate of stretch for a spring when force is applied to it. What is the rate of stretch in centimeters per pound?

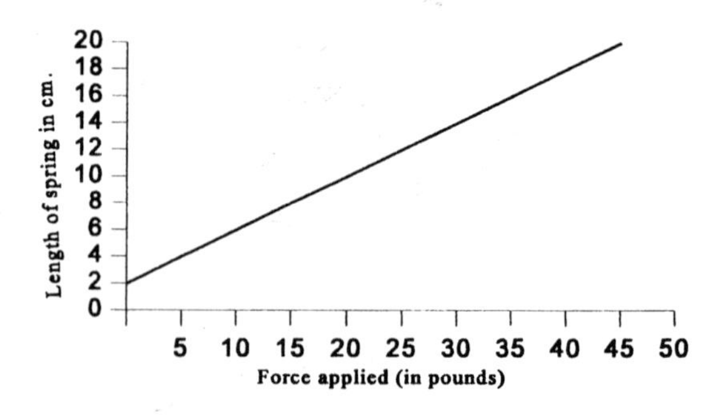

Answer in Journal

This question relates to examples (i) and (ii) on page 38. Please reread them. We noted that it was possible for Tom's scoring rate to be 14 points per game even though he had not scored exactly 14 points in any of the games. Is it possible for Suzy's average driving speed to be 65 miles per hour even though she never drove at exactly 65 miles per hour during the trip? Explain.

Section 13: Meaning of Decimals

The meaning of decimals must be developed from a real understanding of place value and fractions. Decimal notation is an extension of place value based on ten. The decimal point indicates that succeeding digits represent tenths, hundredths, thousandths, etc. So obviously, a person must know what these fractions mean in order to know what the corresponding decimals mean.

The key to understanding decimals and their relationship to fractions and mixed numbers, is **reading** them correctly. Decimals and their fractional or mixed number equivalents are read the same way. For example:

a) .057 and $\frac{57}{1000}$ are both read "fifty-seven thousandths."

b) $2\frac{3}{10}$ and 2.3 are both read "two and three tenths."

c) 1.2053 and $1\frac{2053}{10000}$ are both read "one and two thousand fifty-three ten-thousandths."

Teaching Note:

Elementary school children should not be allowed to use the word "point" in reading decimals. This shortcut is wonderful for people who already understand the meaning of decimals, but it is completely inappropriate in the sixth and seventh grades when students are just beginning to develop an understanding of these numbers.

Pupils must notice two things:

1) The word "and" is very significant in reading both decimals and mixed numbers - it separates the whole number from the fractional part. This is the **only** correct use of "and" in reading numbers.
2) The name of the fractional part of a decimal number (hundredths, ten-thousandths, etc.) is the place value of the last digit. It also indicates the denominator of an equivalent fraction.

If elementary pupils are frequently asked to read decimals, fractions, and mixed numbers, they will discover that the number of digits behind the decimal point corresponds to the number of zeros in the denominator of the equivalent fraction. Let them discover this themselves — it will be much more meaningful than if we give it to them as a rule.

Equivalence and Ordering

Zeros, which are behind a decimal's last non-zero digit, can be added or removed without changing the decimal's numerical value. This can easily be explained by referring to fractions.

For example: .732 is the same number as $\frac{732}{1000}$ and $\frac{732}{1000} = \frac{7320}{10000} = \frac{73200}{100000} = \ldots$

So $.732 = .7320 = .73200 = \ldots$

Note: For decimals which represent **measurements**, these "tail-end" zeros have an important meaning. Although they do not affect the numerical value of the number, they indicate the **precision of the measurement** and cannot be added or deleted. In this context, these zeros are called **significant digits**.

Two decimals are equal only if they are identical or they can be made identical by adding or removing tail-end zeros.

How do we determine which of two unequal decimals is greater?

i) If they are both tenths or both hundredths or both thousandths, etc., then they are the same kind of creatures - and we compare them as we would whole numbers. For example:
   a) $.41 > .39$ because 41 hundredths is greater than 39 hundredths
   b) $5.2468 < 5.2471$ because 2468 ten-thousandths is less than 2471 ten-thousandths

ii) If they have different numbers of decimal digits, we simply add tail-end zeros to the shorter one so that they are the "same kind of creatures." For example:
   a) $.35 > .347$ because 350 thousandths is greater than 347 thousandths
   b) $7.26 < 7.2604$ because 2600 ten-thousandths is less than 2604 ten-thousandths

Relating Fractions and Terminating Decimals

As we noted earlier, writing a decimal as an equivalent fraction requires nothing more than understanding how to read these numbers. However, the reverse process of translating a fraction to a decimal requires some thinking and calculation.

Fractions which can be written with denominator 10, 100, 1000, etc., are easily changed to decimal notation. For example:

a) $\frac{3}{5} = \frac{6}{10} = .6$

b) $\frac{11}{20} = \frac{55}{100} = .55$

c) $\frac{19}{250} = \frac{(4)(19)}{(4)(250)} = \frac{76}{1000} = .076$

d) $\frac{13}{16} = \frac{(625)(13)}{(625)(16)} = \frac{8125}{10000} = .8125$

How can we tell whether a particular fraction in simplest form can be written with denominator, 10, 100, 1000, etc.? We can tell by finding the prime factors of the denominator of the fraction. If the denominator has any prime factor other than 2 and 5, then the fraction cannot be written with denominator 10, 100, 1000, etc.

For example, $\frac{5}{6}$ cannot be written with denominator 10, 100, 1000, etc., because $\frac{5}{6}$ is in simplest form and none of the powers of 10 is a multiple of 6. All multiples of 6 have 3 as a factor — and the only prime factors of powers of ten are 2 and 5.

Similarly, $\frac{18}{35}$ cannot be written with denominator 10, 100, 1000, etc., because $\frac{18}{35}$ is in simplest form and 7 is a factor of 35. Therefore, any fraction which is equivalent to $\frac{18}{35}$ must have a multiple of 7 as its denominator.

Fractions such as $\frac{5}{6}$ and $\frac{18}{35}$ which cannot be written with denominator 10, 100, 1000, etc., are equivalent to infinite decimals. These will be discussed later in the course when we consider the operation of division.

Homework: Section 13

1. Read each of these numbers. (Don't use the word "point".)

   a) $9\frac{5}{23}$ b) 1.073   c) .0089   d) 4827.961

2. Write each decimal as a fraction or mixed number.

   a) 7.93   b) .2051   c) 4.007   d) .69381

3. Put >, <, or = between each pair of numbers.

   a) 2.961   2.97   b) $\frac{14}{25}$ .558   c) 1.02   .985

4. List the prime factors of each number:

   a) 375   b) 128   c) 245   d) 162   e) 55,000

5. Write each fraction as an equivalent fraction with denominator 10,100,1000, etc., if possible.

   a) $\frac{7}{32}$ b) $\frac{18}{75}$ c) $\frac{9}{14}$ d) $\frac{8}{27}$ e) $\frac{21}{875}$

6. List all the fractions which are equivalent to $\frac{12}{21}$.

7. How many factors does 54 have?

8. Give three multiples of 48.

9. List the prime factors of 630.

10. What is the smallest positive number which is a multiple of both 39 and 108?

11. What is the biggest number which is a factor of both 112 and 126?

12. If you want to write $\frac{3}{8}$, $\frac{11}{42}$, and $\frac{31}{36}$ all with the same denominator, what is the smallest denominator you could use? Name two other denominators you could use?

13. Write each number as a decimal: a) $\frac{137}{125}$ b) $\frac{63}{72}$ c) $\frac{171}{40}$ d) $3\frac{38}{95}$

14. For each number, what must it be multiplied by to get a power of ten as a product?
    a) 80 b) 625 c) $2^5 \cdot 5^3$ d) 2^7 e) 5^8

15. Electricity costs 6 cents per kilowatt-hour.
    a) What does the prefix "kilo" mean?
    b) How many watts is 3.87 kilowatts?
    c) How many kilowatts is 76 watts?
    d) What does "kilowatt-hour" mean?
    e) If you burn a 100-watt bulb for 24 hours, how many kilowatt-hours is that? How much will it cost?
    f) How much will it cost to use a 1200-watt hair dryer for 20 minutes?

Answer in Journal

1. Explain what a prime number is.
2. A brownie recipe calls for 3 cups of sugar and 2 eggs. You want to increase the recipe to use 5 eggs. How much sugar should you use? Explain how to find the answer **mentally**.
3. Would it be convenient to increase the recipe by using 4 cups of sugar? Explain.

Section 14: Meaning of Percent

The key to understanding and using percents is knowing that the word **percent** means **hundredths**. (Our familiar percent symbol evolved from "÷ 100".) This establishes a relationship among fractions, decimals, and percents which allows us to write a particular number in several different ways. For example, all of these names and symbols represent the same number:

28% 28 percent 28 hundredths $\frac{28}{100}$ .28

In any given situation, we can choose the representation which is most convenient.

For each of these, notice how various ways of representing the number are related:

a) 100%

100 hundredths

$\frac{100}{100}$

1

b) .73

73 hundredths

$\frac{73}{100}$

73%

c) 1.45

1 and 45 hundredths

$\frac{145}{100}$

100 hundredths and 45 hundredths

145 hundredths

145%

d) $\frac{11}{20}$

$\frac{55}{100}$

55%

.55

e) 12 ½%

12.5%

$\frac{12.5}{100}$

$\frac{125}{1000}$

.125

Note: In translating between decimals and percents, it's helpful to focus on the hundredths represented in the decimal.

f) .762 is 76 (and a little extra) hundredths, or 76 (and a little extra) percent. So we write 76.2%

g) $12\frac{1}{2}$ % or 12.5% is 12 (and a little extra) percent. So the decimal is .12 (and a little extra). We write .125.

These translations are relevant in a variety of very common-place situations, and they emphasize the relationship among fractions, decimals, and percents. However, they are not necessarily the most **efficient** way to find the answer to a question.

Example 1: A Senate committee consists of 11 Democrats and 9 Republicans. What percent of the committee is Republican?

First, notice that $\frac{9}{20}$ of the committee is Republican. So to answer the question, we can easily translate $\frac{9}{20}$ into a percent: $\frac{9}{20} = \frac{45}{100} = 45\%$

Example 2: In one season, Michael Jordan made 572 out of 650 free throws. What percent is this?

This question can be answered by making these translations:

$$\frac{572 \text{ made}}{650 \text{ attempted}} = \frac{22 \text{ made}}{25 \text{ attempted}} = \frac{88 \text{ made}}{100 \text{ attempted}} = 88\%$$

However, it is much more efficient to divide 572 by 650: $572 \div 650 = .88 = 88\%$

Example 3: Sixty percent of the children at the Tennis Camp are boys. There are 16 girls at the Camp. How many campers are boys?

Since 60% of the campers are boys, we know that 40% are girls. We also know that $40\% = \frac{40}{100} = \frac{4}{10} = \frac{2}{5}$. Therefore $\frac{2}{5}$ of the campers are girls. But we are told that there are 16 girls; so $\frac{2}{5}$ of the campers is 16. This means that $\frac{1}{5}$ of them is 8, and $\frac{3}{5}$ of them is 24. So 24 campers are boys.

Example 4: There were 40 students in a class and 7 of them made A`s. What percent of the class made A`s?

The answer can be found by making these translations: $\frac{7}{40}$ of the class made A`s.

$\frac{175}{1000}$ of the class made A`s.

.175 of the class made A`s.

17.5% of the class made A`s.

But again, division is much more efficient: $7 \div 40 = .175 = 17.5\%$

The terms percent increase and percent decrease are commonly used in a variety of real-world situations. These quantities are usually represented as fractions,

$$\text{Percent increase or decrease} = \frac{\text{amount of change}}{\text{original quantity}};$$

and then the fractions are translated into percents.

Example 5: Last year a house was appraised for $95,000, and this year it was appraised for $98,800. What percent increase is this?

$$\text{Percent increase} = \frac{\text{amount of change}}{\text{original quantity}} = \frac{3800}{95000} = .04 = 4\%$$

Example 6: The enrollment at Plato High School dropped from 1600 to 1580. What percent decrease is this?

$$\text{Percent decrease} = \frac{\text{amount of change}}{\text{original quantity}} = \frac{20}{1600} = .0125 = 1.25\%$$

Homework: Section 14

1. Write each number as a percent:
   a) .06 b) 1.35 c) .276 d) .0844 e) 7.3
   f) $\frac{21}{12}$ g) $\frac{7}{8}$ h) $\frac{11}{125}$ i) $6\frac{1}{4}$

2. Write each number as a decimal:
   a) 48% b) 12 ½ % c) 280% d) 15.7% e) .3%

3. Write <, >, or = between each pair of numbers:
   a) 42% $\frac{21}{50}$ b) $\frac{9}{25}$ 38% c) $\frac{2}{3}$ 67%
   d) 255% $\frac{7}{3}$ e) $7\frac{1}{4}\%$ .073

4. Write each of these in words: a) .0083 b) 17.6% c) 9,000,006,005 d) 3.007

5. If a pair of dice is thrown 3000 times, about how many times will the sum be 7?

6. Write a base seventeen numeral to represent one thousand one hundred seventy-one.

7. List all the factors of 144 which are multiples of 3.

8. Write $\frac{21}{75}$ as a decimal.

9. How many things does this numeral represent? 20144_{five}

10. Write $\frac{81}{2^5 \cdot 3^3 \cdot 5^9}$ as a decimal.

11. The ratio of length to width of a rectangle is 5 to 3. The perimeter is 32 cm. What is the area?

12. There were 164 women and 246 men entered in the U.S. Open this year. What percent of the players were women?

13. Of the 104,000 seniors who took the ACT last year, 8½% of them scored above 28. How many seniors scored above 28?

14. Pete scored 70% on a test. He answered 35 questions correctly. How many questions were on the test?

15. Mary runs 7 laps around her yard in 12 minutes.
    a) What is her rate in laps per minute?
    b) What is her rate in laps per hour?
    c) How long will it take Mary to run 21 laps?
    d) How many laps can Mary run in 2 minutes?

16. The grading scale at Jones Junior High is A(93 - 100), B(85 - 92), C(75 - 84), D(70 -74), F(below 70). Final averages in Mrs. Boudreau's class were:

    89, 72, 54, 97, 77, 92, 85, 74, 75, 63, 84, 78, 71, 80, 90.

    What percent of the students received a grade of C?

17. Last month Joe paid $12 finance charge on his $800 unpaid Visa balance. Joe can get an $800 loan from the bank for 12½% annual interest. Should he get the bank loan and pay off his Visa balance? Explain your answer.

18. The gauge on a coffee maker shows there are 35 cups left when the coffee maker is 28% full. How many cups of coffee does it hold when it is full?

19. Suzy made twelve of thirty shots in the first three games of this basketball season. In her next game, she took ten shots and made six of them. What was her shooting average (percentage of shots made) after four games?

20. This year taxpayers are allowed to reduce their income tax by $2\frac{1}{4}\%$. If Ms. Thibodeaux's tax comes to $6,700 before reduction, how much tax must she pay?

21. Andre bought a TV set for $403. This was 65% of the regular price. What was the regular price of the TV set?

22. The enrollment at the University was 15,500 last fall, and it is 16,120 this fall. What percent increase in enrollment is this?

23. A farmer planted 3000 tomato seedlings. Of these 2520 grew to maturity. What was the farmer's percent loss of seedlings?

24. Of all the people who responded to a survey, 390 were under 18 years of age. If 20% were under 18, how many people responded to the survey?

25. The ratio of the length of a side of Square A to the length of a side of Square B is 3 to 2. What is the ratio of the area of Square A to the area of Square B?

26. For each figure, what percent is shaded?

a)

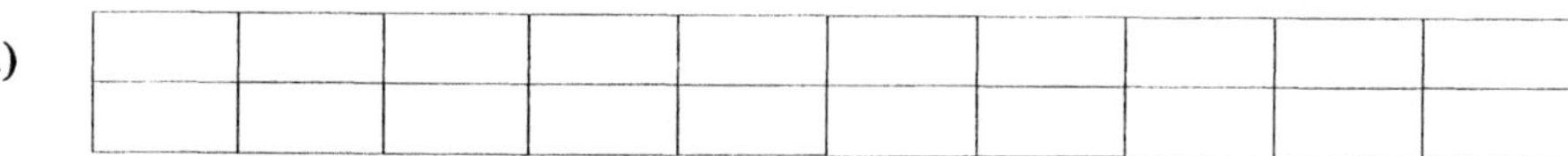

b) c)

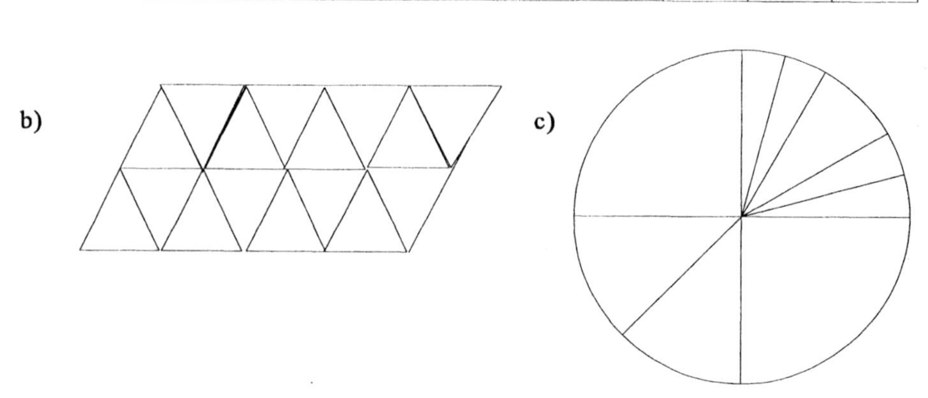

Section 15: Opposites

Concrete Meaning of Opposites

There are many quantitative situations which involve opposites of some kind — above or below zero on a thermometer; assets or debts in bookkeeping; points won or lost in a game; etc. In each of these situations, whole numbers (and fractional or decimal parts of whole numbers), are used to represent one of the quantities. The opposite quantity must be represented by a **different kind of number**. These **opposites,** of whole numbers, fractions, and decimals, are called **negative numbers**. They are denoted by the familiar negative sign, and they are read as "negative 3" or "the opposite of 3" (- 3), "negative 2.7" or "the opposite of 2.7" (- 2.7), etc.

The set of whole numbers and their opposites is called the set of **integers**: ..., -2, -1,0, 1, 2, 3, ...

Example 1:

a) 6.3 and -6.3 are opposites

b) The opposite of -8 is 8. (Notice that this can be written as -(-8) = 8)

Example 2: If the mercury in a thermometer stands at 40 spaces above 0, we say the temperature is 40°; if the mercury stands at 10 spaces below 0, we say the temperature is - 10°.

Example 3: Level 7 of a parking garage is seven floors above ground; Level -2 is two floors below ground.

Example 4: If ABC Company makes a profit of two million dollars on a merger, this is recorded in the Company records as $2,000,000. If the Company loses a half million dollars on a real estate deal, this is recorded as -$500,000.

Example 5: Jim and Mary are playing a game. On Jim's first turn, he loses 7 points; and on Mary's first turn, she loses 10 points. Jim's score is now -7 and Mary's score is -10. Who is ahead in the game? Clearly, Jim is ahead. (Notice that this explains, in a very concrete way, why -7 is greater than -10.)

Opposites and Absolute Value on the Number Line

As we move beyond specific, concrete situations, the concept of opposite is represented in a more general way by using a number line. Each integer, fraction, mixed number, and decimal corresponds to a point on this line.

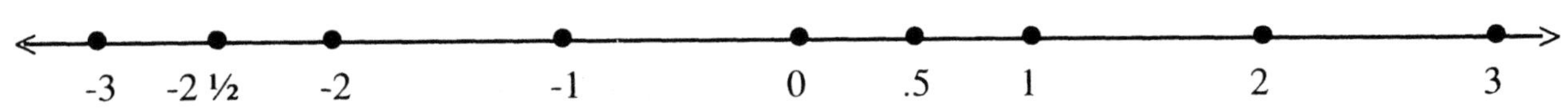

Notice that numbers increase from left to right, so we can tell which of two numbers is greater by comparing their positions on the number line.

The **absolute value** of a number is its distance from zero on the number line. It's obvious that a number and its opposite are the same distance from 0 — hence they have the same absolute value. We indicate absolute value by the symbol, | |.

Example 1:

a) $|6|$ is read "the absolute value of 6," and it indicates how many spaces 6 is from 0 on the number line.

b) $|-5|$ is read "the absolute value of negative five," and it represents the distance between -5 and 0 on the number line.

Example 2: All of the following statements mean the same thing.

a) The distance of -7 from 0 is 7 spaces.

b) The absolute value of -7 is 7.

c) $|-7| = 7$

Homework: Section 15

1. Name the opposite of each number: a) $3\frac{2}{7}$ b) 45% c) -65 d) 1.248 e) 0

2. What two numbers are each 5 spaces from 0 on the number line?

3. What other number has the same absolute value as 2.7?

4. Draw a number line and locate each of these points: - 3, 75%, $2\frac{2}{3}$, 5, $-\frac{3}{2}$, $-\sqrt{4}$

5. Write >, <, = between each pair of numbers:

   a) $-\frac{3}{8}$ $\quad$ $-\frac{1}{3}$ $\qquad$ b) $-\frac{15}{7}$ $\quad$ - 210%

   c) - 9.873 $\quad$ - 9.8729 $\qquad$ d) $-\frac{5}{12}$ $\quad$ $-\frac{3}{7}$

6. If $|j| = 3$, what are all the possible values of j ?

7. Which of these are integers? 0, $\frac{5}{9}$, $|-3|$, 18%, $-(-4)$, $\frac{12}{3}$, $\sqrt{9}$

8. **Read** each of these (don't use the words "minus" or "negative"):

   a) - 3 $\qquad$ b) $-5.2 < 0$ $\qquad$ c) $\left|\frac{3}{5}\right|$ $\qquad$ d) $12\frac{5}{16}\%$ $\qquad$ e) $-8 < -1$

9. What is the smallest whole number? What is the smallest integer?

10. How many spaces is 27 from -8 on the number line?

11. Name an integer between -16 and -13. Name two fractions between $\frac{1}{3}$ and $\frac{1}{2}$.

12. Out of 860 people who completed a survey form, 129 of them said they prefer tea to coffee. What percent of these people prefer tea?

13. When a pair of dice is thrown, what is the probability that the sum will be 5?

14. Measurements of air temperature at various altitudes above New Orleans were all taken at the same time. The results are shown at the right.
   a) Draw a graph to represent this data.
   b) What is the rate of decrease in temperature as altitude increases?
   c) What temperature would have been recorded at 1,500 ft? At 14,000 ft?
   d) At what altitude would the temperature have measured 78°F? At 60°F?
   e) What is the slope of your graph?

Altitude	Temperature
200 ft	86° F
2,800 ft	82° F
6,700 ft	76° F
11,900 ft	68° F
21,000 ft	54° F

15. The ratio of the length to the width of a rectangle is 7 to 4. The area of the rectangle is 252 sq. inches. What is the perimeter of the rectangle?

Answer in Journal

1. Explain in your own words what a negative number is.
2. Give a concrete explanation of why -2 > -3.
3. Explain why the absolute value of a number can never be negative.

Chapter I Review

Part I.

1. Write a base three number to represent four hundred apples.

2. If the ratio of green to red marbles in a bag is 2 to 3, what percent of the marbles are red?

3. Write each number as a fraction with denominator 35, if possible:

   $$\frac{21}{15},\ 40\%,\ 0.6,\ \frac{15}{9},\ \frac{28}{49}$$

4. How many **things** does this numeral represent? 795_{twelve}

5. I have only oak and pecan trees in my yard. The ratio of oaks to pecans is 7 to 5. True or False:

   a) I have $\frac{5}{7}$ as many pecans as oaks.

   b) $\frac{7}{5}$ of my trees are oaks.

   c) If I have 36 trees, 15 are pecans.

   d) I have $1\frac{2}{5}$ as many oaks as pecans.

   e) If I have 21 oaks, then I have 6 more oaks than pecans.

   f) I have 40% more oaks than pecans.

   g) I have 40% fewer pecans than oaks.

   h) If I have 60 trees, then I have 10 fewer pecans than oaks.

   i) If I cut down 1 oak and 1 pecan, the ratio will remain the same.

   j) If I cut down half the oaks and half the pecans, the ratio will remain the same.

   k) If I cut 20% of the pecans, what will be the ratio of oaks to pecans?

6. Write 156 as the product of three whole numbers, all greater than 1.

7. Name four divisors of 96.

8. Is 3774 divisible by 12?

9. Can 3774 be divided by 12? If so, what is the quotient?

10. How many factors of 72 are bigger than 5?

11. Name three multiples of 14.

12. What is the biggest number which is a divisor of both 54 and 90?

13. What is the biggest factor of 2876?

14. What number is a divisor of every whole number?

15. Write 4741 as a product of primes, if possible.

16. Find these products and quotients mentally:

    a) 70×8000 b) $39000 \div 130$ c) $(4000 \times 300) \div 60$

17. Which of these numbers are multiples of 46? 2, 92, 0, 1, 46, 234

18. Which of these are prime numbers? 323, 1061, 2361

19. Farmer Brown brings 1832 eggs to market. When they are put into cartons of one dozen each, how many eggs are left over?

20. What is the smallest non-zero number that has both 14 and 35 as factors?

21. What does $2^2 \times 3 \times 11$ have to be multiplied by to get $2^4 \times 3^2 \times 7 \times 11$?

22. Is $3^2 \times 7^2 \times 19$ a factor of $3^4 \times 7 \times 19^3 \times 23$?

23. Which prime numbers are divisors of 8000?

24. Is 0 even or is it odd? How do you know?

25. Mentally determine which of these numbers are multiples of 36:

    a) 2×3^4 b) $2^3 \times 3^2 \times 7$ c) $2^2 \times 3^2 \times 7$ d) $2^2 \times 3^2 \times 19 \times 41^3$

26. A machine assembles 210 switches per hour. How many of these machines would be needed to assemble 35 switches per minute?

27. From 1:55 am to 2:19 am is what fraction of an hour? What fraction of a day? What fraction of a week?

28. If two dice are rolled, what is the probability that the sum of the numbers is divisible by 3?

Part II. Explain how to find the answer to each of these questions **mentally**.

1. Sue missed 12 out of 80 questions on a test. What percent of the questions did Sue get right?

2. The ratio of Democrats to Republicans on a 30-member committee is to be 3 to 2. How many Republicans must be on the committee?

3. On his motorbike, Mike rides 7 miles in 16 minutes. If he starts out at 9 am, what time will he get to Lake Sunshine, which is 35 miles away?

4. If a contestant on a game show spins the wheel shown at right, what is probability that she will win $500 or more?

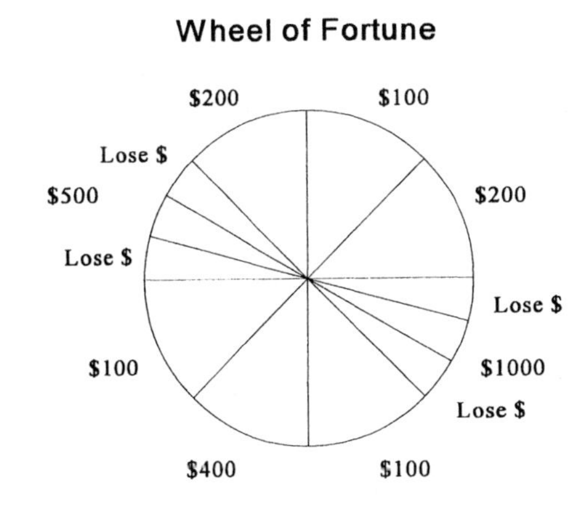

5. Joe made 79, 86, 74, and 77 on his first four Math tests. Determine <u>mentally</u> what he must make on the fifth test in order to have an 80 average.

6. Twenty-five percent of 16 is equivalent to one-half of what number?

Part III. Explain each of these in terms that a sixth-grader could understand.

1. What does $\frac{17}{56}$ mean?

2. What does it mean to say that the ratio of boys to girls in Mrs. Smith's class is 5 to 4?

3. Suppose two fractions have the same numerator but different denominators. Explain how you can tell which fraction is larger.

4. Suppose you need to write a decimal as a mixed number. Explain what to do.

5. Suppose you need to write .237 as a percent. Explain how you know what percent it is without talking about "moving the decimal."

6. Suppose you read that, in Lafayette, the probability that a particular person will be robbed during the next 12 months is $\frac{1}{9}$. What does that mean?

7. Explain how to determine whether two fractions are equal; and if they are not equal, which is greater.

Part IV.

1. Write 8.4% as a fraction in simplest form.

2. A rule-of-thumb is that a one-year-old weighs 400% of its birth weight. Using the rule-of-thumb, what should a one-year-old weigh if its birth weight was $7\frac{1}{2}$ pounds?

3. At the beginning of a trip, the odometer on a car read 63674.2 and the gas tank was full. During the trip, 10.3 and 11.4 gallons of gasoline were purchased. At the end of the trip, the odometer reading was 64283.8 and 6.2 gallons were needed to fill the tank. How many miles per gallon did the car use during the trip? (Round to tenth of a mile.)

4. The average of ten scores is 91. When the lowest score is dropped, the average of the remaining nine scores is 94. What is the value of the score which was dropped?

5. There are 185 women and 65 men working for a certain company. What percent of the employees are men?

6. The scale on a map reads: 1 cm = 60 mi. If the road between City A and City B is $2\frac{1}{4}$ cm on the map, and you are averaging 45 miles per hour, how long will it take you to drive from City A to City B?

7. A work crew of four people requires four hours and forty minutes to do a certain job. How many minutes will it take a crew of seven people to do the same job if everyone works at the same rate?

8. A sofa is advertised at "30% off the regular price." If the sofa is selling for $343, what is its regular price?

9. The Polo Shop is offering a 35% discount on all shirts in stock. In addition to this, 10% is deducted from the sale price if the buyer pays cash. Harry paid cash for a shirt which was marked $30 before the sale. How much did he pay for the shirt?

10. It is estimated that the ratio of New Yorkers who ride the subway to those who don't is 5 to 3. If there are nine million six hundred forty thousand New Yorkers, how many of them ride the subway?

11. Terry bought a car for 15% off the sticker price. If he paid $8,160 for the car, what was the sticker price?

12. It is 220 miles from Lafayette to Shreveport. The trip is all on I-49 except a 20 mile stretch through Alexandria. You know from experience that you can average only 50 mph along this road. What must be your average speed on I-49 in order to make the whole trip in 3½ hours?

13. "Tank" Jackson signed a pro contract for four and a half million dollars over the next five years. What will his monthly salary be if 35% income tax is deducted?

14. A recipe for making 50 cookies calls for 3 eggs and 4 cups of flour. If you use only 2 eggs, how much flour should you use? How many cookies will you get?

15. A large manufacturer offers a ⅔% discount to any customer who pays his bill within 15 days. If a business owes $720,000, how much would it save by paying its bill promptly?

16. The national budget for next year is one trillion eight hundred sixty billion dollars. If one-fifth of the budget is spent on defense, and one-third of all defense spending is for weapon production, how much will be spent next year for weapon production?

17. Over the past five years the school enrollment in the Central School District went from 80,000 to 92,000. What percent increase is this?

18. Eighty percent of the USL faculty are male and ⅔ of the male faculty are married. If there are 570 people on the USL faculty, how many of them are married men?

19. This year Sally's books cost $46.44. This is a 20% increase over last year's cost. How much did she pay last year?

20. Two identical bottles are filled with salad dressing. The ratio of oil to vinegar in one bottle is 3 to 2, and the ratio of oil to vinegar in the other bottle is 2 to 1. If the contents of the two bottles are poured into one big bottle, what is the ratio of oil to vinegar in the final mixture?

21. Only 20% of New York workers drive to work. Of those who do not drive, 3/16 ride bikes. What percent of New York workers ride bikes to work?

22. Between 50 and 100 books are stored on a shelf. Exactly 20 percent of them are textbooks. Exactly one-seventh of them are novels. Can the exact number of books on the shelf now be determined? Explain how you got you answer.

23. After a drop of 15% in the depth of water in the floodway, the depth gauge read 37.5 feet. What was the original depth? (Round to the nearest tenth foot.)

24. Proposed State Budget.
    a) What fraction of the Budget is for Education and Services combined?
    b) If the Budget is 1.5 billion dollars, how much will be spent for Welfare? for Services ? for Highways?
    c) What percent of the Budget is for Environment?
    d) If the Budget is one and a quarter billion dollars, how much more will be spent for Services than for Education?

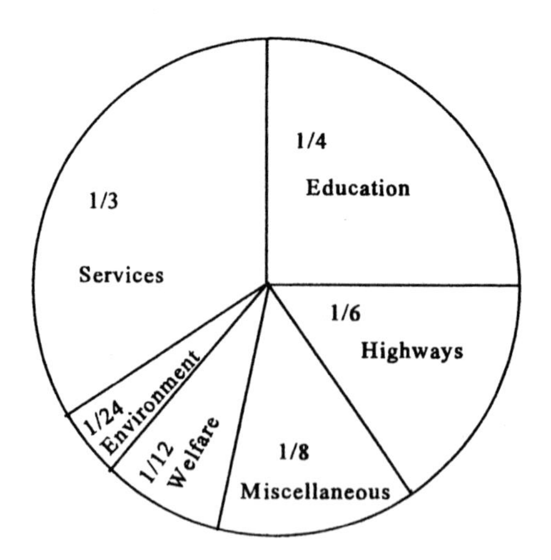

Chapter II

Section 16: Understanding Addition

Combining Things That are Alike

In the primary grades, children learn the meaning of addition in a concrete way by combining groups of things. "Three plus four" is found by putting 3 apples and 4 apples together, then counting to find that the sum is 7 apples. Similarly, if there are 2 pencils in one box and 6 pencils in another box, when I put them all together, I have 8 pencils. If two sets each contain the same kind of things, then I can express the sum as **one number** of that kind of thing. But, suppose I have 4 Snickers and 3 Milky Ways. When I put them all together, what do I have? One obvious answer is that I still have 4 Snickers and 3 Milky Ways. But I can also express the sum as a single number of **candy bars.** In this case, I recognize something common about the two types of things, and I relabel each of them in order to get a combined total. This, of course, is the fundamental idea of addition — for whole numbers, fractions, decimals, radicals, and variables.

When we add whole numbers we combine ones, then we combine tens, then hundreds, etc. For fractions, it's clear that 2 sevenths + 3 sevenths is 5 sevenths. Mathematically, we write: $\frac{2}{7} + \frac{3}{7} = \frac{5}{7}$. What about 2 fifths + 3 fourths? This is like the Snickers and Milky Ways. We show children that 2 fifths and 3 fourths can each be expressed as some number of <u>twentieths</u>; then the sum is <u>twentieths</u>: $\frac{2}{5} + \frac{3}{4} = \frac{8}{20} + \frac{15}{20} = \frac{23}{20}$

For decimals, 13 hundredths + 54 hundredths is 67 hundredths; or .13 + .54 = .67. If the decimals to be added do not have the same names, we can easily give them a common name.

Consider 28 hundredths + 579 thousandths. These can both be expressed as thousandths, and the sum is thousandths: .28 + .570 = .280 + .579 = .859.

This is what is happening when we add decimals by "lining up the decimal points." We are combining digits which have the same place value — just as we do in ordinary addition of whole numbers.

From here, it's an easy step to adding radicals or variables -- combine things which are alike. If the numbers to be added have different names or labels to begin with, express them with a common name in order to find the sum. These examples emphasize the process:

1. $4\sqrt{3} + 2\sqrt{3} = 6\sqrt{3}$
2. $\sqrt{18} + \sqrt{50} = 3\sqrt{2} + 5\sqrt{2} = 8\sqrt{2}$
3. $5j + 3j = 8j$
4. $2(x + 7) + 3(x + 7) = 5(x + 7)$
5. $4mc + 2c^2 = 2m(2c) + c(2c) = (2m + c)(2c)$

Adding Opposites

It's valid to apply this same reasoning when adding two negative numbers: 5 negatives + 2 negatives is 7 negatives; or, (-5) + (-2) = (-7). But the process can't be extended to finding the sum of a positive and a negative number. There is no common label for negative and positive; so these numbers can't be renamed and combined in the usual way. The rationale for adding a negative and a positive number is based on the concept of **opposites**. The **concrete** meaning of opposite relates to the number line: two numbers are opposites if they are on opposite sides of the number line (from 0), and they are the same distance from 0.

Another fundamental property of opposites relates to **addition**: the sum of opposites is 0. This is the property which is relevant to adding a positive and a negative number.

Activity

The class, separated into groups of three, plays this card game:

1. Remove aces and face cards from a standard deck. Deal out ten cards to each person.
2. Black cards win points according to face value; red cards lose points according to face value.
3. Each person determines the score for her hand. One player in each groups records the scores for each person. (Don't use the negative sign in recording scores; and don't use the words "negative" or "minus" in talking about scores.)
4. Repeat three more times.
5. Determine the winner in each group.

Group Discussion:

1. What arithmetic operation is involved in determining a player's score for one hand?
2. What would be the score for each of these hands:

   a) Red 2, 7, 4, 5

      Black 3, 8, 6, 6, 8, 4

   b) Red 3, 6, 2, 2, 9

      Black 5, 7, 8, 7, 2

   c) Red 4, 5, 2, 8, 9, 4

      Black 6, 3, 2, 2

   d) Red 6, 2, 9, 4, 2

      Black 8, 3, 3, 6, 4

3. Discuss various ways of determining scores for the hands given in #2.
4. Which of the hands in #2 has the best score? Why?

This activity provides a concrete basis for addition of integers. It particularly illustrates that a black 7 and a red 7 "neutralize" each other — their combined score is 0. On a more abstract level, we say that 7 + (- 7) = 0.

This relationship gives meaning to our process for finding the sum of a positive and a negative number. Consider these examples:

Example 1: -8 + 3

(-5) + (-3) + 3

-5 + 0

-5

Example 2: -7 + 16

-7 + 7 + 9

0 + 9

9

How the Operation Works: Commutativity, Associativity; Special Role of Zero

Once addition is understood as combining things which are alike, several properties of the operation become obvious.

First, notice that if two sets are being combined, there's no reason to think of one set as "first" and the other as "second." This is the concrete basis for recognizing that 3 + 2 is the same as 2 + 3.

In general, we say that no matter what m and t are, m + t = t + m. This is the **commutative** property of addition. (It may be helpful to remember that one meaning of the word **commute** is to go back and forth or switch positions.)

Next, it's important to recognize that although several sets of things can all be shoved together at once, our processes for addition are limited to two numbers at the time. (Try adding three numbers all at once!) So, in adding three or more numbers, we must select two of them to begin with. Children intuitively understand that when we do this with sets of things, it doesn't matter which two we choose first; but they must **learn** the implication for symbols: (3 + 4) + 5 is the same as 3 + (4 + 5). Or abstractly, no matter what c, k, and n are c + (k + n) = (c + k) + n. This is the **associative** property for addition. (It may be helpful to remember that the word **associate** means come together or pair off.)

This example illustrates the concrete basis for these two properties: Suppose I have several sacks with a few potatoes in each, and I want to combine all the potatoes into one of the sacks. It doesn't matter which two sacks I start with (associative); and after I pick the two to start with, it doesn't which one I pour into the other (commutative).

Teaching Note:

Elementary students need to understand that 5 + 2 = 2 + 5 and (4 + 3) + 2 = 4 + (3 + 2), but they shouldn't be burdened with the formal names or the abstract statements of these properties.

Finally, it's important for children to recognize the special properties of zero **as a number.** Their first experience of zero is as a place holder in numerals. In that context, 0 indicates that a position is empty — nothing is there. In extending this concept of zero to its role in arithmetic, we must help children understand that although zero represents how many things are in an empty set, **zero itself isn't nothing** — it's a perfectly good number.

For addition, the special property of 0 is that, no matter what number j is, 0 + j = j.

Homework: Section 16

1. What is the opposite of $-3\frac{1}{2}$?

2. How many spaces is 6 from -18 on the number line?

3. What is the absolute value of -5.3?

4. **Read** each of these (don't use the words "minus" or "negative" or "point"):

   a) $|7|$ b) -5 c) -(-2) d) $12 > 0$ e) $-11 < -6$ f) 1.073 g) $6\frac{3}{8}\%$

5. Write each number as a decimal: a) $2\frac{7}{16}$ b) 13.9% c) $\frac{51}{34}$

6. Which of these are integers? -14, 2.6, $|-8|$, -(-3), 0, $\frac{5}{12}$

7. Find all the factors of 52.

8. Which of these are prime numbers? -385, 9462, 83, 3723, 167, 323, .7

9. What is the smallest positive number which is a multiple of both 140 and 168?

10. Write each number as a percent: $\frac{39}{65}$, 3.4, .006, $\frac{153}{68}$, .8253, $1\frac{1}{10}$

11. Write True or False for each statement:

    a) $-2 > -7$

    b) 14 is to the right of 9 on the number line

    c) 18 is a divisor of 416

    d) When 972 is divided by 45, the remainder is 6

    e) No matter what t is, $-t$ is negative

    f) If m is to the left of j on the number line, then $j > m$

    g) The sum of two odd numbers is odd

12. For each statement, what does k have to be in order for the statement to be true?

    a) $-k = 8$ b) $|k| = 17$ c) $-(-k) = 3$ d) $k + 12 = 21$

    e) $k + 12 = 5$ f) $k + 12 = -11$ g) $k = \sqrt{25}$ h) $16 = k^2$

13. Find the sums **mentally:**

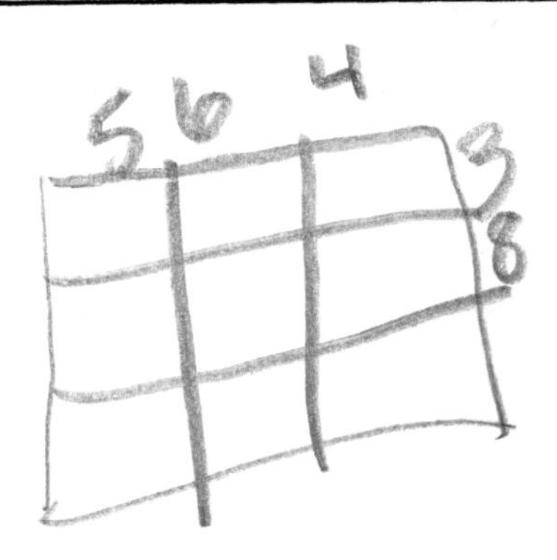

a) $3 + (-4) + (-11) + 2 + (6) + (-1) + (-8)$

b) $-6 + (-2) + 7 + (-8) + 3 + (-14)$

c) $\frac{5}{8} + (-\frac{3}{8}) + (-\frac{1}{4}) + (\frac{1}{16})$

d) $-.7 + .3 + (-.8) + (-.4) + .9$

e) $2j + (-7j) + (-13j) + j + (-4j) + 8j$

f) $3\sqrt{17} + (-4\sqrt{17}) + (-\sqrt{17}) + (6\sqrt{17})$

g) $-8t + (-3t) + (-5m) + 12t + (-m)$

14. The temperature in Detroit yesterday morning was -18°F. By that afternoon, the temperature was 17°F. How many degrees did the weather warm up during the day?

15. Harry got a notice from the bank that his account was $87 overdrawn. He deposited $153. What was his balance then?

16. On the number line shown below, locate the point which represents each of these numbers:

$-j,\ -m,\ j + k,\ j + m,\ m + k$

```
<--------------------------------•-----------•----•------------------•----------------------->
                                 m           0    j                  k
```

Answer in Journal

1. What does the word **sum** mean?
2. Explain what **opposites** are.
3. What does it mean to say that "addition is commutative?"
4. What does it mean to say that "addition is associative?"
5. Explain why adding a positive number and a negative number is fundamentally different from addition in other contexts.

Section 17: Understanding Subtraction

Concrete Meaning; Inverse Relationship with Addition

The concrete basis for subtraction is the concept of take-away — removing things from a set. If I have a bag of 11 apples, and I take 4 of them out, then there are 7 apples left in the bag. Written without labels, this becomes 11 - 4 = 7.

Using the same example, if I put the four apples back into the bag with the 7 that were left, I get the original 11 apples. It's obvious that in dealing with sets of **things**, subtraction and addition are reverse operations. In doing arithmetic, this relationship between addition and subtraction explains the way we "check" subtraction by adding:

Subtract	**Check**	**Subtract**	**Check**
$\begin{array}{r} 25.7 \\ -\underline{8.4} \\ 17.3 \end{array}$	$\begin{array}{r} 17.3 \\ +\underline{8.4} \\ 25.7 \end{array}$	$\begin{array}{r} 6\frac{7}{8} \\ -\underline{2\frac{3}{4}} \\ 4\frac{1}{8} \end{array}$	$\begin{array}{r} 4\frac{1}{8} \\ +\underline{2\frac{3}{4}} \\ 6\frac{7}{8} \end{array}$

The inverse relationship between addition and subtraction can be expressed in a completely abstract way: no matter what numbers the variables represent, m - c = k **means the same thing as** k + c = m.

For sets of things, it's clear that subtraction is not commutative. Taking 3 cookies out of a box of 8 cookies, leaves 5 cookies; but taking 8 cookies out of a box of 3 cookies is meaningless. However, there are many contexts in which 3 - 8 does have meaning — and it is different from 8-3.

Example 1: If Joey has a bank balance of $3, and he writes a check for $8, then his account is overdrawn by $5. This is the **opposite** of having a balance of $5; so we represent the situation with a negative number. His account total is now -$5. This is a concrete illustration that 3 - 8 = -5.

Example 2: Referring again to the card game, notice that removing a black 8 from a hand has the same effect as putting in a red 8. This observation provides another concrete basis for the definition of subtraction: **subtracting a number is equivalent to adding its opposite**.

Example 3: Remember that addition and subtraction are inverse operations, so

3 - 8 = ? **means the same thing as** 3 = ? + 8

Therefore the question, "What is 3 - 8," can be rephrased as, "What must be added to 8 to get 3?" And the answer, of course, is -5.

All of these examples lead to the purely symbolic definition of subtraction:

t - v = t + (-v)

In writing sums and differences which involve negative numbers, parentheses are used for clarification whenever two signs (+ or -) must be written next to each other. For example: 3 + (-5); 2 - (-7); etc.

Teaching Note:

Elementary school children must understand that distance, on a ruler or thermometer or any numbered line, is related to subtraction. They should associate the words 'distance' and 'difference'. It is also important for them to recognize that the distance from position A to position B is the same as the distance from B to A; so they can develop the convenient habit of always subtracting the smaller number from the larger.

Homework: Section 17

1. What must be added to -58 to get the sum 0?

2. What is the distance between 14 and -19 on the number line?

3. Write each difference as a sum:

   a) (-11) - 6 b) 5 - (-13) c) (-7) - (-2)

   d) 8 - 19 e) $2\frac{1}{4} - 5\frac{1}{6}$ f) (9.8) - (-13.4)

4. No matter what *m* is, $-m + m =$ ______.

5. Fill-in each blank with **positive** or **negative** :

   a) If m > 3, then m is ________________

   b) If x is negative, then -x is ________________

   c) If -t > 0, then t is ________________

   d) The sum of two negative numbers is ________________

   e) If k is any number except 0, then |k| is ________________

   f) If a > 5, then a - 5 is ________________

   g) If m < 3, then m - 3 is ________________

6. Find these sums mentally:

   a) (-6) + (-4) + 7 + 2 + (-11) + (-6) + 4 + (-1) + 20 + (-2) + 1 + (-6)

   b) 10 + (-14) + (-9) + 3 + (-5) + 1 + (-8) + 12 + (-4) + (-2) + 7 + 9

   c) -17 + (-5) + 14 + (-7) + (-2) + 8 + (-4) + (-9) + 20 + 7 + (-4)

7. Perform these operations:

   a) 3 - 11 + (-6) - (-8) - 9 + (-2) - 7

   b) -4 - (-3) + (-11) - 8 - 2 - (-7) + (-1)

   c) 16 - 21 - (-10) + (-8) - 11 + 16

8. Joe averages 18 miles per hour on his bike. How long will it take Joe to go 45 miles on his bike?

9. Write 750 as a product of prime numbers.

10. Which of these are primes? 3772, 3729, 4895, 323, 113

11. Read each of the following out loud:

a) $7 < t$ b) $|m|$ c) $-(-a)$ d) 207,000,400,093

12. Find all the divisors of 54.

13. Write symbolically:

a) *j* is greater than 26
b) the opposite of c
c) the sum of *m* and the opposite of 8
d) the sum of $2\frac{1}{7}$ and k
e) 7 more than y
f) the difference between 19 and 16

14. Write each number as a fraction:

a) 1.734 b) 28.35% c) $4\frac{11}{17}$ d) .8%

15. Write each number as a fraction with denominator 24, if possible:

a) $\frac{11}{8}$ b) $\frac{1}{96}$ c) $\frac{21}{36}$ d) $\frac{39}{51}$ e) 2 f) 0

16. The whole numbers 1 through 59 are written on separate slips and one slip is drawn at random.

a) What is the probability that the number drawn is odd?
b) What is the probability that it is a multiple of 5?
c) Which is more likely, that it is greater than 26 or that it is less than 33?
d) What is the probability that the number drawn is prime?
e) What is the probability that it is a negative number?
f) What is the probability that it is less than 82?

17. Forty percent of the people in this class are freshman, twenty-five percent are sophomores, and fifteen percent are juniors. The remaining 8 people are seniors. How many juniors are in the class?

18. On the number line shown below, locate the point that represents these numbers:

$-d$, $c + d$, $b - a$, $b + a$, $b - c$

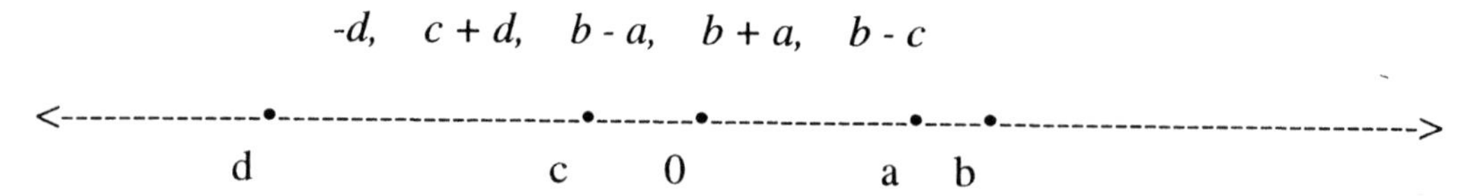

Answer in Journal

1. Explain why $-(-7)$ is 7.
2. What does it mean to say that the **difference** between two numbers is 4?
3. Is subtraction commutative? **Explain** your answer.
4. Is subtraction associative? How do you know?

Section 18: Understanding Multiplication

Concrete Models: Repeated Addition; Area

The first concrete models used to explain multiplication of whole numbers are based on the concept of **repeated addition**: 5 bags of cookies with 3 cookies in each bag (5 x 3), 4 rows of chairs with 7 chairs in each row (4 x 7), 6 baseball teams with 9 players on each team (6 x 9), etc. Because the process in each of these examples is addition, it's clear that the product represents whatever kind of things were being added. This model for multiplication is only useful in situations where a quantity can be considered as equal groups or equal portions — then the product is found by multiplying the number of groups or portions by the amount in each group.

Example 1:

a) 5 pieces of string, each piece is 18.3 cm long
total length is (5 x 18.3) cm

b) 4 crackers; each weighs $\frac{1}{5}$ ounce
total weight of crackers is $(4 \times \frac{1}{5})$ ounces

c) 3 round trips to Baton Rouge; each trip is 108.6 miles
total distance is (3 x 108.6) miles

d) 8.5 years; investment earns $796.40 per year
total earnings are (8.5 x 796.40) dollars

e) 1 hour 20 minutes of driving; average rate is 54.3 miles per hour
total distance is $(1\frac{1}{3} \times 54.3)$ miles

In the third or fourth grade, children learn to find the area of rectangles. The concept must be carefully developed in a meaningful sequence:

1. They develop an understanding of **what area is** (a measurement of an amount of surface — something which could be **painted**).
2. They learn that area is commonly measured in **squares** (square inches, square centimeters, square feet, square meters, etc.), and they learn **what these units look like.**

Square Centimeter

Square Inch

3. They use grids to find the area of figures. For irregular figures, this is a tedious process of counting; for rectangles, the process is much easier because multiplication (repeated addition) can be used.

   This rectangle consists of 3 rows of square centimeters with 5 squares in each row, therefore the area is 3 x 5 square centimeters.

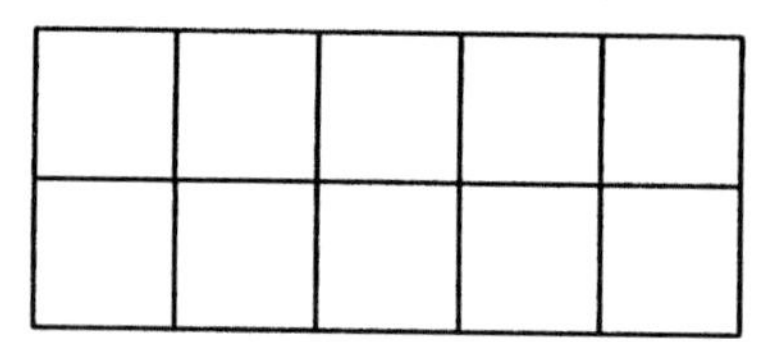

4. Finally, children learn to calculate the area of a rectangle from the measurements of its length and width.

   2 **inches** x 5 **inches** = 10 **square inches**

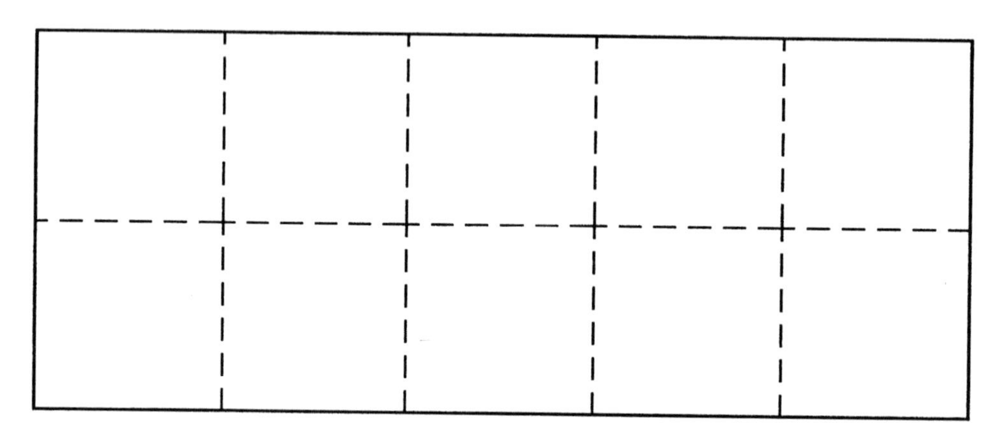

This is an important transition from the concept of multiplication as repeated addition. In this process, the product (square inches) is a completely **different kind of creature** from the quantities which were multiplied (inches).

Example 2: A rectangular ranch is 3.2 miles long and 1.7 miles wide. What is its area?
(3.2 **miles**) x (1.7 **miles**) = 5.44 **square miles**

Example 3: A 2.4 kilowatt oven operates for 3 hours 15 minutes. How much electricity is used?
(2.4 **kilowatts**) x (3.25 **hours**) = 7.8 **kilowatt-hours**

Example 4: What is the product of 7j and 3j?
$(7j)(3j) = 21j^2$

Teaching Note:
In beginning algebra, it's very common for students to mistakenly write a single number to represent a sum such as $2x^2 + 3x$. This error is rooted in their failure to understand that the product of x times x results in a completely **different kind of creature** named x^2.

Our familiar processes for multiplying whole numbers, fractions, and decimals are all based on the understanding of products as **different kinds of things** from the quantities which were multiplied.

Example 5: 3 **tens** x 2 **hundreds** = 6 **thousands**

Using symbols instead of words, we write: 30 x 200 = 6000.

For paper and pencil multiplication (which should **not** be necessary in this case), correct position of digits in the product is assumed by shifting each successive partial product one space to the left.

$$\begin{array}{r} 200 \\ \underline{\times\ 30} \\ 000 \\ \underline{600\ \ } \\ 6000 \end{array}$$

Example 6: An area model provides a clear, concrete illustration for the process of multiplying fractions.

$\frac{4}{5}$ of length $\times$ $\frac{2}{3}$ of width is $\frac{8}{15}$ of area

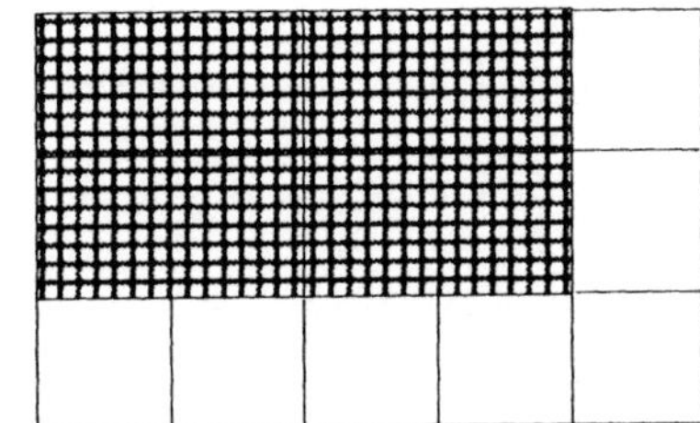

2 **thirds** x 4 **fifths** = 8 **fifteenths**

With symbols instead of words, we write: $\frac{2}{3} \times \frac{4}{5} = \frac{8}{15}$

Our abstract definition for multiplication of fractions is a generalization of this model:

$$\frac{j}{k} \bullet \frac{m}{v} = \frac{jm}{kv}$$

Notice that the short-cut for reducing a fraction, by "cancelling" common factors in the numerator and denominator, can be applied before multiplying.

whole process: $$\frac{7}{25} \bullet \frac{5}{14} \bullet \frac{15}{9} = \frac{\overset{1}{(\cancel{7})}\overset{1}{(\cancel{5})}\overset{1}{(\cancel{15})}}{\underset{\cancel{5}\,1}{(\cancel{25})}\underset{2}{(\cancel{14})}\underset{3}{(\cancel{9})}} = \frac{1}{6}$$

short-cut: $$\frac{\overset{1}{\cancel{7}}}{\underset{\cancel{5}\,1}{\cancel{25}}} \bullet \frac{\overset{1}{\cancel{5}}}{\underset{2}{\cancel{14}}} \bullet \frac{\overset{\cancel{3}\,1}{\cancel{15}}}{\underset{3}{\cancel{9}}} = \frac{1}{6}$$

Example 7: **7 tenths** x 8 **tenths** = 56 **hundredths**

Written symbolically as fractions, we have: $\frac{7}{10} \times \frac{8}{10} = \frac{56}{100}$.

Therefore, for decimals, the product must be: (.7)(.8) = .56

It is important to notice that products are represented in several different ways:

1. the familiar "times" symbol: 7 x 3
2. a raised dot: $\frac{2}{3} \bullet \frac{5}{7}$
3. parentheses: (8.3)(2.7)(5.4) or $3(j + 4)$
4. exponents: 3^5
5. no symbol, when factors are variables: $2xy$
6. combinations of these: $2^4 \bullet 3^2$; $(-3)^2(-5)$; $2j(k + 1)^2$
7. the word "of": $\frac{2}{3}$ of 78; 48% of 200

Homework: Section 18

1. Find sums, differences, and products **mentally**.

   a) $-3-(-7)+(-4)+2-(-6)-5+(-4)-8-(-9)$ b) 8000 x 700

   c) $\frac{1}{2}-\frac{1}{3}+\frac{1}{4}-\frac{1}{6}$ d) 5 - 1.67 e) 16 x 25 f) .2 x .5 x.7 g) 10 - 5.29

2. Write each number as a fraction:

   a) $2\frac{7}{17}$ b) $17\frac{2}{5}\%$ c) 4.097 d) .03%

3. Write each number as a decimal:

   a) $\frac{21}{30}$ b) $\frac{42}{175}$ c) 265% d) $\frac{3}{4}\%$

4. At a conference, the ratio of smokers to non-smokers is 3 to 5. What percent of the participants smoke?

5. Twenty-eight percent of USL students have a GPA above 3.0. If a student is chosen at random, what is the probability that her/his GPA is above 3.0?

6. Suzy can mow Farmer Brown's field in 5 hours.
   a) If she mows for 2 hours, what portion of the field will be done?
   b) If she mows for 1 hour 20 minutes, what portion will be done?
   c) How long will she have to work in order to mow 70% of the field?

7. Find answer **mentally**:

   On her diet, Janie is allowed an average of 20 grams of fat per day. For the first 4 days of this week, she averaged 18 grams per day. For the past 2 days, she had 23 grams each day. What is the maximum grams Janie can have today in order to stay within her allowed amount for the week?

8. Seven robots assemble 840 picture tubes in an 8-hour day. How long will it take 12 of these robots to assemble 450 picture tubes?

9. Find **mentally:** How many factors does 10,000 have?

10. Name an even multiple of 19 which is between 300 and 325.

11. How many **things** does this base seven numeral represent? 20065_{seven}

12. Tom completes a lap around the training track in 3 minutes 20 seconds, and Harry completes a lap in 4 minutes flat. If they begin running at the same time, how long will it be before they again meet at the starting line?

13. Write a **base twenty** numeral to represent five hundred thousand.

14. I want to pack 20 dozen peaches in bags so that each bag has an equal number of peaches. I have only 9 bags, and each bag holds a maximum of 35 peaches. How must they be packed?

15. What number must be multiplied **times itself** to get the given product? If there is no such number, explain why.
    a) 49 b) 144 c) .81 d) $\frac{25}{64}$ e) - 9 f) 14

Answer in Journal

1. What is a **product**?
2. Use a rectangle model to illustrate why $\frac{3}{7} \times \frac{5}{8} = \frac{15}{56}$.
3. **Explain how** you found the answers to these Homework Exercises: 1(c), 7, 9, and 14.

Section 19: Properties of Multiplication

In situations where multiplication can be considered as repeated addition, it's clear that the operation is commutative and associative. Concrete examples are easily found:

a) 12 bags of M&Ms with 7 candies in each (7×12)

is the same number of M&Ms as

7 bags of M&Ms with 12 candies in each (12×7)

b) 5 layers of bricks, with each layer having 4 rows of 3 bricks per row $5 \times (4 \times 3)$

is the same number of bricks as

3 layers of bricks, with each layer having 5 rows of 4 bricks per row $(5 \times 4) \times 3$

Moving beyond concrete examples, these properties of multiplication are stated in a general way using symbols:

Commutativity: $j \times k = k \times j$

Associativity: $m \times (c \times v) = (m \times c) \times v$

Notice that if we have 1 bag with 13 M&Ms in it, then we obviously have 13 candies. This illustrates the special multiplicative property of 1: **No matter what k is,** $1 \times k = k$

Now suppose we have j empty bags (j bags with 0 candies in each). Then, of course, we have 0 candies. In symbols, this translates into: **No matter what j is,** $j \bullet 0 = 0$

Zero has another very important multiplication property: if the product of two numbers is zero, then one of the numbers must be zero.

Written symbolically: **If** $j\,k = 0$, **then** $j = 0$ **or** $k = 0$

For a concrete illustration of this property, we can refer again to the bags of candy:

If Suzy has no M&Ms, then she either has no bags **or** she has some bags but they are all empty.

Linking Addition and Multiplication: The Distributive Property

The close relationship between addition and multiplication has been emphasized throughout all of our discussion of multiplication. This relationship deserves very special attention because it's the cornerstone of arithmetic and algebra.

Example 1: As always, we begin with **things.**
Suppose Harry has 3 bags of marbles with 5 marbles in each. Then, his friend gives him 4 more bags with 5 marbles in each. How many marbles does Harry have now?

We can describe the situation mathematically in two different ways:

(3	×	5)	+	(4	×	5)	**or**	(3 + 4)	×	5
Bags		Marbles in each		Bags		Marbles in each		Bags		Marbles in each

Example 2: Consider these illustrations:

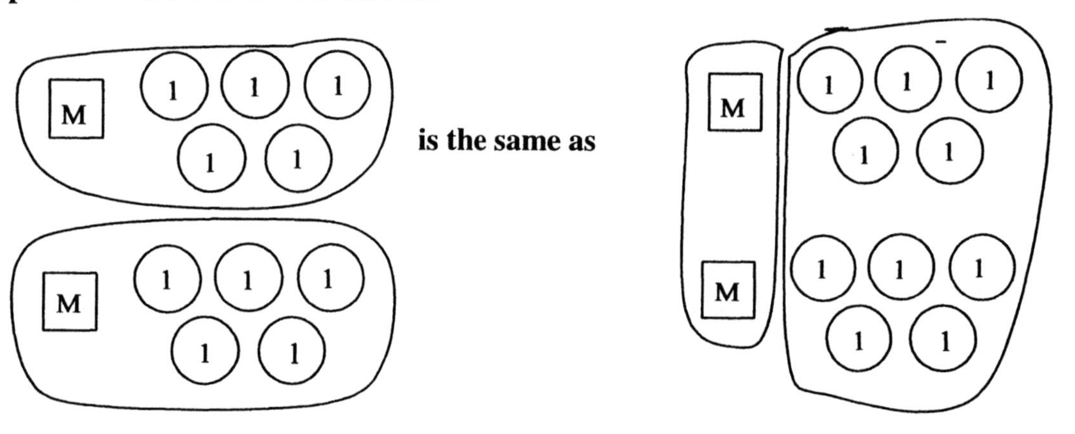

This represents 2 groups with (M + 5) in each group.
$2 \times (M + 5)$

This represents 1 group with two Ms in it, **and** 1 group with two 5s in it.
$(2 \times M) + (2 \times 5)$

Example 3: Notice these pictures:

☆☆☆ ☺☺ **is the same as** ☆☆☆ ☺☺

☆☆☆ ☺☺ ☆☆☆ ☺☺

A + B (A + B)

☆☆☆ ☺☺ **is the same as** ☆☆☆ ☺☺

$\frac{1}{2}A + \frac{1}{2}B$ $\frac{1}{2}(A + B)$

$$\mathbf{\frac{1}{2}A + \frac{1}{2}B = \frac{1}{2}(A + B)}$$

This relationship can be described, clearly and concisely, for **all** situations and **all** numbers. It's called the **Distributive Property:**

$j \times (m + t) = jm + jt$ **or** $ck + yk = (c + y) \times k$

Notice that commutativity and associativity refer to a single operation (either addition or multiplication); the Distributive Property involves **both** operations.

Our familiar algorithm for multiplication is an efficient way to apply the Distributive Property in a place value numeration system.

This is what's "going on":

243 × 57

(2 hundreds + 4 tens + 3 ones) × (5 tens + 7 ones)

(2 hundreds × 5 tens) + (4 tens × 5 tens) + (3 ones × 5 tens) + (2 hundreds × 7 ones) + (4 tens × 7 ones) + (3 ones × 7 ones)

(10 thousands) + (20 hundreds) + (15 tens) + (14 hundreds) + (28 tens) + (21 ones)

(1 ten thousand) + 2(thousands) + (1 hundred + 5 tens) + (1 thousand + 4 hundreds) + (2 hundreds + 8 tens) + (2 tens + 1 one)

(1 ten thousand) + (3 thousands) + (7 hundreds) + (15 tens) + (1 one)

(1 ten thousand) + (3 thousands) + (7 hundreds) + (1 hundred + 5 tens) + (1 one)

(1 ten thousand) + (3 thousands) + (8 hundreds) + (5 tens) + (1 one)

13851

This can be abbreviated:

```
   243
  x 57
    21
   28
  14
   15
  20
 10
 13851
```

Notice that shifting positions preserves place value for all these partial products

```
   243
  x 57
  1701
 1215
 13851
```

Our familiar algorithm shortens the process even more by "carrying":

Our multiplication algorithm is just one of many that have been used throughout history. This "galley" method is another example:

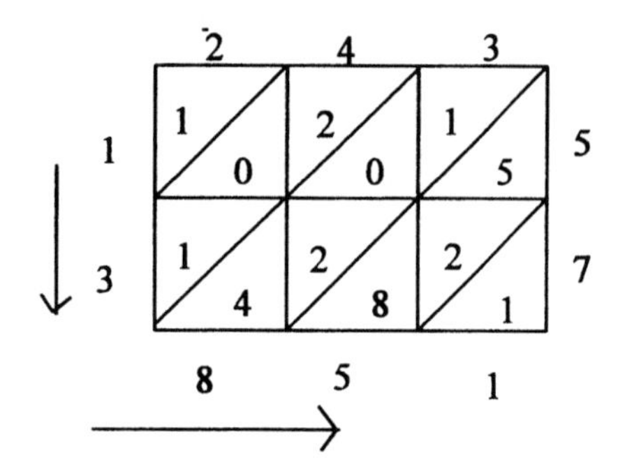

Partial sums are added along the diagonals (with "carrying" if necessary) beginning at lower right. The product is read from upper left.

Multiplication of mixed numbers, without translating them into fractions, would be a direct application of the Distributive Property. For example:

$$3\frac{3}{4} \times 1\frac{3}{5}$$

$$(3 + \frac{3}{4}) \times (1 + \frac{3}{5})$$

$$(3 + \frac{3}{4})(1) + (3 + \frac{3}{4})(\frac{3}{5})$$

$$3 + \frac{3}{4} + (3)(\frac{3}{5}) + (\frac{3}{4})(\frac{3}{5})$$

$$3 + \frac{3}{4} + \frac{9}{5} + \frac{9}{20}$$

$$\frac{60}{20} + \frac{15}{20} + \frac{36}{20} + \frac{9}{20}$$

$$\frac{120}{20} = 6$$

It is usually more efficient to translate the mixed numbers into fractions:

$$3\frac{3}{4} \times 1\frac{3}{5}$$

$$\frac{15}{4} \times \frac{8}{5}$$

$$\frac{15 \cdot 8}{4 \cdot 5} = \frac{3 \cdot 2}{1} = 6$$

Algebraically, the Distributive Property is applied in all factoring and multiplication of polynomials and other expressions. For example:

a) $jm + j^2 + 3j = j(m + j + 3)$

b) 6% of $(4t + v) = .06(4t) + .06v$

c) $\sqrt{20k^2} + \sqrt{45} = 2k\sqrt{5} + 3\sqrt{5} = (2k + 3)\sqrt{5}$

Homework: Section 19

1. Write "seven times three" in three different ways using mathematical symbols.
2. What value(s) of k will make each statement true? Find answer **mentally**.

   a) $k + 2\frac{1}{2} = 5\frac{7}{8}$ b) $12 - k = 14$ c) $-k = 3.07$

   d) $|k| = \frac{4}{11}$ e) $k^2 = 121$ f) $\frac{2}{7}k = 16$

   g) $3k + 1 = 13$ h) $2.3k = 0$ i) $k + .73 = 4$

   j) $2(k + 3) = 18$ k) $k - 7 = -15$ l) $k^3 = -8$

   m) $-(-k) = 5$

3. What is the smallest positive multiple of 295?
4. **Find** the area of the rectangle at the right. (Use a ruler!)
5. Read each of these aloud. (Don't say "point"; "minus"; "negative", "absolute value" or "over".)

   a) $-j > 0$ b) 1.007 c) $\frac{5}{267}$ d) $|t|$

6. The result of addition is a ________________.
7. 3 hundreds x 2 thousands is 6 ________________.
8. 60 is what percent of 24?
9. 7 fifths x 3 eighths is 21 ________________.
10. If all the letters of the alphabet are written on separate cards, and one is drawn at random, what is the probability that it will be one of the letters in "Louisiana"?
11. 35% of Mrs. Thibodeaux's class are girls. What is the ratio of boys to girls in the class?
12. The result of subtraction is a ________________.
13. 4 tenths x 2 hundredths is 8 ________________.
14. The result of multiplication is a ________________.
15. Which property is illustrated in each:

    a) $21k = 3(7k)$ b) $(2m + 5)t = 2mt + 5t$ c) $(5x^2 - 2) + 3a = 3a + (5x^2 - 2)$

Challenge:

Find the sum, difference, and product **in the bases given.** Don't translate to base ten. **THINK** about what carrying and borrowing **mean!**

a) Base seven

$$\begin{array}{r} 56_{seven} \\ +\ 45_{seven} \\ \hline \end{array}$$

b) Base eight

$$\begin{array}{r} 241_{eight} \\ -\ 26_{eight} \\ \hline \end{array}$$

c) Base five

$$\begin{array}{r} 42_{five} \\ \times\ 13_{five} \\ \hline \end{array}$$

Section 20: Product of Two Negative Numbers

The product of a positive number and a negative number can be represented as repeated addition: $5 \times (-3) = (-3) + (-3) + (-3) + (-3) + (-3) = -15$.

So it is clear that these products are negative.

There is, however, no simple way to illustrate why $(-4) \times (-3)$ must be 12. The explanation is based on the fundamental properties of equality, addition, and multiplication; hence it is completely abstract.

$-3 + 3 = 0$	This is the **meaning of opposite**.
$(-4)(-3 + 3) = (-4)(0)$	Two equal quantities, (-3 + 3) and 0, have been multiplied by the same number, -4. So the products are obviously equal.
$(-4)(-3 + 3) = 0$	Any number multiplied by 0 gives the product 0.
$(-4)(-3) + (-4)(3) = 0$	This is the Distributive Property.
$(-4)(-3) + (-12) = 0$	We explained above why (-4)(3) has to be -12.

Now notice that (-4)(-3) is added to -12 to get 0.

What is the **only** number that, when added to -12, gives the sum 0?

Right! 12 is the only number. Therefore, (-4)(-3) **has to be** 12.

Homework: Section 20

1. Do this arithmetic **mentally.**

a) $-7 + (-5) - (-11) + (-4) - 8 - (-5) - 4 - (-6) + (-10)$ b) (2000) x (-60) x (-500)

c) 10 - 4.89 d) (-4)(-3)(-2)(-10)(-2) e) 20% of 43 f) 150% of 18

g) $\frac{5}{8}$ of 56 h) 2.67 x 1000 i) $2.65 \div 100$

2. Mark each statement true or false.

a) 3.7 = 3.700 b) No matter what k is, $-k$ is negative.

c) If $m > j$, then $-m < -j$. d) If $t = v$, then $t + y = v + y$, no matter what number of y represents.

e) Every fraction can be written as a fraction with positive denominator.

f) If $c^2 = 64$, then c must be 8.

g) -11 is the **only number which can be added to** 11 to give the sum 0.

h) Between any two fractions, there is always another fraction.

i) Zero is nothing.

j) $\frac{5}{12}$ can be written as a terminating decimal.

k) If j is a multiple of 3 and k is even, then 6 is a factor of jk.

l) If I select a number from this set, {2, 4, 6, 8, ...}, the probability that it will be odd is 0.

3. What is the ratio of consonants to vowels in the word LOUISIANA?

4. Write a base three numeral to represent five hundred **things.**

5. Name 3 factors of 48.

6. What does m have to be to make each statement true?

a) $17m = 0$ b) $m^2 = 20$ c) $3m + 2 = 2 + 3m$ d) $6m = 5m$ e) $|m| = \frac{5}{7}$

7. Locate the points on the number line: $-t$, $j + t$, $t - j$, $j + k$, $k - t$, $t - k$, $-k + j$

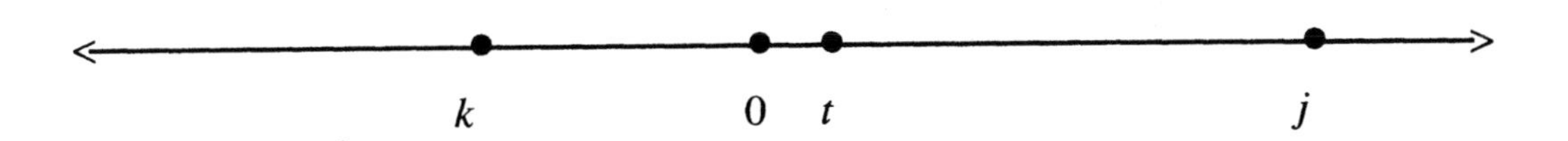

8. Write each number as a decimal: $5\frac{1}{4}\%$, $\frac{15}{24}$, $\frac{9}{400}$, $\frac{7}{2^5 5^7}$

9. Five-twelfths of Wisdom College students are women. What is the ratio of men to women students?

10. Find the sum in **base twelve**

$$\begin{array}{r} 9T6_{twelve} \\ 85E_{twelve} \\ +\ 477_{twelve} \\ \hline \end{array}$$

11. Find the difference in **base eight**

$$\begin{array}{r} 6013_{eight} \\ -\ 2254_{eight} \\ \hline \end{array}$$

12. Find the product in **base five**

$$\begin{array}{r} 423_{five} \\ \times\ 34_{five} \\ \hline \end{array}$$

13. What is the next **integer** to the left of $-7\frac{5}{6}$ on the number line?

14. How many factors does $13^5 17^4$ have?

Section 21: Meaning of Division

Concrete Models: Repeated Subtraction; Separating a Set into Equal Groups

Children progress in their understanding of division from very concrete situations to very abstract concepts. But from the beginning, it's clear that **division is the reverse of multiplication.**

The meaning of division is developed in several stages:

a) Separate 20 things into groups with 4 things in each group. How many groups are there?

b) What must be multiplied by 4 to get the product 20? $(4 \times ? = 20)$

c) What is 20 **divided by** 4? $(20 \div 4 = ?)$

Notice that b) and c) clearly show that multiplication and division are inverse operations:

$4 \times ? = 20$ **means the same thing as** $20 \div 4 = ?$

d) What is 4 divided into 20? $(4\overline{)20})$

It's very important for children to understand that this "long division" notation is read in two different ways – 4 **divided into** 20 or 20 **divided by** 4.

To "check" long division (when there is no remainder) we simply multiply quotient times divisor to see if their product is the dividend. The process is a direct application of the definition of division as the inverse of multiplication.

e) What is $20 \div 8$?

To a child, $8\overline{)20}$ doesn't seem much different from $4\overline{)20}$. but of course there is a very significant difference: 8 is not a factor of 20; so the division process doesn't provide a nice whole number "answer." Elementary students learn to deal with this situation in three different ways.

1. **remainder**

$$\begin{array}{r} 2 \\ 8\overline{)20} \\ \underline{16} \\ 4 \end{array}$$

answer: 2 r 4

check: $2\frac{1}{2} \times 8 = 20$

2. **mixed number quotient**

$$\begin{array}{r} 2 \\ 8\overline{)20} \\ \underline{16} \\ 4 \end{array}$$

answer: $2\ \frac{4}{8}$ or $2\ \frac{1}{2}$

check: $(2 \times 8) + 4 = 20$

3. **decimal quotient**

```
   2.5
8|20.0
  16 x
   40
   40
```

answer: 2.5

check: 2.5 x 8 = 20

f) What is 20 ÷ 6?

A fifth or sixth grader will expect 20 ÷ 6 to be the same kind of "problem" as 20 ÷ 8. But again, there is a **very important difference**. Finding a decimal quotient for 20 ÷ 6 brings children face-to-face with the concept of infinite decimals.

```
   3.33
6|20.000
  18 xxx
   20
   18
    20
    18
     20
```

It soon becomes apparent that this process will never end — the quotient is an infinite sequence of 3's

answer: 3.333.... or $3.\overline{3}$

(Notice the two different ways of representing this number.)

check: impossible

(We can't multiply 6 by the quotient because multiplication begins with the **last** digit of each number and an infinite decimal doesn't have a last digit!)

calculator answer: 3.333333333

calculator check:

3.333333333 x 6 = 19.99999999

Teaching Note:

In all of this, elementary pupils are confronted with several new and difficult ideas: the concept of infinity, the process of rounding, and the subtle differences among the numbers

$3\frac{1}{3}$, $3.\overline{3}$, 3.33, 3.3333, etc.

g) $\frac{2}{7}$ means $2 \div 7$

This definition involves another difficult transition from concrete to abstract: it introduces a completely new concept of fractions. As we have emphasized, children develop an understanding of fractions as parts of things: $\frac{2}{3}$ of the class had measles; this is $\frac{1}{6}$ of the pie; $\frac{3}{4}$ of an inch is less than $\frac{7}{8}$ of an inch; the probability of getting a green ball is $\frac{7}{12}$, etc. In these contexts, although the numerator and denominator are separate numbers (each with a specific meaning), the fraction itself is also considered as a **number.**

When we say that $\frac{2}{7}$ means $2 \div 7$, we are using fractional notation to represent an **operation**. In this case , the long division process results in a decimal quotient with a block of six repeating digits.

```
     .285714
  7 |2.000000
     14xxxxx
      60
      56
       40
       35
        50
        49
         10
          7
          30
          28
           2
```

at this point, we realize that "we're back where we started" — so this whole cycle will repeat forever.

answer: .285714285714.... or $.\overline{285714}$

check: impossible

calculator answer: .2857142
OR .2857143
OR .285714286
OR others, depending on the calculator used

calculator check: .2857142 x 7 = 1.9999994

Teaching Note:

Have elementary school students do this division on their calculators and compare answers; talk about how some calculators round and others do not, and even those that do round, do so differently. Talk about the number of places displayed on calculator, and what is a reasonable place to round.

Recap

1. All of these mean the same thing:

   $20 \div 4$ 20 divided by 4 $4\overline{)20}$ 4 divided into 20 $\frac{20}{4}$

2. Every fraction can be written as a decimal which either

   i) **terminates** ($\frac{3}{4} = .75$; $\frac{61}{200} = .305$; $\frac{13}{25} = .52$, etc.)

   or

   ii) **repeats** ($\frac{5}{6} = .8\overline{3}$; $\frac{4}{9} = .444...$; $\frac{2}{7} = .\overline{285714}$; etc.)

3. For a fraction in simplest form:
   a) if the only prime factors of the denominators are 2 and 5, then its decimal equivalent **terminates**.
   b) if the denominator has other prime factors, then the fraction is equivalent to an infinite repeating decimal.

4. An infinite decimal must be rounded off (or written in fractional form) in order to do arithmetic with it. Calculators round off infinite decimals automatically — so for such a number, the digits shown in the display are **always an approximation.**

Homework: Section 21

1. Mental arithmetic

   a) 50% of 870 b) $3.9 \times 10{,}000$ c) $\frac{7000 \times 90{,}000}{300}$

   d) $\frac{5}{6}$ of 42 e) 15% of 62.40 f) $\frac{21}{23} \div 3$

   g) $78 \div 1000$ h) $4\frac{1}{3} - 1\frac{1}{6}$ i) 25% of 28

   j) $2\frac{2}{3} \div 4$ k) $20 - 17.63$ l) $2^4 - 3^2 - |-1|$

2. Write each number as a fraction: $-3\frac{1}{8}$, 17%, .093, 41.47, 102%, $\frac{1}{2}$%

3. What integer is closest to each of these? $7.8\overline{3}$, 148%, $\frac{39}{11}$, $-\frac{3}{8}$, 13.57, $2.4\overline{96}$

4. Write each number as a decimal: $2\frac{1}{4}$, $9\frac{1}{2}$%, $\frac{7}{8}$, $\frac{17}{12}$, 420%, $11\frac{2}{3}$

5. Write each number as a percent: .182, $\frac{3}{8}$, .0075, $1\frac{1}{4}$, $\frac{12}{17}$, $\frac{31}{18}$

6. Round:

   a) 8573 to hundreds b) .072 to tenths c) 4.675 to hundredths

   d) 9.618 to a whole number e) 7,582,496 to thousands

7. Put >, < or = between each pair of numbers:

   a) $-\frac{3}{15}$ $\frac{4}{-20}$ b) $\frac{7}{3}$ $\frac{9}{4}$ c) $4\frac{6}{11}$ $\frac{40}{9}$

   d) 18% $\frac{9}{51}$ e) $.\overline{67}$ .677 f) $\frac{250}{503}$ $\frac{18}{35}$

8. When a pair of dice are thrown, what is the probability that the sum will be more than 3 and less than 10?

9. Sue walked 14 miles in 5 hours and Joe walked 8 miles in 3 hours. Who had the greater average rate? How much greater?

10. Twenty ounces of water must be mixed with $8\frac{1}{3}$ ounces of canned milk to make a baby's "formula." How many ounces of water should be mixed with $18\frac{3}{4}$ ounces of milk?

11. List all the common factors of 198 and 220.

12. The sum of two numbers is $2\frac{5}{6}$. If one of the numbers is 7, what is the other number?

13. Harry spent $\frac{4}{7}$ of his savings on a baseball glove. If the glove cost $48, how much money did he have left?

14. What must be added to $\frac{5}{8}$ to get the sum $\frac{11}{12}$?

15. What fraction of a yard is 20 inches?

16. One hundred forty of the 500 employees at TopCo Corporation are women. What is the ratio of men to women employees at TopCo?

17. If you pick a card from a full deck, what is the probability that it will be a spade or an ace?

18. What is the smallest positive whole number that has both 78 and 130 as factors?

19. What is the distance between $-\frac{21}{5}$ and $\frac{14}{3}$ on the number line?

20. Dividing a number by 12 has the same effect as multiplying the number by ___?

21. Write each fraction with denominator 28, if possible. $\frac{9}{7}, \frac{22}{56}, \frac{15}{12}, \frac{8}{21}, \frac{25}{35}$.

22. Joe got 19 hits in 53 times at bat. What percent of his at-bat were hits?

Answer in Journal

1. Explain how to get the answer to #13 of the Homework Exercises **mentally**.
2. This is a question in a 6th grade math book: What is 473 ÷ 18? The answer in the back of the book is $26\frac{5}{18}$. Suzy got 26 r 5. Joe used his calculator and got 26.277778. Both students want to know "What is the right answer?" What would you tell them?
3. Is division commutative? Justify your answer.
4. Is division associative? Justify your answer.

Section 22: Making Sense of How We "Do" Division

The process of long division is an application of the inverse relationship between multiplication and division. We "guess" what has to be multiplied by the divisor to get the dividend, test to see if it's correct and make adjustments if it isn't. Throughout the process, it's important for children to remember that the product of quotient times divisor cannot be **more** than the dividend.

There are **several** techniques for doing all of this, but the underlying inverse relationship between multiplication and division is the basis for all of them.

Example 1: $73\overline{)2044}$

It is easier to guess what must be multiplied by 7 to get 20, than to guess what must be multiplied by 73 to get 2044. So, the simplest process for long division is to guess the quotient one digit at the time, trying to get as close to a factor of the dividend as possible without going over it.

First digit guess:

```
     2
73)2044
   146
    58
```

Test: Since 58 ≥ 0 and 58 < 73, the first digit is correct.

Second digit guess:

```
     28
73)2044
   146
    584
    584
```

Check: 28 x 73 = 2044, therefore 2044 ÷ 73 = 28.

Example 2: $47\overline{)2632}$

First digit guess:
$$\begin{array}{r} 6 \\ 47\overline{)2632} \\ \underline{282} \end{array}$$

Test: Since $263 - 282 < 0$, 6 is too big.

First digit, new guess:
$$\begin{array}{r} 5 \\ 47\overline{)2632} \\ \underline{235} \\ 28 \end{array}$$

Test: Since $28 \geq 0$ and $28 < 47$, the first digit is correct.

Second digit guess:
$$\begin{array}{r} 57 \\ 47\overline{)2632} \\ \underline{235} \\ 282 \\ \underline{329} \end{array}$$

Test: Its clear that 7 is too big.

Teaching Note:

After a little experience, children will realize from the beginning that 7 is too big in this situation.

Second digit, new guess:
$$\begin{array}{r} 56 \\ 47\overline{)2632} \\ \underline{235x} \\ 282 \\ \underline{282} \end{array}$$

Check: $56 \times 47 = 2632$, therefore $2632 \div 47 = 56$

To extend this process to include long division with decimals, the first step is to re-write the divisor and quotient so that the divisor is a whole number. Then, place value is preserved by putting the decimal in the quotient directly above the decimal in the dividend.

Example 3: $9.3\overline{)52.08}$

First, notice that $52.08 \div 9.3$ **is the same as** $520.8 \div 93$

$$\left(\text{because } 52.08 \div 9.3 = \frac{52.08}{9.3} = \frac{52.08 \text{ x } 10}{9.3 \text{ x } 10} = \frac{520.8}{93} = 520.8 \div 93\right).$$

So we can write $93\overline{)520.8}$. Next we place the decimal in the quotient directly above the decimal in the dividend, and proceed as with whole numbers:

$$\begin{array}{r} \bullet \\ 93\overline{)520.8} \end{array}$$

Example 4: $0.007\overline{)2.1}$

$2.1 \div 0.007$ **is the same as** $2100 \div 7$. So we write $7\overline{)2100}$ and divide as usual.

Division of Fractions

In order to understand our process for division of fractions, children must first understand these two things:

1) No matter what j is, $j \div 0$ **is not a number**.
2) The fractions $\frac{a}{b}$ and $\frac{b}{a}$ are called reciprocals; every fraction except 0 has a reciprocal; and, the product of reciprocals is 1.

Math textbooks usually say that "division by zero is undefined." This doesn't mean anything to most students (even college students!). It's much better to say that $\frac{5}{0}, \frac{2}{0}, \frac{11}{0}$, etc. **are not numbers** —and explain why.

This is the explanation:

Let's suppose, for example, that $5 \div 0$ is a number. We'll call it k.

Then $\begin{array}{r} k \\ 0\overline{)5} \end{array}$, and $k \cdot 0 = 5$. But this is impossible because, no matter what k is, $k \cdot 0 = 0$.

So, our supposition that $5 \div 0$ is a number cannot be true. And, of course, there's nothing special about 5; we could have chosen any non-zero number, and the same contradiction would have resulted.

So, if m is any non-zero number, $\frac{m}{0}$ is not a number.

This argument doesn't show that $0 \div 0$ isn't a number — because $\begin{array}{r} 0 \\ 0\overline{)0} \end{array}$ doesn't give the same impossible situation when we "check:" $0 \times 0 = 0$ is true! But $0 \div 0$ isn't a number for a different reason.

If $0 \div 0$ were a number, then we could make these translations:

$0 \cdot 0 = 0$ means $0 \div 0 = 0$ **and** $1 \cdot 0 = 0$ means $0 \div 0 = 1$

Obviously, something is very wrong here — $0 \div 0 = 0$ and $0 \div 0 = 1$ imply that $0 = 1$! So, our supposition, that $0 \div 0$ is a number, leads to something impossible (a contradiction). Therefore, it isn't true.

Now back to division of fractions. For example, what is $\frac{3}{4} \div \frac{6}{7}$?

$$\frac{3}{4} \div \frac{6}{7} = \frac{\frac{3}{4}}{\frac{6}{7}} = \frac{\frac{3}{4} \times \frac{7}{6}}{\frac{6}{7} \times \frac{7}{6}} = \frac{\frac{3}{4} \times \frac{7}{6}}{1} = \frac{3}{4} \times \frac{7}{6}.$$

Therefore, $\frac{3}{4} \div \frac{6}{7}$ **is the same as** $\frac{3}{4} \times \frac{7}{6}$.

There isn't anything special about this particular example. The process illustrated for $\frac{3}{4} \div \frac{6}{7}$ is valid for any fractions as long as the divisor isn't 0. In general, we say:

$$\text{If } t \neq 0, \text{ then } \frac{a}{b} \div \frac{t}{k} = \frac{a}{b} \times \frac{k}{t}$$

The inverse relationship of multiplication and division is the basis for our arithmetic algorithm: dividing by a fraction is the same thing as multiplying by its reciprocal. (Notice the analogy to subtraction: subtracting a number is the same thing as adding its opposite.)

There is no efficient procedure for dividing mixed numbers, so to find the quotient of mixed numbers we translate them into improper fractions.

Now, a final observation. One of the important aspects of number sense is to recognize whether an answer is **reasonable**. In many situations this involves a judgement about results from finding a product or a quotient: what can we predict, in advance, about its size?

For positive numbers, we should notice:

i) If j is multiplied by a proper fraction (or its decimal or percent equivalent), then the product will be smaller than j.
Examples: $3\frac{1}{2} \times \frac{4}{7} = \frac{7}{2} \times \frac{4}{7} = 2$ and $28 \text{ x } 15\% = 28 \text{ x } .15 = 4.2$

ii) If j is divided by a proper fraction (or its equivalent), then the product will be bigger than j.
Examples: $12 \div \frac{2}{3} = 12 \text{ x } \frac{3}{2} = 18$ and $2.6 \div .13 = 260 \div 13 = 20$

Homework: Section 22

1. Perform these operations **mentally**.

a) 20% of 55 b) $3.9 \times 10{,}000$ c) $\frac{5}{6}$ of 54 d) $26 \div \frac{1}{2}$

e) 75% of 28 f) $\frac{3000 \times 8000}{600}$ g) $7.8 \div 1000$ h) $\frac{21}{23} \div 3$

ii) $\frac{35}{36} \div \frac{1}{9}$ j) $\frac{45 \times 28}{35}$ k) $7\frac{3}{8} - 2\frac{1}{2}$ l) 15% of 120

m) $\left(\frac{3}{5}\right)^2(50)$ n) $6t \div \frac{2}{3}$ o) $20 - 14.71$ p) $\frac{1}{4}$ of $\frac{8}{17}$

q) (.3)(.5)(.4) r) 250% of 12

2. Write each number as a percent: 1.3, $\frac{11}{12}$, .0847, $\frac{28}{5}$

3. Write >, <, or = between each pair of numbers:

a) 42% $\frac{21}{51}$ b) $.\overline{38}$ .383 c) $\frac{2}{3}$ 67% d) $\frac{8}{15}$ .53

4. Round 1.749 to the nearest tenth.

5. Tom bought some stock this week for $17 per share. Last week the stock was $25 per share. What percent decrease was this?

6. The ratio of men to women at last years' Sugar Bowl was 13 to 7. What percent of people at the game were men?

7. Which of these are prime? 9757, 437, 3597, 797.

8. If you draw a card from a full deck, what is the probability that it will be:

   a) the 7 of clubs? b) a heart? c) a black face card (J, Q, K)? d) a diamond or a 6?

9. From a factory assembly line, 3 out of every 500 toasters are defective. What percent loss does the company have from defects?

10. List the factors of 504.

11. Do arithmetic in the base given:

   a) $234_{six} + 244_{six} + 525_{six}$ b) $7042_{twelve} - 3853_{twelve}$ c) $534_{seven} \times 26_{seven}$ d) $43_{five} \overline{)24232_{five}}$

12. Write each fraction with denominator 18, if possible:

   a) $\frac{42}{27}$ b) $\frac{16}{72}$ c) $\frac{7}{28}$ d) $\frac{0}{5}$

13. Name three common multiples of 56 and 98.

14. What integer is closet to each of these:?

   a) $6\frac{1}{2}\%$ of 804 b) $\left(-\frac{5}{2}\right)^2$ c) $400\% - 3^2 + \frac{2}{7}$ d) $5\frac{3}{8} - |-7|$

   e) $-62\frac{1}{3}\%$ f) $3.\overline{49}$ g) 26.3888 ⋯

15. Over the past five years, the enrollment in Jackson Parish schools went from 80,000 to 92,000. What percent increase is this?

16. Which of these represent zero?

   a) $\frac{0}{2}$ b) $j + (-j)$ c) $|-3| - (-3)$

17. Locate these points on the number line shown below: $k - j$, $-3j$, k^2, jk, $|j|$, $k \div 4$, $j \div \frac{2}{3}$

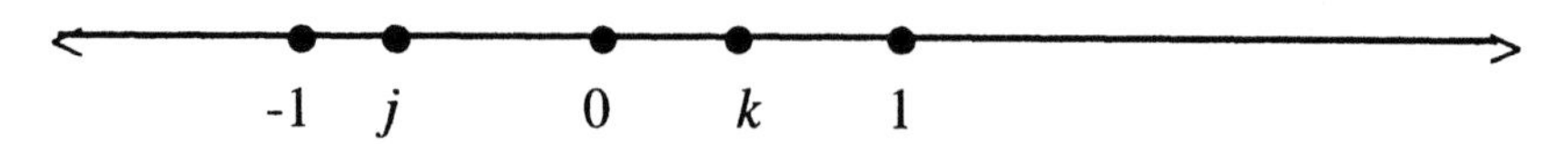

18. An IV tube drips 8 times per minute. Ten drips is one teaspoon of fluid.

    a) What is the rate in drips per hour?

    b) What is the rate in teaspoon per minute?

    c) How long will it take to empty a one-quart bag?

    d) If a new one-quart bag is connected, how many drips per minute would empty it in 2 hrs 40 minutes?

Cumulative Review

1. This numeral is written in base thirteen: 1097_{thirteen}. **Explain** how to determine how many things it represents.

2. Smooth-Write pencils are packed 15 to a box. One day, 11377 pencils needed to be packed. Explain how to find the answer to this question using a calculator: After as many boxes as possible are filled, how many pencils will be left over?

3. Explain what an **even number** is. What is an **odd number**?

4. What does it mean to say that 12 is a divisor of *j*?

5. What is a prime number? For a specific whole number, what must be done to determine whether it is prime?

6. If you know that 15 is a multiple of *k*, what are the possible values of *k*?

7. a) Write a question in which the word **factor** is used as a noun.
   b) Write a question in which **factor** is used as a verb.

8. a) Explain what the greatest common factor of two numbers is.
   b) Describe an efficient process for finding GCF.

9. a) Explain what the least common multiple of two numbers is.
   b) Describe an efficient process for finding LCM.

10. What is the **concrete** meaning of fraction?

11. a) Explain what it means to say that the ratio of peanuts to pecans in a can is 17 to 3.
    b) Does this ratio tell you how many peanuts are in the can?
    c) Could there be exactly 10 pecans? Explain.

12. Suppose you read that for women between 40 and 50 years old, the probability of living to age 80 is $\frac{5}{9}$. What does this **mean** to you?

13. A bag contains red and blue chips, and nothing else. I tell you that the probability of drawing a red chip is $\frac{3}{7}$.
    a) What is the probability of drawing a blue chip? Explain.
    b) What is the probability of drawing a green chip?
    c) Do you know how many chips are in the bag? Explain.
    d) What is the ratio of red to blue chips? Explain how you know this.

14. For each of the three kinds of rates, describe a situation which involves that kind of rate.

15. Suppose a machine assembles 7 toys in 15 minutes. Explain how to find the answers to these questions **mentally**:

    a) How many toys does the machine assemble in 1½ hours?

    b) How long will it take to assemble 280 toys?

    c) What is the machine's rate in toys per hour?

16. For each concept, make up a question which pertains to that concept and has as its answer: ratio, rate, probability, percent.

17. ABC Co. has 30 employees, and their average salary is $19,000. After an additional manager is hired, the average salary goes up to 20,000. Explain how to find the new manager's salary **mentally**.

18. Explain (as you would to a 5th grader) why $\frac{3}{4}$ and $\frac{9}{12}$ are equal fractions.

19. Explain how to find all the positive fractions which are equal to $\frac{15}{21}$.

20. Suppose there are 5 red and 6 blue balls in a bag (and nothing else). In relation to this situation, explain what each of these fractions represents: $\frac{5}{6}$, $\frac{6}{11}$

21. What determines whether or not a particular fraction is equivalent to a terminating decimal? Explain.

22. Explain how to translate a decimal into a percent, and vice-versa, without talking about "moving the decimal."

23. Explain what opposites are in these three contexts:
    a) concrete real-world situations; b) number line; c) addition.

24. What is the concrete meaning of addition? How is this concept extended to addition of fractions, decimals, and variables? How is addition of a positive and a negative number basically different from all these other situations?

25. What is the concrete meaning of subtraction? What is the abstract definition of subtraction?

26. Explain the area model for multiplication of fractions.

27. Explain what the commutative, associative, and distributive properties are.

28. What is the concrete meaning of division?

29. What does it mean to say that multiplication and division are **inverse operations**?

30. Explain why $\frac{3}{0}$ is not a number.

31. Describe the various stages in which children learn about division.

32. If 17 *r* 6 and 17.75 are both correct answers to "What is $a \div b$?", what are a and b? Explain.

33. What number divided by 7 gives a quotient of 652 with remainder 3?

34. What is the largest quotient that can be formed using two numbers chosen from the set $\{-24, -3, -2, 1, 2, 8\}$?

35. A recipe calls for 3 ½ cups of sugar and 2 eggs. If you use 3 eggs, how many cups of sugar should you use?

36. Math 80 is so popular that there has been a 150% increase in enrollment this year. If there were 180 Math 80 students last year, how many are enrolled this year?

37. Maxima Rental Cars charges $55 per day plus 30¢ per mile for each mile over 125 miles per day. What would be the cost for a two day rental if the car was driven 335 miles during the two days?

38. The pediatrician tells little Suzy's mother to give the child two baby aspirins every four hours. In the middle of the night, the mother realizes she has run out of baby aspirin and has only adult tablets. According to the labels, adult aspirin are 325 mg each and baby aspirin are 80 mg each. What fraction of an adult tablet should she give little Suzy?

39. A machine can print 4020 bumper stickers in an hour. How long will it take the machine to print 4690 bumper stickers?

40. Mary bought a radio on sale for 18% off the regular price. If she paid $28.70 for the radio, how much did she save by buying it during the sale?

41. Last spring it rained in Kansas City on March 24 and the next rain was on May 5. How many dry days did Kansas City have between those two rains?

42. A car goes 180 miles on 12 gallons of gas. At this rate of consumption, how much would the gas cost for a 1620-mile trip if gas costs 88.9 cents per gallon? (Round answer to nearest cent.)

43. Out of the USL graduating class of 1040 students last May, 156 entered medical-related graduate programs. What percent of the graduates entered such programs?

44. A few months ago, Leah purchased stock at $16 a share. Last week, the same stock sold for $20 a share. What was the percent increase in the price of the stock?

45. You purchase a home with no insulation. An energy consultant claims that you will save 1/5 of your electric costs by installing insulation and ¼ of your electric costs by keeping the thermostat at 65° in the winter and 78° in the summer. If your electric bill is now $120 per month, how much should you save per year by adding insulation and by keeping the thermostat at the recommended setting?

46. The latest census reports that there are 40 million American families, and a survey shows that 17 out of every 25 American families own a VCR. Based on the census and the survey, how many American families own a VCR?

47. Find the smallest product that can be obtained by multiplying three different numbers from the set $\{-10, -5, -3, -1, 0, 3, 6, 10\}$.

48. The ratio of boys to girls in a class is five to six. If there are fifty-five pupils in the class, how many are boys?

49. If six men build six houses in six days, and nine women build nine houses in nine days, how many houses can eighteen men and eighteen women build in eighteen days?

50. Recently, Frank took a one-hundred question aptitude test where each correct answer scored 5 points, each incorrect answer scored -2 points, and each question not answered scored zero points. Frank answered 80 questions and scored 232 points. How many questions did he answer correctly?

51. AMC Long Distance Co. has a special rate on calls between 11:00 p.m. and 5:00 a.m., to anywhere in the U.S. The charge is 8.5 cents for the first two minutes, and 16 cents for each additional minute.
    a) Mrs. Hebert talked to her son in Chicago from 11:26 p.m. to 12:08 a.m. How much will the call cost?
    b) If Tom calls his girlfriend in New York between 11 and 5, how long can he talk for $10?

52. The average height of the boys in Mr. Brown's class is 5'8", and the average height for the whole class if 5'5". If there are twice as many girls as boys in the class, what is the average height of the girls?

53. 117 out of 180 people passed Math 113 last semester. What percent failed the course?

54. I have some change (pennies, nickels, and dimes) in my pocket. Fourteen of the coins are dimes, 25% of them are nickels, and five-ninths of them are pennies. What is the total value of my change?

55. Three-sevenths of the pupils in Mrs. Landry's class have had chicken pox. If twelve of the pupils have had chicken pox, how many are in the class?

56. The average of four numbers is 85. If the largest of these numbers is 97, then the average of the remaining three numbers is __________ .

57. As part of a study on weather conditions, the temperature in the air at certain altitudes above Denver was measured. These are the results:

ALTITUDE	CELSIUS TEMPERATURE
200 FT	27°
800 FT	24°
2000 FT	18°
9000 FT	-17°

a) What was the temperature at 1000 FT? at 1 mile?

b) At what altitude was the temperature at the freezing point?

c) Was it summer or winter in Denver?

58. The graph relates the distance traveled (in miles) to the time elapsed (in hours) on a trip taken by an experimental airplane. During which hour was the average speed of this airplane the largest?

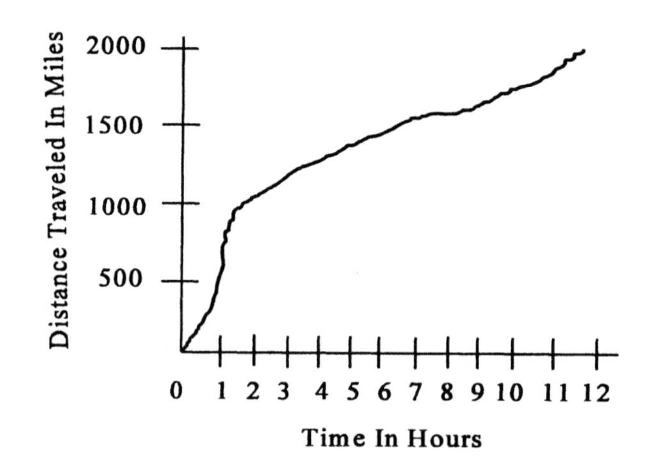

59. Weekday Activity Chart
This pie chart describes the average time that Donna spends on various activities in one day.

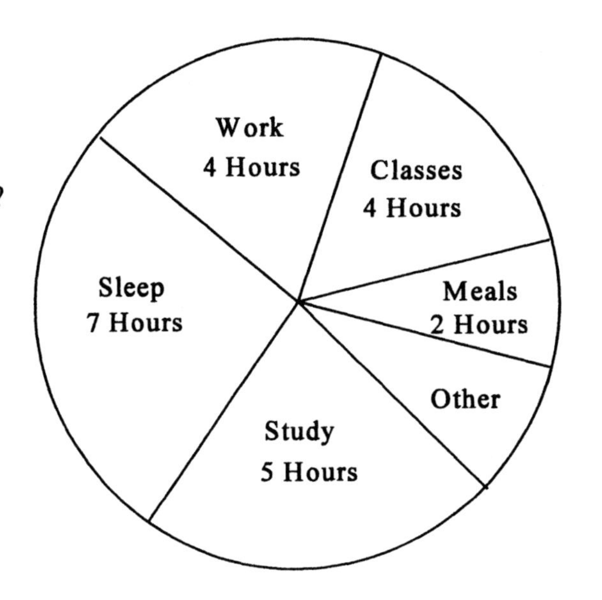

a) How much time does Donna have for activities other than those specifically listed?

b) What percent of her time does Donna spend sleeping? (Round to tenth of a percent.)

c) What fraction of her time does she spend at work?

d) If she sleeps only 6 hours, what percent decrease would this represent? (Round to tenth of a percent.)

e) What fraction of her time does she spend in classes or studying?

60. Deficit in Billions of Dollars.

a) What was the trade deficit in 1985?

b) What was the budget deficit in 1987?

c) How much more was the budget deficit in 1986 than in 1984?

d) What was the percent decrease in budget deficit from 1987 to 1988?

e) The budget deficit exceeded the trade deficit in 1986 by how many dollars? By what percent?

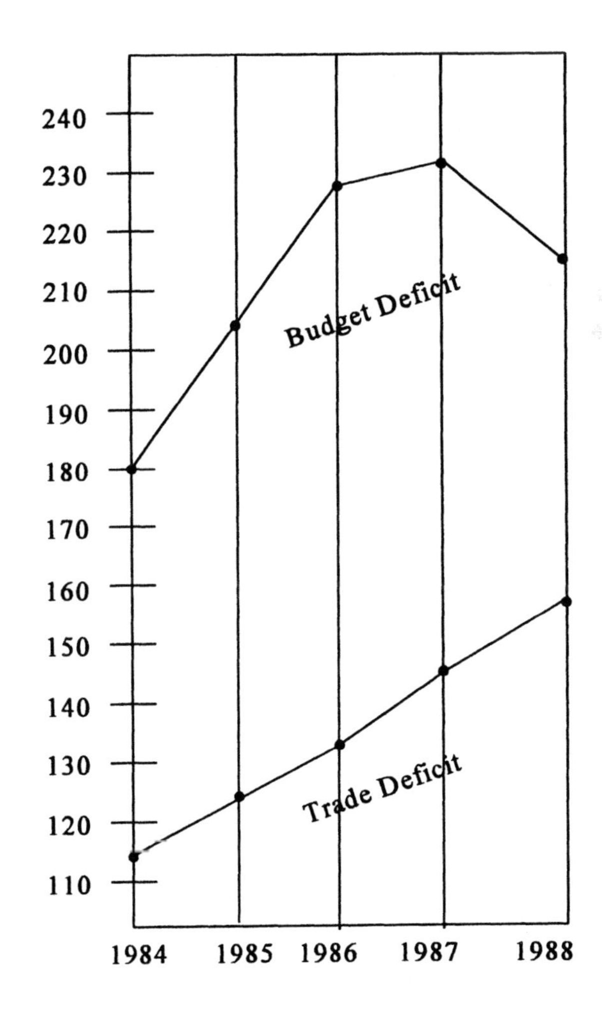

Chapter III

Section 23: What is a Measurement?

Pure or abstract numbers such as 19, 3 ¼, .86, are seldom found anywhere except in textbooks. Every number which is related to something in the real world has a description or label associated with it: 500 pounds, 6 dozen eggs, 32 square inches, 12 ½ % interest, 19.74 gallons, etc. The number indicates how many, and the label indicates how many whats. In order for "16.3 miks" to be meaningful to me, I not only must understand what 16.3 represents, I must also understand what a *mik* is.

If a number is the result of counting, the label identifies the objects which were counted — trees, pencils, dollars, whatever. If a number is the result of measuring, the label identifies some fixed quantity (unit) which was matched with the object being measured. Therefore, an understanding of units is the key to making sense of measurements.

We must begin, however, with an even more fundamental concept: What is a measurement?

Objects can be described in a variety of ways — color, length, temperature, beauty, texture, weight, etc.; and for any of these attributes, descriptions can range from very vague to very specific. For example, consider these descriptions of a steel rod:

i) very long
ii) longer than a kitchen sink
iii) about twice as long as a baseball bat
iv) 6 times as long as a sheet of ordinary notebook paper.
v) 65 inches

Obviously, all of these descriptions pertain to the length of the rod. (It's important to notice that we recognize that the last description pertains to length only because we already understand what an inch is.)

The first description is completely useless. "Very long" could range from a few centimeters, for a thin rod in a clock, to 10 or 12 yards for a supporting rod in a bridge.

The second description is vague, but it gives some information. Since the rod is compared to the length of a sink, we know that it's at least 1 ½ feet long, and probably not more than about 5 feet.

Description (iii) is much more specific than the first two. Baseball bats aren't all exactly the same length, but they vary just a little bit. So "twice as long as a baseball bat" gives a good approximation of the length of the rod.

Description (iv) is even better. Ordinary notebook paper is a standard length, so on the basis of this information, I could determine the actual length of the rod by marking off 6 adjacent sheet-lengths along a line.

Description (v) is the most specific and most useful. It's a measurement whose standard of comparison is a carefully-defined, universally-recognized unit of length. Rulers and tape measures marked in inches are readily available, so the length of the rod can easily be replicated.

Hopefully, this example has helped to clarify what a measurement is: a numerical description of some aspect of an object. A measurement always consists of a number and a standard of comparison (unit); it specifies how many of these units correspond to the object.

Some attributes of objects, such as color, texture, and beauty, are difficult to quantify; so these aren't usually described as measurements. Attributes which are commonly measured are length, weight, temperature, area, and volume or capacity.

Length, or distance, is the attribute which pertains to how far it is from one point to another along a certain path. These are some common things whose most prominent attribute is length: line segments, thread or string, roads or highways, edges or rims of objects.

Weight describes how much force an object exerts on whatever is supporting it. Two identical boxes, one empty and the other filled with iron bolts, are easily distinguishable by holding each of them in your hand. The attribute which makes them very different is their weight.

Temperature is a measure of the hotness or coldness of an object. It is the attribute which distinguishes tap water from boiling water; or the feel of the forehead of a quiet, healthy child from that of a child with a high fever.

Area is the attribute which pertains to the surface of an object — something which might be painted or covered. These are some common things whose most prominent measurable attribute is area: parcels of land, counter tops in a kitchen, tarpaulins, television screens.

Volume is a description of the amount of space an object occupies. A sofa takes up a lot more space in a room than a chair; a watermelon takes up a lot more space in the refrigerator than an apple; a big gumbo pot takes up a lot more space in a cabinet than a little sauce pan. In all of these examples, the difference between the objects in each pair is their volume.

Notice that we could also say that the space inside the gumbo pot is much greater than the space inside the sauce pan. From this point of view, when we consider the objects as containers, we often refer to how much they can hold as capacity rather than as volume.

Homework: Section 23

1. Find the answer to each question **mentally**:

   a) Suppose the five-digit number 91k85 is divisible by 3. What are all the possible values of k?

   b) The ratio of boys to girls in Mr. Stelly's class is 2 to 3. What percent of the class are boys?

   c) If $(2^4)(3^5)m = 6^7$, what is m?

   d) When a certain number is divided by 4, the quotient is 21 and the remainder is 3. What is the number?

   e) Joe scored 84, 88, and 86 on three 100-point math tests. How many points must he score on the 200-point exam in order to get an A (90 average) in the course?

   f) Find the smallest product that can be obtained by multiplying three different numbers from this set: {-10, -6, -3, 0, 4, 5}.

2. Four hundred people are lined up waiting to ride on the new Death-Defying roller coaster. An attendant walks down the line and gives every sixth person a free drink and every tenth person a free candy bar. How many people get both a drink and a candy bar?

3. Each five-digit numeral which has digit-sum of 43 is written on a slip of paper. One slip is drawn at random.

   a) What is the probability that it is divisible by 5?

   b) What is the probability that it is divisible by 11?

4. Cut out 5 examples of measurements from magazines, newspapers, or labels. For each, identify the attribute which was measured and the unit used. Bring your examples to our next class.

Section 24: Measurements on Labels

Fill-in this chart for the items displayed:

Item	Measurement on Label	Attribute Measured	Unit of Measurement
Scotch Tape			
Vanilla			
Soap			
Pain Reliever			
Chocolate Pudding			
Screws			
Iced Tea			
Wrapping Tissue			
Plant Food			
Cinnamon			
Food Storage Bags			
Super Glue			
Paint			
Cloth Tape			
Glue Stick			
Aluminum Foil			
Thread			

These are some important things to notice about the measurements you entered in the chart on the previous page:

Scotch Tape

1. These two sets of measures each give the width of the tape and the length of the roll. No one is interested in the area of the roll of tape. (We wouldn't want to cover anything with Scotch tape!) So the multiplication sign is used as an abbreviation for the word "by" — it doesn't suggest that the two numbers should be multiplied.

2. The metric measurements of width and length have different labels — and they seem "strange." The unit given for the length of the roll is tenths of a meter, which is a very uncommon unit. Also, it's not realistic that the width of the tape was measured in tenths of a millimeter; a tenth of a millimeter is too small to see. So where did these numbers come from? They were almost certainly calculated using the first set of measurements.

Vanilla

The measurements on this label give us a very convenient comparison of English and metric units of liquid capacity (volume): 1 fluid ounce is about the same amount as 30 ml. The small bottle is labeled 1 fl oz — 29 ml, while the large bottle is labeled 2 fl oz – 59 ml. How do you explain the difference?

Soap

We see from the soap label that 1 ounce is about 28 grams. This is another very useful relationship between English and metric units. The two measurements of weight emphasize how little a gram really is.

Pain Reliever

The unit for this measurement is milligram. A milligram is an extremely small amount of weight — one thousandths of a gram. (And remember, from the soap, that a gram itself is a tiny amount.) 200 mg is the weight of the actual painkiller in each tablet; it doesn't represent the weight of a whole tablet.

Thread

Since 35 yd is the same as 32 meters, we notice that there isn't much difference between a yard and a meter — a meter is a little longer than a yard (about 3 inches longer).

Chocolate Pudding

1. The measurements given in terms of ounces and grams describe the weight of the powdered mix. These reinforce the comparison we have already seen between these two units: an ounce is about 28 grams.

2. "Four ½ cups" is a description of the volume of the prepared pudding.

Screws

The label #6 x $1\frac{5}{8}$" indicates that the screws are $1\frac{5}{8}$ inches long. #6 refers to how many threads per inch, which is a count rather than a measurement. Here again, the multiplication sign has nothing to do with multiplication.

Iced Tea

All measurements on this label describe the volume of the tea. They emphasize two very convenient comparisons:

i) A pint is 16 fluid ounces.

ii) Since 16 fl oz is 473 ml, we can calculate that 1 fl oz is about 30 ml.

Wrapping Tissue

1. The label 18 sq ft is a measurement of the area of the paper — how much surface it can cover. (Note: any measurement whose unit is some kind of square is a description of area.)

2. 20 in x 26 in gives the length and width of each sheet of paper. In this case (unlike the Scotch tape), it makes sense to multiply length times width because area is a significant attribute of wrapping paper.

3. Notice that inches multiplied by inches results in a completely different kind of unit — square inches. Similarly, 7 ft x 5 ft is 35 sq ft; 4 cm x 3 cm is 12 sq cm; 10 kilowatts x 6 hours is 60 kilowatt-hours; etc.

Miracle Gro

The measurements on this label call our attention to the fact that ounces and fluid ounces are not the same — they don't even describe the same attribute of the plant food. The contents of the bottle weigh 9.8 oz and have volume 8 fl oz.

Cinnamon

Using the measurements on the soap wrapper, we calculate that 1 oz is approximately 28 grams. Here we have that comparison given directly on the label.

Sandwich Bags

1. The label quart describes the volume or capacity of a bag; whereas 7 in x 8 in gives the length and width of a bag — not a measure of its volume or capacity.

2. 17.8 cm x 20.3 cm is another example of dimensions which were probably calculated rather than measured.

Paint

1. One quart is 32 fl oz and 946 ml is almost a liter. So this label emphasizes an important relationship between English and metric units of capacity: a quart and a liter are approximately the same.

2. There is a description given on the back of the can which we didn't enter in the chart. It says: "One gallon of this paint will cover approximately 400 sq feet." This is a measure of how much surface area the paint will cover — it isn't a measurement of the amount of paint. (Remember, squares are units of area.)

Cloth Tape

The descriptions of length emphasize again that a meter is a little longer than a yard.

From the measurements of the width of the tape we get another important comparison between English and metric units of length: an inch is about the same as 2 ½ centimeters.

Glue Stick

1. This bottle contains just a little bit of glue, so the units used to make the measurements are extremely small — hundredths of an ounce and hundredths of a gram.

2. The measurement .20 oz is particularly significant. Why is there a zero after .2? It's there to say that the weight of the glue was measured in hundredths of an ounce, not tenths of an ounce; and by coincidence, there were two tenths and no hundredths. (This is comparable to writing $.20 to represent twenty cents. We describe money in terms of hundredths of a dollar — so we write .20 to signify two dimes and no cents.)

Homework: Section 24

1. In each situation described below, what attribute of an object must be measured (length, weight, capacity, etc.) ?

   a) To buy carpet to cover a floor

   b) To buy materials for building a fence

   c) To determine the value of a farm

   d) To buy punch to fill a certain thermos jug

   e) To buy ground beef for a casserole

   f) To describe the amount of flour in a cake

   g) To buy ribbon for the hem of a skirt

   h) To order gravel for a driveway

2. What attribute of an object is described by each of these measurements?

a) 18 yd	b) 2 ½ lb	c) 3 cups	d) 5.4 cm	e) 24 sq ft
f) 30 gal	g) 3 ¾ in	h) 158 g	i) 100 m	j) 5 qt
k) 14 oz	l) 81 sq m	m) 32 cu ft	n) 43.7 mi	o) 3 TBS
p) 2 liter	q) 12 fl oz			

3. Each of these represents a measurement: 5.8 glumps; 11 sq miks; 2 cu tibs
   a) Do you know what attribute of an object has been measured in each case? Explain.
   b) From these measurements, can you tell anything about how big or how heavy the object is? Explain.

4. Explain what a measurement is, and describe the process for making a measurement.

5. Explain what this means: $57_{\text{base } m}$.

6. Mrs. Broussard bought a new car in Lafayette where the sales tax is 7%. Her total cost (price of car plus tax) was $19,260.00. If she had bought the car in Abbeville where the sales tax is 6 ½% , how much would she have saved?

7. Suzy bought two boxes of assorted red and green ornaments for her Christmas tree. Each box contained the same number of ornaments. In one box, the ratio of red to green was 4 to 3 and in the other box the ratio was 2 to 3. When all the ornaments were put on the tree, what was the ratio of red to green ornaments?

Section 25: Measurement of Length

The purpose of the activity in Section 24 was to emphasize that an understanding of commonly-used units is the key to making sense of measurements.

The unit specified in a measurement identifies two essential things: (i) what attribute of the object was measured, and (ii) what fixed quantity was the standard for comparison. So in order for measurements to be meaningful, a person must recognize both the attribute and the amount represented by various units.

We begin with a review of the names and relative sizes of units for ordinary measurable attributes. Recall that there are two sets of standard units: the English system (used only in the United States and Great Britain) and the more universal metric system. The metric system is used everywhere for science and technology, and it is becoming common in the U.S. in many non-technical situations.

Measurement of Length

Length, or distance, is the attribute which pertains to "how far" it is from one point to another along a certain path.

These are the common units of length and the relationships among them within each system.

English	
inch (in)	
foot (ft)	12 in
yard (yd)	3 ft or 36 in
mile (mi)	5280 ft

Metric	
millimeter (mm)	
centimeter (cm)	10 mm
meter (m)	100 cm
kilometer (km)	1000 m

The metric system has two very "nice" features:

i) For a particular attribute, the various units are numerically related by powers of ten. Consequently, measurements given in one unit can be mentally converted to any other metric unit.

ii) For each attribute there is only one basic name, and different units are specified by attaching a prefix to that root name. Moreover, the prefixes identify the power-of-ten relationship of that unit to the basic unit. For length, the basic unit is the meter. Prefixes, for all basic units, are milli (which means "one-thousandth"), centi (which means "one-hundredth"), and kilo (which means "one thousand").

Since both the English and metric systems are now widely used in the U.S., it's important for school children to become comfortable with both sets of units. It's also helpful for a person to have some references for comparing units between systems. For length, these three comparisons are particularly useful:

* An inch is about 2 ½ centimeters.
* A meter is a little longer than a yard. (A meter is about 39 inches.)
* 5 miles is about 8 kilometers.

Lengths can be estimated using hands and feet:

* centimeter — width of little finger
* inch — middle joint of index finger
* foot — length of man's foot
* yard — from center of chest to tip of outstretched arm
* yard — long stride

There are a variety of tools for making accurate measurements of length. If the length to be measured is relatively short, a ruler or tape measure can be used. For longer distances, mechanical devices such as pedometers and odometers are available. For extremely long distances, measurements are made by finding how long it takes sound or light (which both have constant speeds in a given medium) to travel from one point to the other.

Group Activity

Find the measurements required to fill-in these charts.

	width		length
front desk		**classroom**	

	circumference	diameter	c + d	c - d	cd	c ÷ d
hula-hoop						
basket bottom						
basketball						

Some Things To Notice:

1. It's much more difficult to measure the length of the classroom than the width of the desk; so a wider range of answers is expected among the measurements of the room. For the desk, the width can be accurately determined to the nearest fourth-inch; so results from all the groups should be within a fourth-inch range. For the room, accuracy to the nearest inch seems reasonable; so answers among all the groups should be within a one-inch range.

2. The units given in the entries for c + d, c - d, cd, and c ÷ d deserve special attention. If we assume that the measurements for c and d have the label inches, then c + d and c - d will also have that label. But the label for cd is square inches, and c ÷ d has no label.

3. If we look at the entries for c + d, we find a wide variation among the values for the three circular objects. Similarly, there is a wide range among the three entries for c - d and for cd. but the three values for c ÷ d are approximately the same! This suggests a very significant relationship: The circumference of a circle is a little more than 3 times its diameter.

Homework: Section 25

1. Estimate each using English units:

   a) length of this segment ______________________________

   b) length of a new pencil

   c) diameter of a quarter

   d) distance from Lafayette to Baton Rouge

   e) height of President Clinton

2. Estimate each using metric units:

   a) length of this segment ______________________________

   b) distance from Lafayette to Houston

   c) length of a baseball bat

   d) diameter of a tennis ball

   e) height of an ordinary room

3. Find the perimeter of each figure using the given unit:

   a) millimeter

   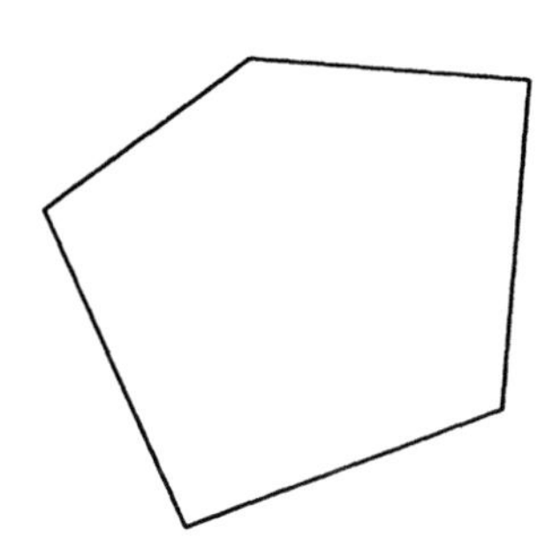

   b) centimeter

   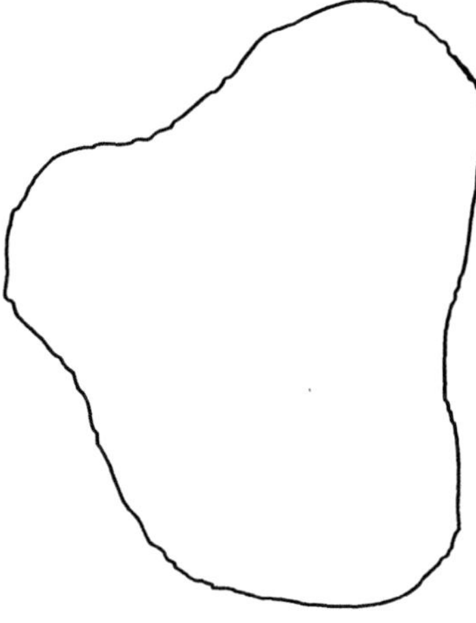

4. Find the answers mentally:

   a) How many centimeters are equivalent to 7 kilometers?

   b) $2\frac{1}{4}$ ft is equivalent to how many inches?

   c) 24 inches is what part of a yard?

   d) 21.6 centimeters is the same as _______ meters and ________ millimeters.

   e) $5\frac{2}{3}$ yd is equivalent to how many feet?

5. How many seconds are in a centihour?

6. Is 8 mm film closer to $\frac{1}{2}$ inch or to $\frac{1}{3}$ inch?

7. As you drive along I-10, you notice that the speed limit is 65 mph. The speedometer on your BMW reads 100 km-per-hr. Are you speeding?

8. How many kiloinches are in a mile?

9. Which is longer, the 100-meter dash or the 100-yard dash? About how much longer?

10. In the Olympic Program, the height of a Russian basketball player is given as 215 cm, and the height of an American player is given as 6 ft 10 in. Which player is taller?

Section 26: More about Circumference

Activity

1) Draw a segment which is 4.0 cm long.
2) Set your compass opening at the length of the segment.
3) In the space below, construct a circle using this compass opening.

4) What is the radius of your circle?
5) What is the diameter of your circle?
6) Mark point A anywhere on the circle.
7) Using point A as a starting point, and your compass set at the same opening, mark off six equally spaced points on the circle.
8) Draw segments to connect adjacent points.
9) What is the perimeter of the hexagon formed by the segments your drew?
10) How is the perimeter of the hexagon related to the diameter of the circle?

Now notice that the circumference of the circle is just a little longer than the perimeter of the hexagon. So we see that the circumference of a circle is just a little more than three times the diameter.

This is a graphic demonstration of the relationship you noticed for the circular objects in the last activity — for each object, $c \div d$ was a little more than 3.

The actual number, $c \div d$ or $\frac{c}{d}$, is the same for all circles. It's an infinite, non-repeating decimal which is very close to 3.14. (Remember that fractions are equivalent to repeating decimals; so this number cannot be written exactly as a fraction. But it's very close to $\frac{22}{7}$.) As you know, the name of this very important number is pi (π).

Homework: Section 26

1. The distance around a circular lake is about a mile. About how far is it across the lake? (Give your answer in yards.)

2. Construct a hexagon which has all sides equal and perimeter of 9 inches.

3. The spokes of Suzy's bicycle wheels are 25 cm long. About how many revolutions do the wheels make when Suzy rides 2 kilometers?

4. Suppose a rectangle has these properties:
   * The lengths of all the sides are a whole number of inches.
   * The perimeter is 14 inches.

   a) Could the rectangle be a square? Explain.

   b) From the description given, can you determine the area of the rectangle? If so, what is it? If not, explain why it can't be determined.

5. Mr. Soileau has a circular rose garden which is 14 ft in diameter. He wants to put a fence, consisting of 4 strands of wire, around the garden. Wire is sold in 15-yard spools. How many spools does he need to build the fence?

6. Draw a circle whose circumference is approximately 19 cm.

7. How many hours are in a centiyear?

8. Find the answers **mentally**:

   a) How many inches are equivalent to $1\frac{1}{4}$ yd?

   b) How many meters are equivalent to 857 cm?

   c) How many yards are equivalent to 17 ft?

   d) How many meters are equivalent to 16.3 km?

   e) How many meters are equivalent to 8 mm?

   f) How many feet are equivalent to $4\frac{1}{2}$ yd?

   g) 35 miles is about ______ kilometers.

   h) 2 meters is about ______ inches.

   i) 10 cm is about ______ inches.

   j) 10 meters is about ______ feet.

   k) 4 yards is about ______ cm.

   l) 1 foot is about ______ centimeters.

   m) Azaleas are to be planted about four feet apart around a circular patio. If the radius of the patio is 10 ft, how many azaleas will be needed?

   n) If you draw a card at random from a full deck, what is the probability that it will be a spade or a red queen?

   o) List these numbers in order from smallest to largest:

   1) $41 + \frac{1}{9683}$ 2) $41 - \frac{1}{9683}$ 3) $41 \times \frac{1}{9683}$ 4) $41 \div \frac{1}{9683}$ 5) 41.9683

   p) At a "20% off" sale, how much would you pay for a coat that was originally selling for $65?

   q) A Senate committee of 21 members (all either Republican or Democrat) is to be selected so that the ratio of Republicans to Democrats is 4 to 3. How many Democrats will be on the committee?

Section 27: Measurement of Weight

Remember that weight is the force which an object exerts on whatever is supporting it. These are the common units of weight and the relationships within each system:

English	Metric
ounce	milligram
pound (16 ounces)	gram (1000 milligrams)
ton (2000 pounds)	kilogram (1000 grams)

Two comparisons between the system are worth remembering:
* An ounce is about 28 grams.
* A kilogram is about 2.2 pounds.

Activity

1. Estimate the weight of each object listed in the chart.
2. Measure the weight of each object in both English and metric units.

Object	Estimated Weight(English)	Actual Weight (English)	Estimated Weight (metric)	Actual Weight (metric)

Homework: Section 27

1. Make a list of eight non-food items which are commonly measured by weight.

2. Estimate the weight of each of these:

   a) gallon of milk
   b) ping-pong ball
   c) grapefruit
   d) a cat
   e) compact car
   f) 2-liter bottle of coke
   g) average-sized 12 year old
   h) a brick
   i) an encyclopedia
   j) a slice of bread
   k) pair of men's tennis shoes
   l) a large refrigerator

3. Find these answers **mentally**:

   a) 3/4 yd is the same as _____ inches.
   b) What is the radius of a circle whose diameter is $10\frac{5}{8}$ inches.
   c) 6.3 kilograms is equivalent to ____ grams.
   d) How many ounces are in a milli-ton?
   e) 4 meters is almost the same as ____ feet.
   f) If all sides of an octagon are $1\frac{1}{4}$ inches long, what is the perimeter of the octagon?
   g) 3.5 tons is ____ pounds.
   h) 38 millimeters is _____ centimeters and _____ meters.
   i) A 100-foot garden hose is lying in a circle on the ground. Approximately what is the diameter of the circle?
   j) 2 ounces is about _____ grams.

4. Which of these are measurements? For each measurement, what attribute of the object does it describe?

   a) 4.83 lb
   b) 4.7 cm
   c) 1.5 dozen
   d) 240 mg
   e) size $9\frac{1}{2}$ shoe
   f) 18 sq yd
   g) 5 kg
   h) 15.89 dollars
   i) 14 oz
   j) 21.7 cu mi

5. Every time these two wheels are spun, two numbers are selected by the pointers. What is the probability that the sum of the two selected numbers is even?

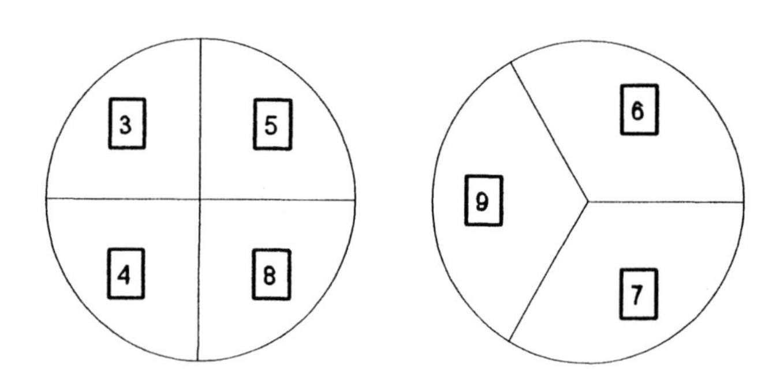

6. The average height of the boys in Mr. Brown's class is 5'8", and the average height for the whole class is 5'5". If there are twice as many girls as boys in the class, what is the average height of the girls?

7. Ann, Bob, and Cathy were planning a camping trip. They agreed to equally share the cost of food and supplies. Cathy had to work the day before the trip, so Ann and Bob bought everything. Ann spent 25% more than Bob. Cathy's share was $18. How should the $18 be split between Ann and Bob?

8. Make a transparency of the next page and bring it to class.

Make an overhead transparency of this page.

Section 28: Measurement of Area

Area is a measure of surface — something which might be covered or painted.

Units of area are almost all squares.

English

sq in
sq ft (144 sq in)
sq yd (9 sq ft)
sq mi (5280 x 5280 sq ft)

Metric

sq cm
sq m (10,000 sq cm)
sq km (1,000,000 sq m)

The most common non-square unit of area is the acre. An acre is about 44,000 sq ft, and it has no particular shape.

It's clear that the area of a rectangle can be calculated from measurements of length and width.

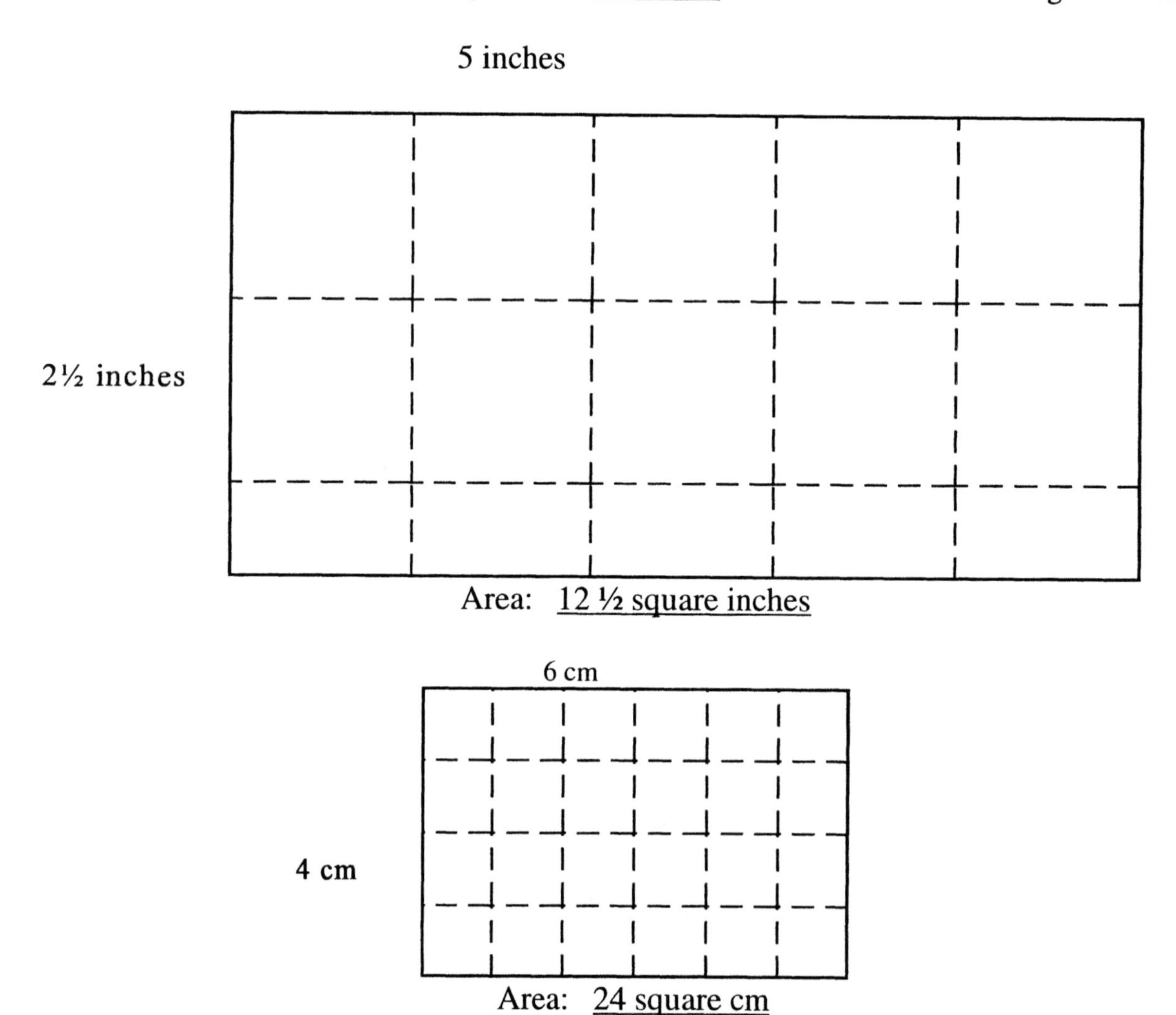

The area of circles, triangles, and other polygons can also be calculated from linear measures. We will discuss these in detail in later sections.

For irregular figures, area can be approximated using grids. The area of the figure shown below is approximately 52 square centimeters.

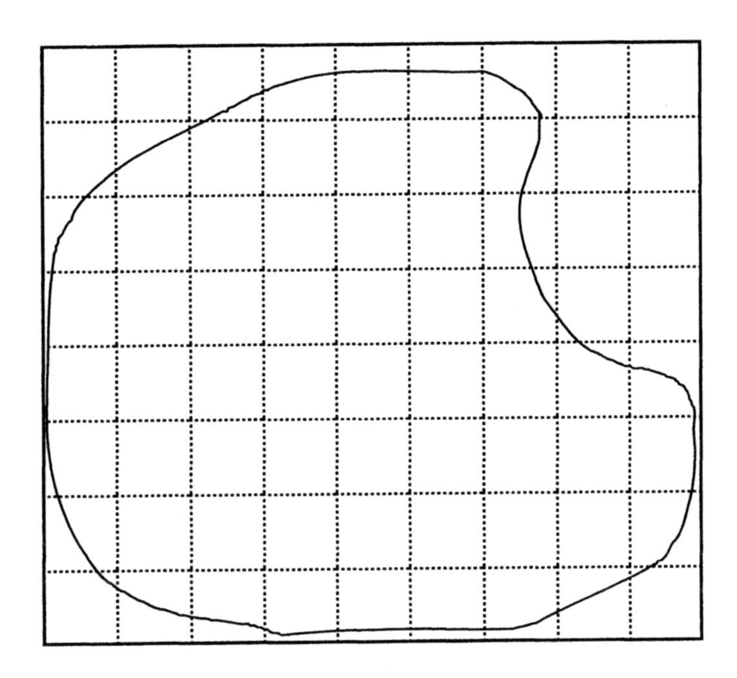

Homework: Section 28

I. Find the approximate area of each figure with a square centimeter grid. What is a reasonable range of answers for each figure?

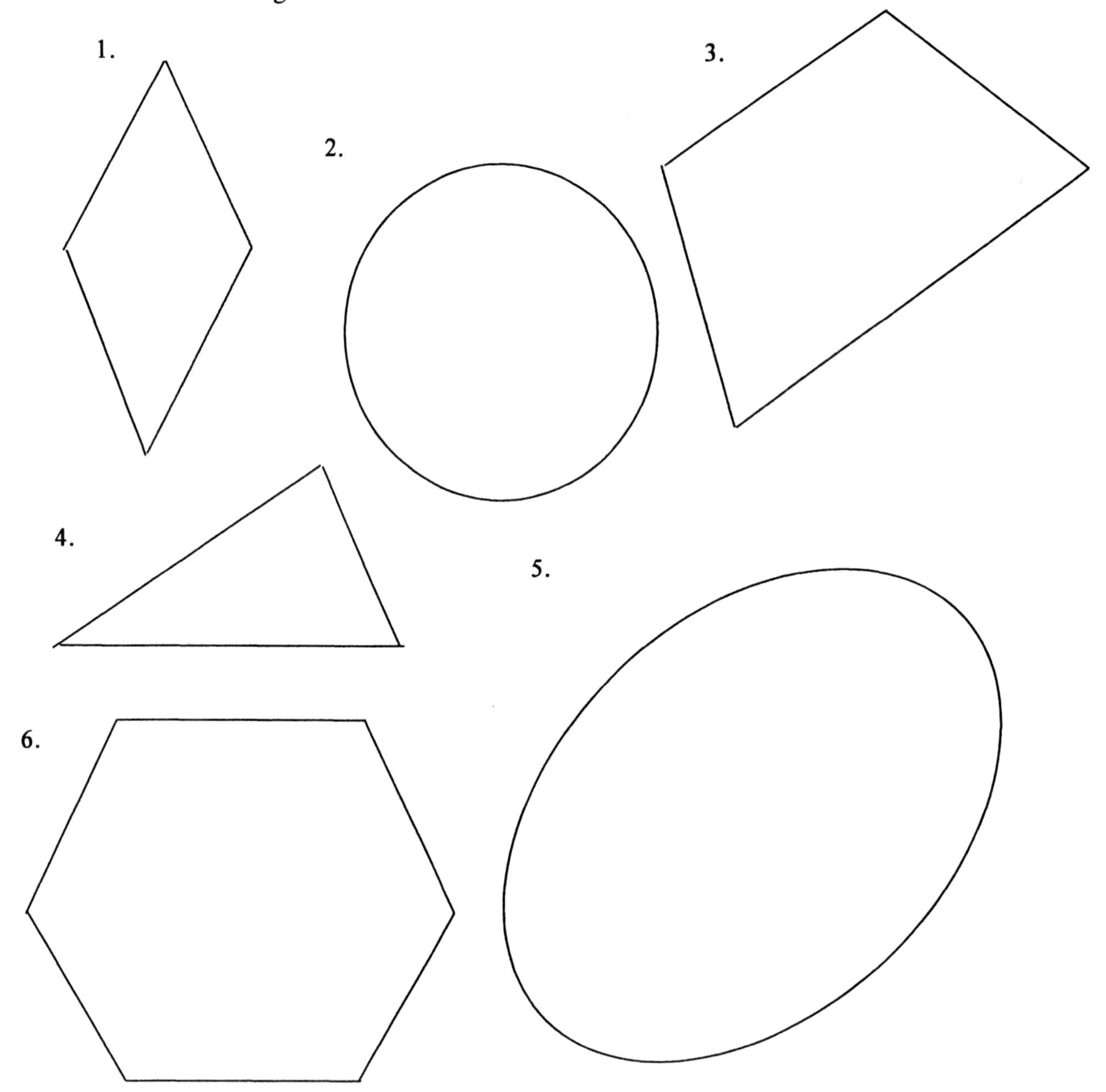

II. 1. For each figure, what percent of the grid area is shaded? Do you expect everyone to get the same answers? Explain.

a)

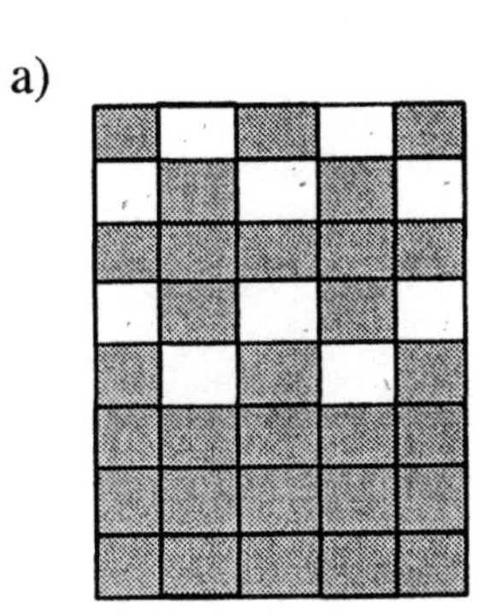

b)

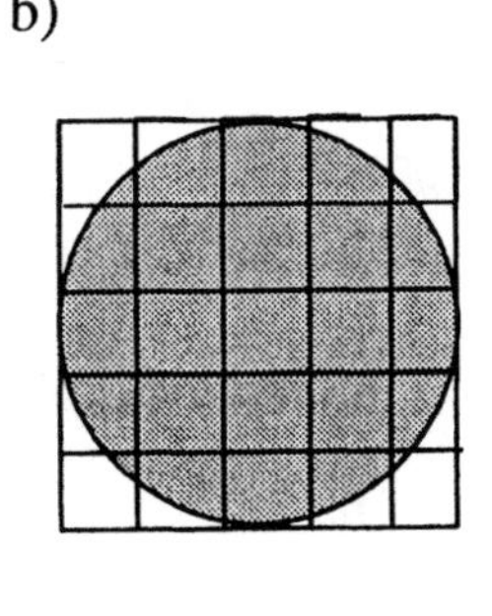

c)

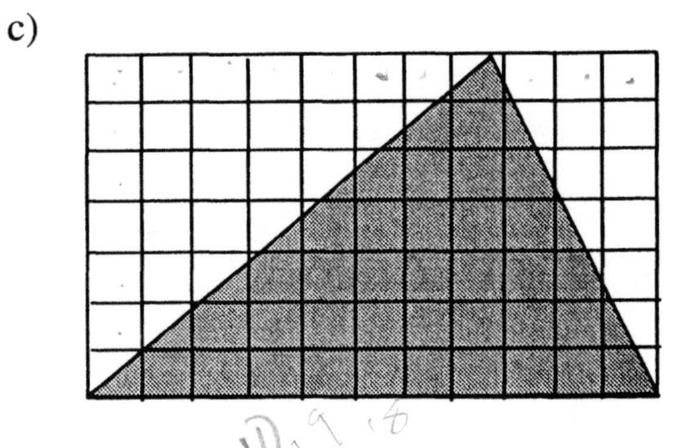

2. On the grid to the right, what is the ratio of shaded to unshaded area?

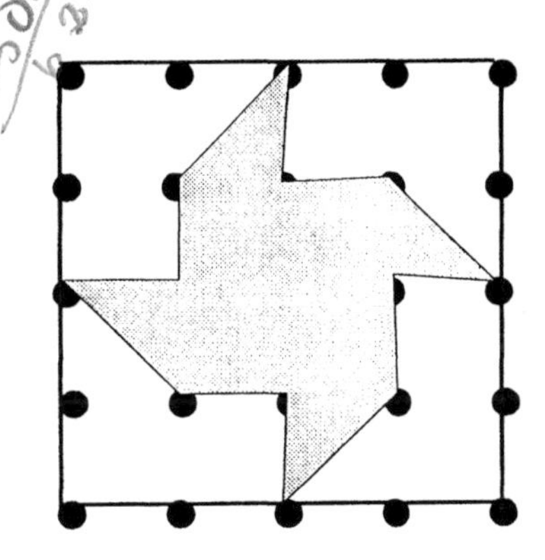

3. a) $3\frac{1}{2}$ sq ft is _______ sq inches.

b) 9.6 sq m is equivalent to _______ sq cm.

4. A football field is 100 yd long and $53\frac{1}{3}$ yd wide. Approximately how many acres is it?

5. About how many square feet are in one square meter?

6. The area of the figure represented at the right is 180 square centimeters. What is the perimeter of the figure?

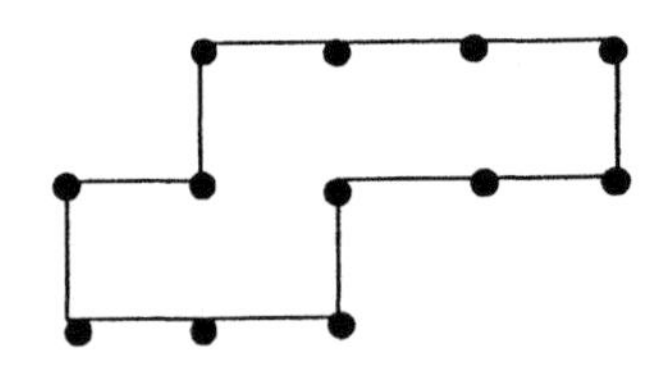

Section 29: Measurement of Volume

Volume describes how much space an object occupies, or how much a container can hold. There are two different sets of units for measuring this attribute.

Cubic units are used to measure the volume of appliances (washers, dryers, refrigerators), construction materials (concrete, sand, gravel), engines, and cooling capacity of air conditioners. Notice that none of the items in the chart on page 96 are labeled in cubic units. This is because these units are not commonly used for portable household things.

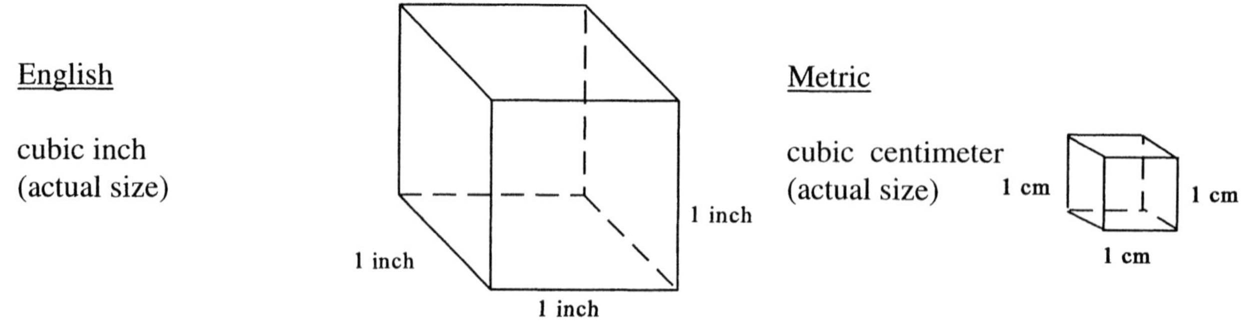

cubic foot: 12 x 12 x 12 cu in

cubic yard: 3 x 3 x 3 cu ft

cubic meter: 100 x 100 x 100 cu cm

To measure the volume of some object using cubic units, we must determine how many cubes, one unit on each edge, would be required to construct the object (if solid) or fill it up (if hollow).

If the object is rectangular (a box), its volume can be calculated from measures of its length, width, and height. The block shown at right clearly consists of 120 cubic centimeters. This can be calculated by multiplying the measures of length (5 cm), width (4 cm), and height (6 cm).

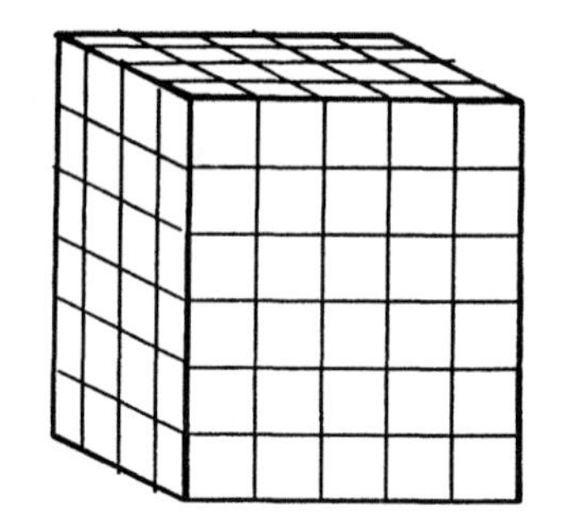

Volume for many objects which are not rectangular can also be calculated from linear measures. We will consider these in later sections.

For most household items, volume is measured in units of liquid capacity. Despite the name, these units aren't restricted to measurement of liquids. These are the common units:

English	Metric
teaspoon	milliliter
tablespoon (3 tsp)	liter (1000 ml)
fluid ounce (2 TBS)	
cup (8 fl oz)	
pint (2 cups)	
quart (2 pts)	
gallon (4 qts)	

There are two very useful comparisons between these English and metric units of liquid capacity:

* A liter is just a little more than a quart.
* A fluid ounce is about 30 milliliters.

Note: In the English system there is no direct link between cubic measure and units of liquid capacity. For example, it can be demonstrated that a cubic foot is comparable to about 7 ½ gallons. But this is only an approximation.

$$7\frac{13}{16}\text{ in} \times 3\frac{13}{16}\text{ in} \times 3\frac{13}{16}\text{ in}$$

7.8 in x 3.8 in x 3.8 in

volume of half-gallon: $\approx$ 113 cu in

volume of gallon: $\approx$ 226 cu in

cubic foot = 12 x 12 x 12 cu in
= 1728 cu in

$1728 \div 226 \approx 7.6$

Therefore, a cubic foot is approximately 7 ½ gallons.

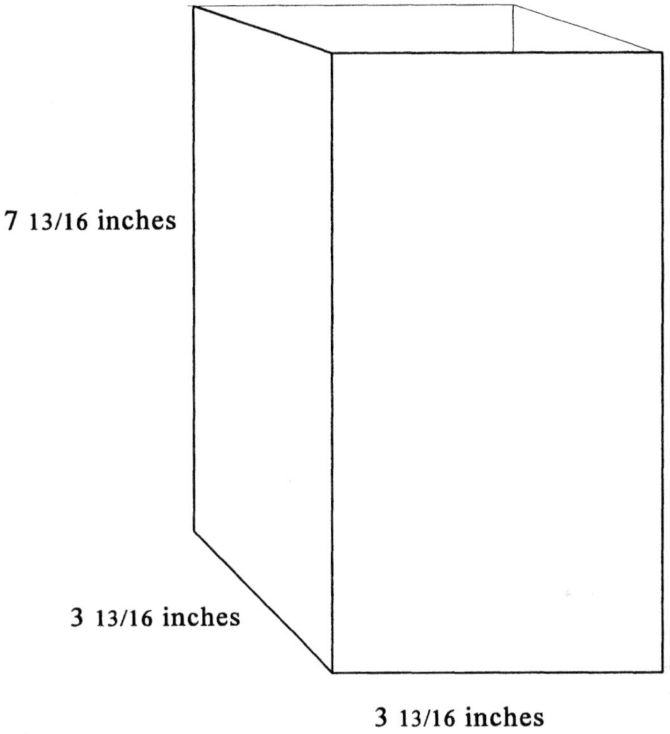

7 13/16 inches

3 13/16 inches

3 13/16 inches

Half-gallon milk carton

In the metric system, however, we have an equivalence which allows us to convert directly between cubic measure and units of liquid capacity:

A milliliter is exactly the same as a cubic centimeter.

Homework: Section 29

1. Find these answers **mentally**:
   a) 3.17 liters is the same as _____ ml.
   b) 263 ml is the same as _____ cu cm.
   c) A 20-qt pot holds _____ gallons of gumbo.
   d) A pint is _____ fluid ounces.
   e) 40 miles is about _____ km.
   f) 2 teas is about _____ ml.
   g) 78.9 kg is the same as _____ grams.
   h) 3 kg is about _____ lbs.
   i) A cubic yard is the same as _____ cu ft.
   j) A liter is about ___ cups and about ___ fl oz.
   k) A cup is about _____ ml.
   l) A square meter is about _____ Sq ft.
   m) 2 feet is about _____ cm.
   n) 17 mm is the same as ___ cm and ___ m.

2. For each of these measurements, what attribute of the object is being described:
   a) 21.4 cm b) 280 ml c) 4.3 acres d) 9.6 kg
   e) 2 TBS f) 3 ½ oz g) 14 fl oz h) 250 mg
   i) 1 ½ cu yd j) 75 sq ft k) 2 liters l) 1 pint
   m) 21 g n) 1.6 cu nibs o) 4630 km p) 5.4 sq pacas

3. For each pair of measurements, which represents the greater quantity?
   a) 18 inches, 60 cm b) 3.4 kg , 5 lb 12 oz
   c) 28 sq in, 50 sq cm d) 5 gallons, 20 liters
   e) 1 cup, 400 ml f) 3. fl oz, 5 TBS

4. A swimming pool is 18 ft long, 14 ft wide, and 5 ft deep. About how many gallons of water does it hold?

5. A bottle of vaccine contains 86 ml. How many 1 ½ cc "shots" can be given from the bottle?

6. Two hundred cubes are linked together as shown. How many square units of surface does this object have?

7. A jar of peanut butter is labeled "20 oz; 14 fl oz". Another jar of the same brand is labeled "10 fl oz". What is the weight of the peanut butter in the smaller jar?

8. The length and width of the rectangle shown are whole numbers of centimeters, and its perimeter is less than 100 cm. The ratio of length to width is 3 to 2. What is the maximum possible area of the rectangle?

Section 30: Measurement of Temperature

Remember that temperature is a description of the hotness or coldness of an object.

On the Fahrenheit thermometer to the right, there are 180 spaces between the boiling and freezing points of water. On the Celsius thermometer, there are only 100 spaces between these two fixed temperatures. So, 9 Fahrenheit degrees correspond to 5 Celsius degrees.

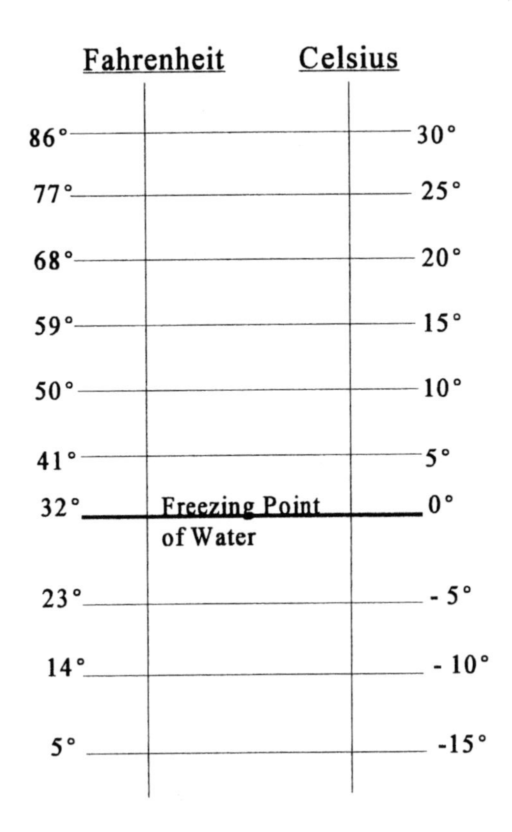

Some often-measured temperatures are shown on the scales at right:

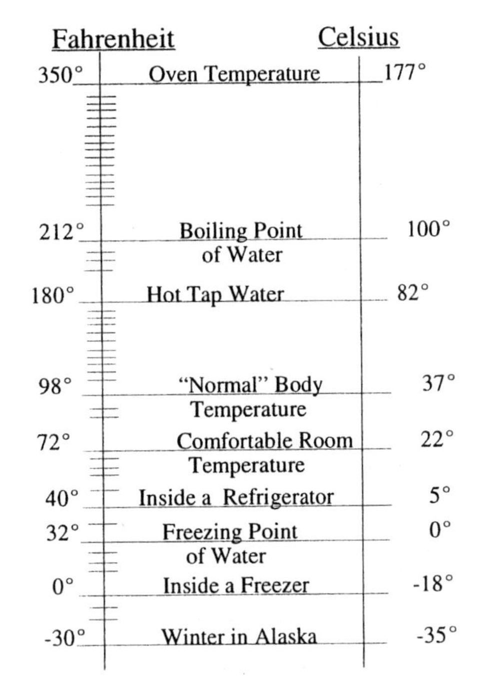

Chapter III Review

I. To describe each of these, what attribute is usually measured?

1. amount of flour in a cake
2. amount of oranges in a sack
3. size of a garbage can
4. size of a house
5. amount of caffeine in a cup of coffee
6. amount of butter in a stick
7. dose of cough medicine
8. size of a farm
9. amount of aspirin in a tablet
10. amount of cereal in a box
11. amount of 7-up in a can
12. amount of sirloin steak in a package
13. amount of vaccine in an injection
14. size of a new-born baby
15. size of a paper cup

II. Specify which English and which metric label is commonly used for each of these measurements.

1. length of a room
2. weight of a Boeing 747
3. distance from Boston to Dallas
4. weight of a T-bone steak
5. distance from Martin Hall to Griffin
6. length of a pencil
7. thickness of a penny
8. distance from the earth to the sun
9. area of a room
10. volume of a warehouse
11. area of a large city
12. volume of a shoe box
13. area of a sheet of paper
14. volume of a gasoline tank
15. area of a farm
16. volume of a swimming pool
17. volume of a big gumbo pot
18. surface area of the earth
19. volume of a thermos jug
20. volume of a drinking glass
21. area of a postage stamp
22. volume of a refrigerator.

III. **Estimate** each of the following in metric units:

1. area of the top of a slice of bread
2. weight of a large apple
3. distance from New Orleans to New York
4. capacity of a small microwave oven
5. width of a compact car
6. volume of a basketball
7. length of a sofa
8. weight of a backpack with 3 large notebooks and 2 textbooks
9. area of a King-size sheet
10. capacity of a large mixing bowl

IV.

1. A highway sign says that the speed limit is 72 km/hr. You are going 45 mi/hr. Are you speeding?

2. One event in the Olympics is the 60 meter run. About how many yards is that?

3. Joe is 4 ft 11 in tall; his friend is 142 cm tall. Who is taller?

4. About how many liters of gasoline can you buy for $20 if gas costs $1.25 per gallon?

5. A piece of imported cheese is marked 2.6 kg. If it sells for $5 per pound, about how much will it cost?

6. A rectangular farm is .35 miles in length and .26 miles in width. About how many acres is it?

7. A platinum ring weighs 18 grams. If platinum is worth $6700 per ounce, how much is the ring worth?

8. The concrete slab for a house will have a surface area of 2200 sq ft. There is to be a 4 inch layer of sand below the slab. How many cubic yards of sand will be needed?

9. Carpet costs $24 per square yard. How much will it cost (approximately) to put wall-to-wall carpet in a room which is 18 ft long and 14 ft 9 in wide?

10. What kind of clothing would be comfortable in each of these temperatures?
    28° C; 5°C; 20° C; 20°F; 45° F; 79° F.

11. A brick is 8 in long, 4 in wide, and 2 in thick. A wooden crate is 48 in long, 26 in wide, and 18 in high. How many bricks will fit inside the crate?

12. Find the required measurements in metric units.

    a) Find the length :

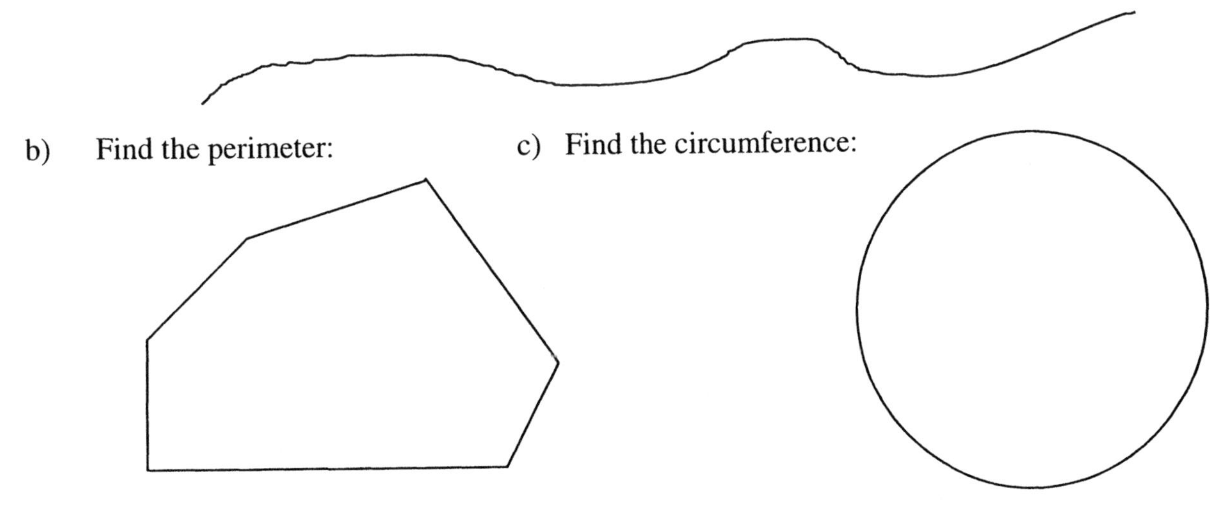

13. A package of salami weighs 734 grams and sells for $3.95 per pound. How much will it cost?

14. A six-pack of 12-fl ounce cans of American beer sells for $2.79. An eight-pack of 245 ml bottles of imported beer sells for $2.15. Which is the more expensive beer?

15. Make as close an approximation as you can of these areas (in sq cm).

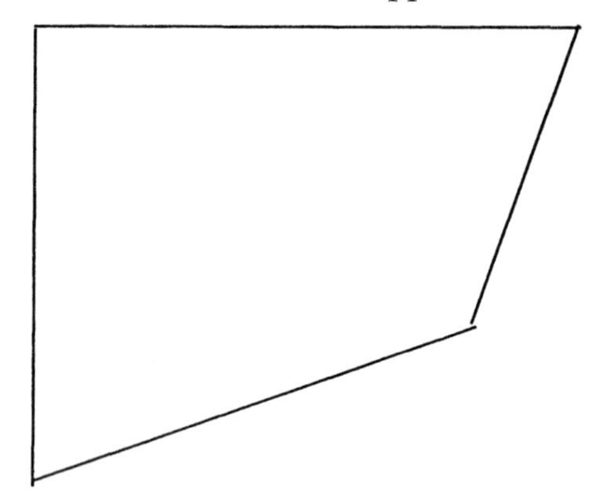

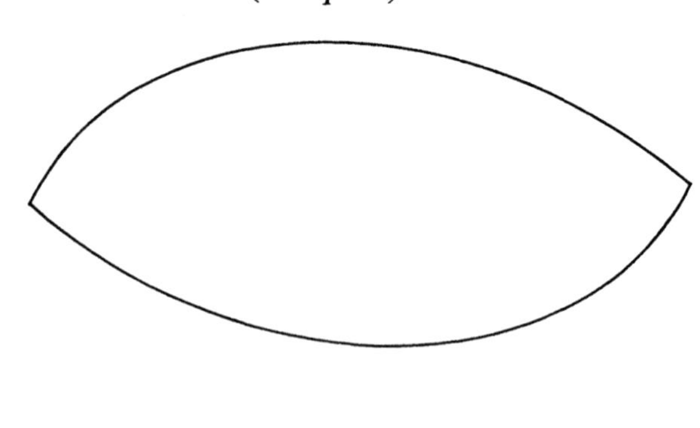

V. Optional (if additional reinforcement is needed)

A. Fill in the equivalent amounts:

1. 7 ⅔ yd = ________ ft
2. 0.06 m = ________ mm
3. 75 g = ________ kg
4. 52 oz = ________ lb
5. 8.4 mi = ________ yd
6. 3.042 m = ________ cm
7. 0.063 ton = ________ lb
8. 3.74 km = ________ cm
9. 76 in = ________ ft
10. .043 kg = ________ mg
11. .28 mg = ________ g
12. 6 ⅜ lb = ________ oz
13. 2112 ft = ________ mi
14. 21.3 g = ________ mg
15. 160 lb = ________ ton
16. 873 cm = ________ m
17. 99 in = ________ yd
18. 8 ⅔ ft = ________ in
19. 92 cm = ________ km
20. 0.007 ton = ________ oz
21. 14 sq yd = ________ sq ft
22. 3780 sq cm = ________ sq m
23. 14.8 cu cm = ________ ml
24. 6 ½ gal = ________ qt
25. 7 teas = ________ Tbs
26. 13 cups = ________ pt
27. 12.4 l = ________ ml
28. 108 cu ft = ________ cu yd
29. .24 sq mi = ________ sq ft
30. 7296 cu cm = ________ cu m
31. 2 ⅓ sq ft = ________ sq in
32. 12.7 sq cm = ________ sq mm
33. .38 l = ________ ml
34. 168 fl oz = ________ qt
35. 504 sq in = ________ sq ft
36. 1080 cu in = ________ cu ft
37. 4970 ml = ________ l
38. ⅓ sq yd = ________ sq in
39. 3.1 sq m = ________ sq cm
40. 3 ¼ cup = ________ fl oz

B. Fill in the approximate amounts.

1. 16 ½ in ≈ ________ cm
2. 2.09 lb ≈ ________ kg
3. 385 mi ≈ ________ km
4. 87 g ≈ ________ oz
5. 16 ¾ yd ≈ ________ m
6. ¾ oz ≈ ________ mg
7. 9385 kg ≈ ________ lb
8. 5.8 m ≈ ________ in
9. 350 mg ≈ ________ oz
10. 14.6 km ≈ ________ yd
11. 98 ft ≈ ________ m
12. 8 3/5 oz ≈ ________ g
13. 7 ⅓ yd ≈ ________ cm
14. 24 cm ≈ ________ in
15. 1 ⅜ in ≈ ________ mm
16. 0.88 kg ≈ ________ oz
17. 2 ½ ft ≈ ________ cm
18. 13,000 ft ≈ ________ km
19. 3.5 mi ≈ ________ m
20. .79 lb ≈ ________ g
21. 8 fl oz ≈ ________ ml
22. 15 acres ≈ ________ sq ft
23. 8 cu m ≈ ________ cu ft
24. 12 ½ gal ≈ ________ l
25. 24 cu in ≈ ________ cu cm
26. 48 sq in ≈ ________ sq cm
27. 5 acres ≈ ________ sq m
28. 1½ pt ≈ ________ ml
29. 2¼ cups ≈ ________ ml
30. 0.56 sq mi ≈ ________ acre
31. 68°F ≈ ________ °C
32. 95°F≈ ________ °C
33. 350°F≈ ________ °C
34. - 4°F≈ ________ °C
35. 31°C≈ ________ °F
36. - 10°C ≈ ________ °F
37. 62°C ≈ ________ °F
38. 20°F ≈ ________ °C

Chapter IV

Introduction:

Geometry developed as practical, experimental mathematics. As early as five or six thousand years ago, people were measuring objects which they called line segments, angles, circles, etc. They thought of these things as sets of points — tiny dots similar to the point of a pen. From their measurements, these early geometors discovered almost all of the geometric relationships which we know today.

In time, just as man abstracted the idea of "four" from sets of four objects, he began to abstract the idea of purely mathematical figures from physical models of points, circles, triangles, etc. The study of geometry moved from the real world to the mind, where there are none of the limitations of physical reality — mathematical points have no size at all; segments can have exact lengths; circles are always perfect. Abstract geometric relationships were proved by logical reasoning from a given set of definitions and axioms; physical geometric relationships were approximated from measurements.

We will try to deal with this dual nature of geometry without constantly referring to it. It will usually be clear from the context whether "point" means mathematical point or a point in the real world, and whether "circle" means a perfect abstract circle or a circular object.

Section 31: Straightness, Flatness, and Dimension

The figure at the right shows several paths between A and B. If you were traveling from A to B along one of these paths, your distance would be the length of that path.

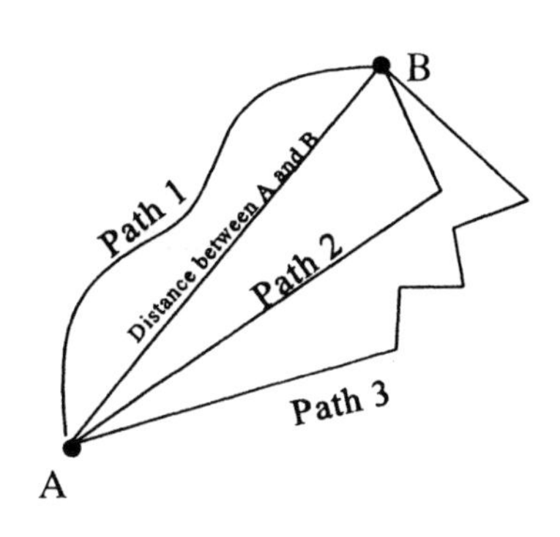

In mathematics, we use the term <u>distance</u> to specify the length of the shortest path (or straight path) between the points. This length, or distance, is designated as AB. The straight path between A and B is called a <u>segment</u>, and it is denoted as $\overline{AB}$ or as $\overline{BA}$. Notice that $\overline{AB}$ (with a bar over the letters) refers to the path or segment, whereas AB (with no bar) refers to the length of the segment. $\overline{AB}$ is a <u>figure</u> and AB is a <u>number</u>.

A and B are the <u>endpoints</u> of $\overline{AB}$. The rest of the segment is made up of all the points which are between A and B along the straight path. In this context, the word <u>between</u> has a very specific meaning which defines "straightness."

Consider the points M, P, and Q as the location of three houses along a country road. It would be natural, in everyday conversation, to say that P is between M and Q. In geometry, however, the word <u>between</u> implies "on a straight path." So in this sense, P is not between M and Q.

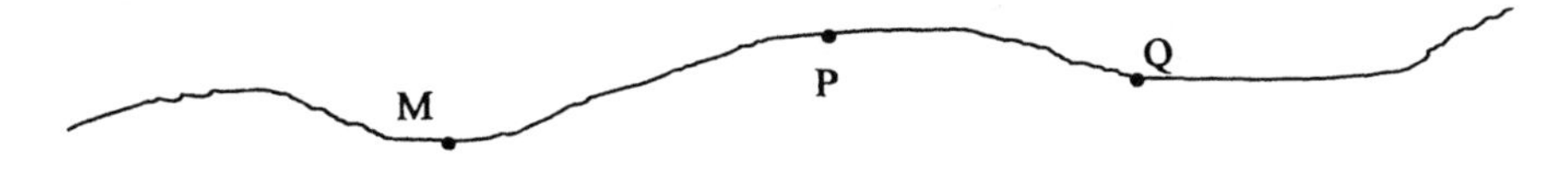

In the figure below, K is between T and V because TK + KV = TV; this is the relationship that defines "straightness," and assures us that K is on segment $\overline{TV}$.

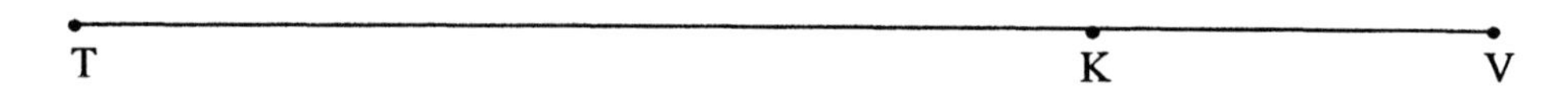

Now suppose $\overline{AB}$ is extended (along a straight path) infinitely far beyond B:

The resulting figure is a ray. A is its endpoint, and it is named using A (first) and any other point on the ray (second). This ray could be designated as $\overrightarrow{AB}$, $\overrightarrow{AM}$, $\overrightarrow{AK}$,etc.

The best physical model for $\overrightarrow{AB}$ is a beam of light originating at A and shining through B. It's easy to see that the same ray will be produced if the beam is focused on M or K or V instead of B.

It's also clear that if a beam originates at B and is focused on A, a completely different ray will result. This is $\overrightarrow{BA}$ or $\overrightarrow{BC}$ or $\overrightarrow{BP}$

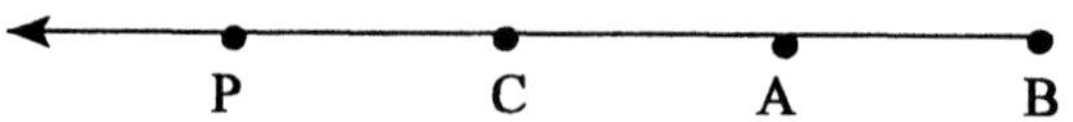

These two opposite rays, together, form a line. It can be named using any pair of its points: $\overleftrightarrow{AB}$, $\overleftrightarrow{VC}$, $\overleftrightarrow{KB}$, etc.

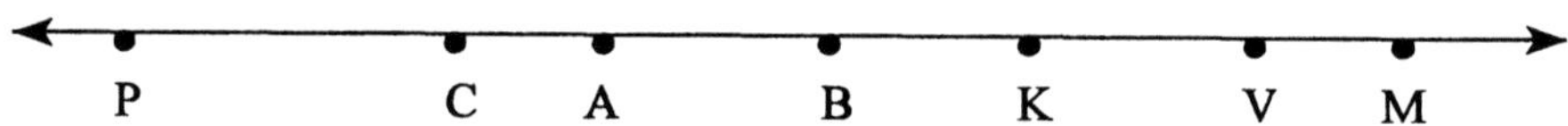

It's important to note that for any two points, there is exactly one line which contains them.

In the picture at right, M, P, and Q are non-collinear points because there is no line which contains all three of them.

Any set of three non-collinear points are the vertices of a triangle. For M, P, and Q, the triangle consists of segments $\overline{MP}$, $\overline{PQ}$ and $\overline{MQ}$, which are called sides of the triangle. It is named $\triangle$PMQ or $\triangle$MQP or any other permutation of the three points.

The lengths of the sides of a triangle are related by the Triangle Inequality: the sum of the lengths of any two sides of a triangle is greater than the length of the third side. (Notice that this is actually a definition of non-collinear points - because if M, P, and Q were all on a line, with P between M and Q, then MP + PQ would be equal to MQ.)

Three non-collinear points, and the triangle which they define, lie in exactly one plane. A plane is often described as an infinite flat surface. However, flatness, like straightness, is one of those things that we recognize when we see it; but it isn't easily defined. This is the criteria for flatness: an infinite surface is flat (which means it's a plane) if, for every two points on the surface, the entire line which contains those two points is on the surface.

Segments, rays, and lines have only one attribute — length. Hence, they are said to be <u>one-dimensional</u> figures. The length of a segment is <u>measurable</u>; length for rays and lines is not measurable because it's infinite. (In order to determine a length, there must be a beginning point and an ending point; but a ray has only one end point and a line has none.)

Planes, triangles, and other non-collinear figures which lie in a plane, are <u>two-dimensional</u>. In order to locate a point on a surface, two coordinates are required. For example, locations on a map or globe must include both longitude and latitude. The measurable attribute of finite, closed two dimensional figures is <u>area</u>.

Activity

1. Equal Measures and Congruent Figures

   A ________________________________ B

   a) Trace $\overline{AB}$ on a sheet of patty paper.
   b) On another sheet of patty paper, trace the line segment and label its endpoints C and D.
   c) Record the length of $\overline{AB}$ by making marks on the edge of a sheet of patty paper. Compare this with the length of $\overline{CD}$.
   d) The two line segments, $\overline{AB}$ and $\overline{CD}$, are different figures; but they have the same length. So AB = CD, and the two segments $\overline{AB}$ and $\overline{CD}$ are congruent. Two figures are congruent if they are <u>copies</u> or <u>duplicates</u> of each other.

2. Midpoint of a Segment
   a) Find the **midpoint** of $\overline{AB}$ by folding the paper until A lies on top of B and one half of the segment lies on top of the other half. Pinch a crease at this point, then open the page and label the midpoint M.
   b) What is the relationship between MA and MB?
   c) Name a pair of congruent figures.
   d) Since M separates $\overline{AB}$ into two congruent segments, we say that M **bisects** $\overline{AB}$.

Homework: Section 31

1. What do we mean by the <u>distance</u> between two points?

2. What kind of figure does each picture represent?

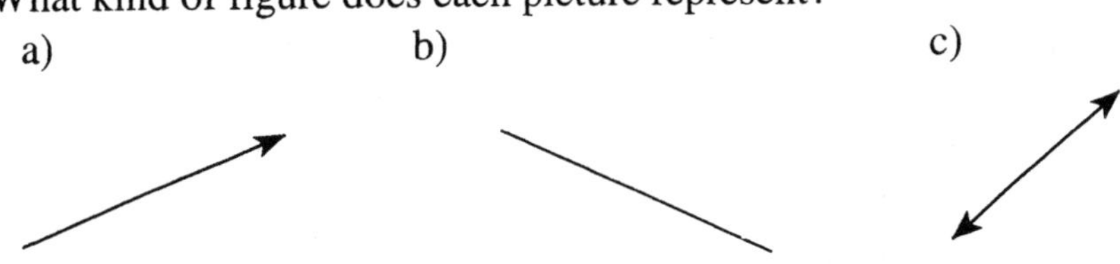

3. What does each of these symbols represent?

   a) $\overline{TC}$ b) $\triangle MVP$ c) $\overleftrightarrow{AK}$ d) RM e) $\overrightarrow{CD}$

4. If V is between M and Q, then _____ + _____ = _____.

5. If A, T, and P are non-collinear points, then they lie in exactly one __________, and they are the vertices of a ____________.

6. Describe a concrete model of each of these:

   a) segment b) plane c) ray

7. Explain the criteria we use to judge:

   a) whether three points are collinear

   b) whether an infinite surface is a plane

8. a) How many different lines contain point T?

   b) How many different lines contain points A and B?

   c) Suppose A, K, and Q are the vertices of a triangle. How many lines contain points A, K, and Q?

   d) How many planes contain points A and B?

   e) How many planes contain $\triangle RCV$?

9. Name three one-dimensional figures. What does "one-dimensional" mean?

10. Name three two-dimensional figures. What does "two dimensional" mean?

11. a) Suppose you are at Blackham Coliseum on Johnston Street, and you want to know the location of the Amtrak Station, which is also on Johnston Street. What information do you need?

    b) Suppose you want to know the location of Hurricane Patricia. What information do you need?

    c) Suppose a pilot, en route from New York City to New Orleans, must be given the location of a very small, but severe, area of turbulence which she needs to avoid. What information must by given?

12. You have three plastic straws which measure 14.6 cm, 8.7 cm, and 5.5 cm, respectively. Can they be connected to form a triangle? Explain.

13. Assume that AK = 5 units and MA = 11 units. If A, K, and M all lie on a line, what are the possible values of MK?

14. a) Draw a point P in the space at right.

    b) Draw ten points which are each 3 cm from P.

    c) If you drew <u>all the points on this page which are 3 cm from P</u>, what kind of figure would they form?

    d) Draw this figure with your compass.

    e) Are there any points which are 3 cm from P that are not on this page? If so, where are they? What kind of figure do they form?

15. A truck carrying nuclear waste materials overturned on a straight stretch of I-10 in New Mexico. Several canisters split open and contaminated the surroundings with dangerous radioactivity for a distance of 20 miles from the accident site. What geometric figure is related to each of these questions?
    a) How many miles of I-10 were contaminated?
    b) How much land was contaminated?
    c) What was the total air contamination?

16. TRUE OR FALSE:

    a) If the units digit of a number is even, then the number is a multiple of 2.
    b) If a number is a multiple of 2 , then its units digits is an even number.
    c) If a number is a multiple of 15, then it is a multiple of 5.
    d) If a number is a multiple of 5, then it is a multiple of 15.
    e) A number is divisible by 3 if and only if the sum of its digits is divisible by 3.
    f) The product of two numbers is even if and only if the two numbers are even.

17. A metric clock has 10 <u>metric hours</u> in a day (24 hours), and 100 <u>metric minutes</u> in one metric hour. If this clock reads 8:40, what time is it according to a standard clock? (Indicate AM or PM.)

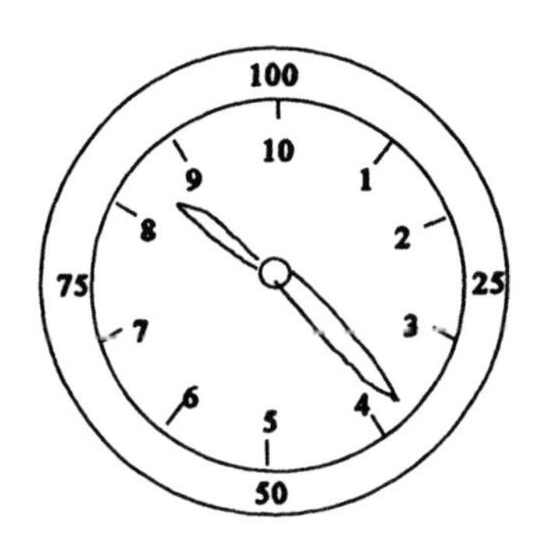

Section 32: Angles and Parallels

Consider the picture on the right:

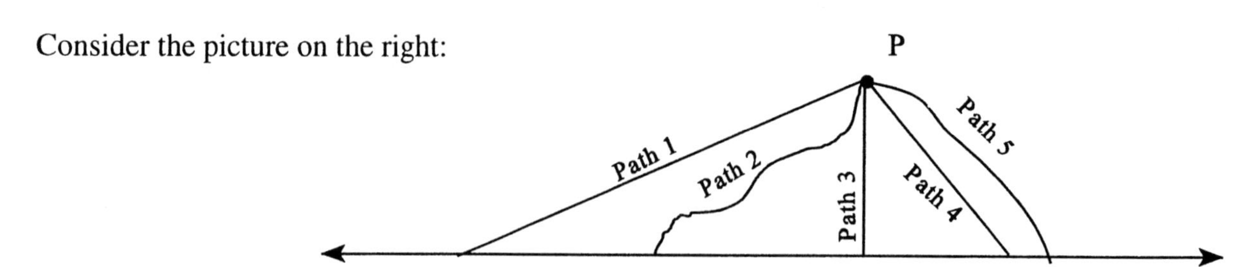

If you are at P, and you travel along one of these paths to the line, the distance you travel will vary according to your choice of path.

In this situation, as in the case of distance between two points, we specify that the distance from a point to a line is the length of the shortest path from the point to the line.

In the picture above, it's easy to see that Path 3 is the shortest; but in order to explain what distinguishes Path 3 from the others, and makes it the shortest, we must consider angles and their measurement. (It's not enough to say "straight" path, because Paths 1, 3, and 4 are all straight.)

An angle consists of two rays which have the same end point. This common end point is the vertex of the angle.

Notice in the figures to the right that an angle can be named in three different ways: by a number placed within the rays ($\angle$ 5 or $\angle$ 3); by the vertex ($\angle$ T or $\angle$ C); or by three points — the vertex in the middle and a point on each ray ($\angle$ACK).

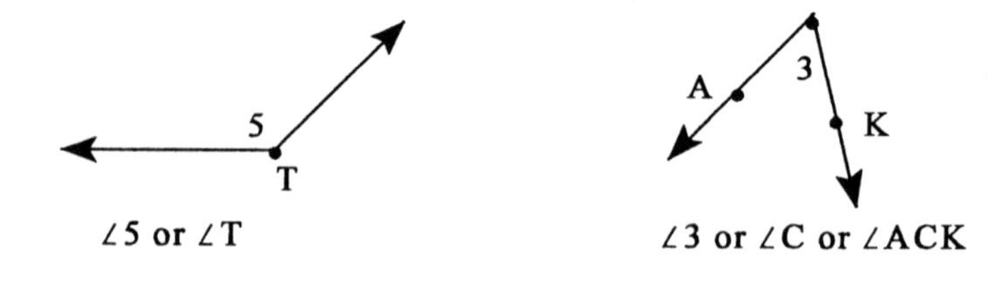

In the figure at right, three angles are defined — all having vertex P. $\overrightarrow{PB}$ cuts $\angle$APC into two parts (which can be named $\angle$1 and $\angle$ 2 or $\angle$APB and $\angle$BPC). These parts which, together, form $\angle$APC are adjacent angles. (The name is very descriptive since the word adjacent means "next to" or "side-by-side.")

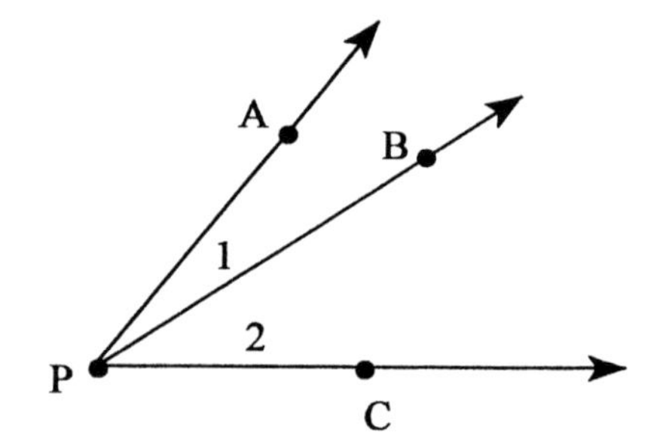

In the examples below, the angles have increasing openings between rays as we move from left to right. This "opening" can be thought of as one ray remaining stationary, as the other ray rotates away from it.

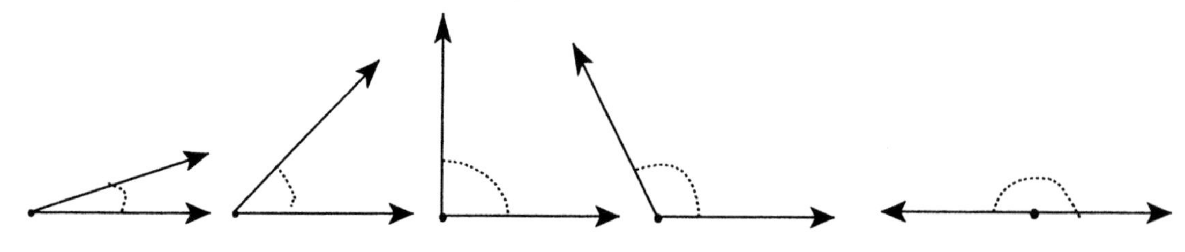

This "opening between rays," or "rotation of one ray away from the other," is the measurable attribute of angles. It is measured in units called degrees, where a degree is $\frac{1}{360}$ of a complete rotation.

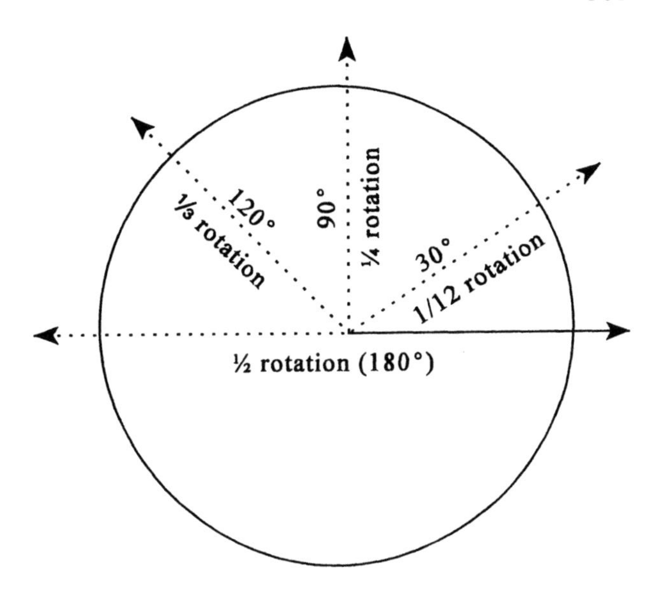

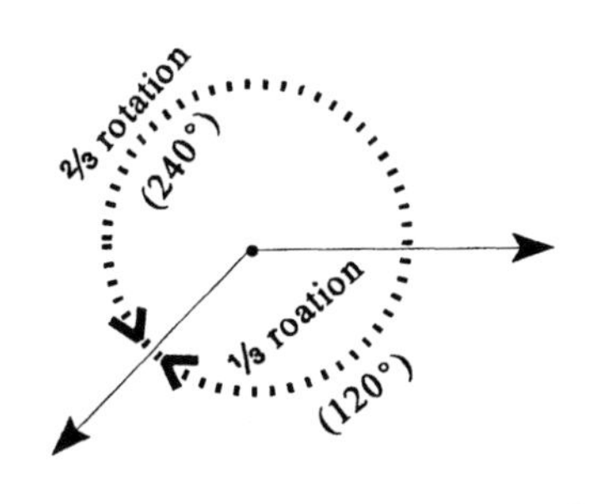

In the picture at right, notice that the "opening" between the rays can be looked at from two different points of view. If the rotating ray has moved counter-clockwise, then it has made two-thirds of a rotation, and the angle measure would be 240°. If, on the other hand, the rotating ray has moved clockwise, then it has made one-third of a rotation and its measure is 120°. In trigonometry and physics, both of these angle measures are possible; and the appropriate one must be determined from the context of the situation. In geometry, our focus is on the relative positioning of the two rays, not on how one of them is rotated. So to avoid ambiguity, we will agree to always use the smaller of the two possible measures. With this agreement, 180° (one-half rotation) is the greatest possible angle measure. Notice that a 180° angle is formed by two opposite rays; hence it's a line:

An angle whose measure is 90° (one-fourth rotation) is called a right angle, and its rays are said to be perpendicular. Similarly, lines or segments which form right angles are also perpendicular. Examples:

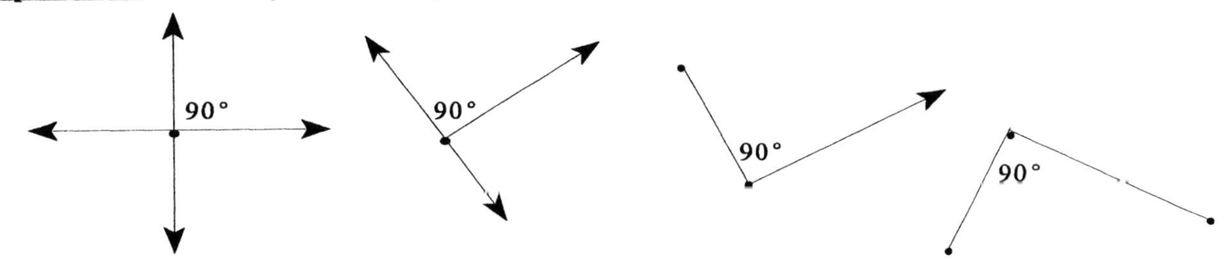

To measure angles we use a protractor, which is a semi-circle marked in degrees. The angle shown in the picture below measures 54°.

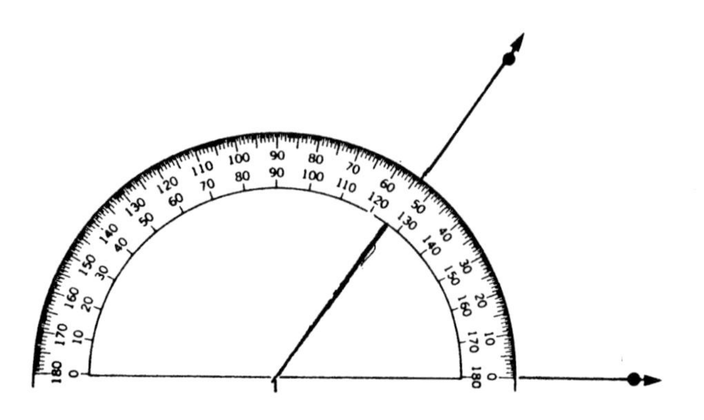

Notice that the sum of the measures of all the adjacent angles around a point is 360°:

$m\angle 1 + m\angle 2 + m\angle 3 + m\angle 4 + m\angle 5 = 360°.$

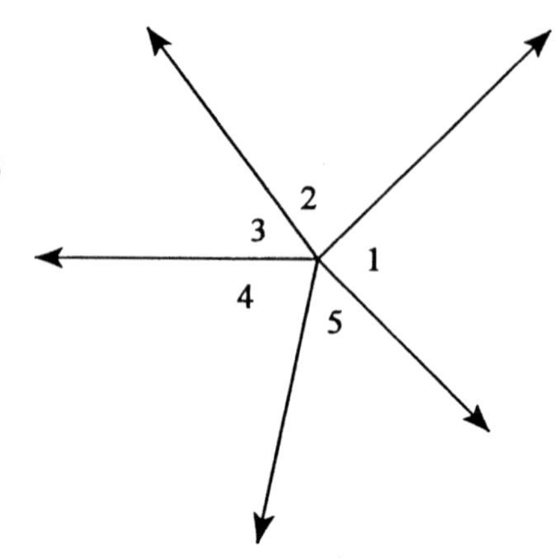

Now we are ready to return to the concept of distance from a point to a line. It is the length of the perpendicular segment from the point to the line. In the picture below, the distance from A to $\overleftrightarrow{KC}$ is the length of $\overline{AT}$.

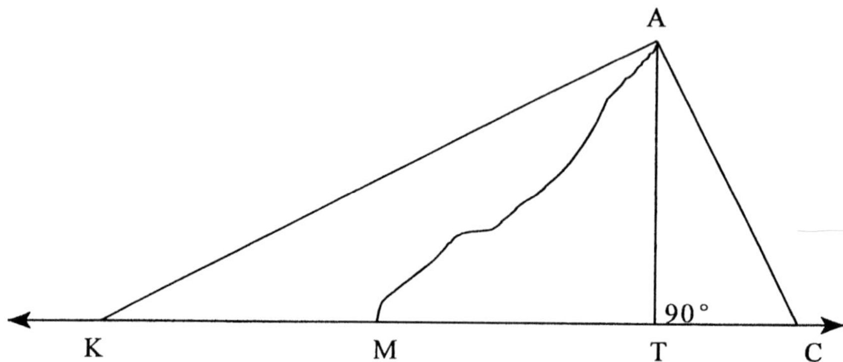

Finally, we consider the distance between two lines.

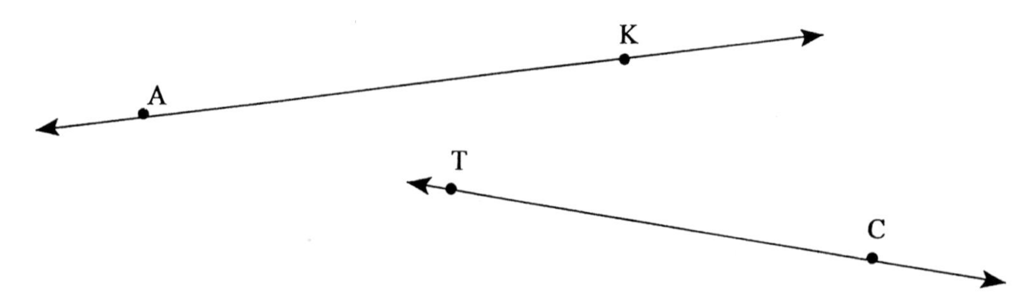

In the picture above it is clear that there is no such thing as the distance between the two lines -- it depends on "where you are". From point T, you are closer to $\overleftrightarrow{AK}$ than from point C. So we cannot talk about the distance between the two lines; we can only talk about the distance from a particular point on one line to the other line.

There is, however, a situation in which "the distance between two lines" is meaningful:

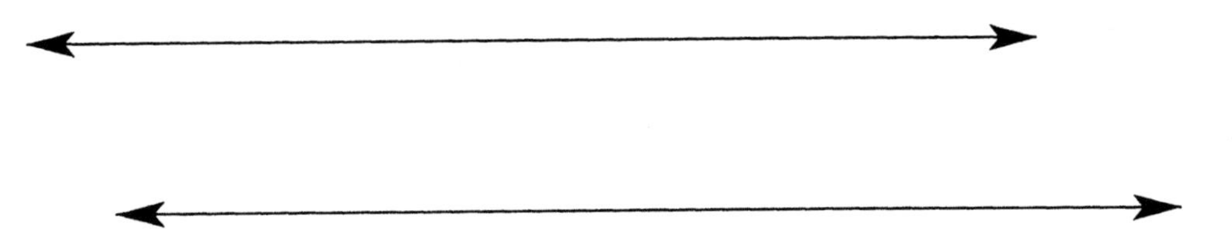

In this picture, the two lines are the same distance apart from any point on either line. As you know, these lines are said to be parallel.

In the primary grades, children learn about parallel lines from models such as railroad tracks or opposite sides of a picture frame. They are described as lines on a flat surface which never intersect. In about the fifth grade, parallel lines are defined in terms of distance.

Activity

1. To find the distance between a line and a point using patty paper:
   a) Fold the patty paper to create a crease across the sheet and then open it up so it is flat on the desk. We can imagine that this crease is a line that extends infinitely far in both directions. Draw the line and label two points on it as V and W.

   b) Draw a dot not on the line to represent the point P.

   c) Pick up another sheet of patty paper and notice that a corner represents an angle of 90°. Use this corner and an edge to create a segment which is perpendicular to the line and passes through P.

   d) Let S be the point where this segment crosses the line. Mark the length of $\overline{PS}$ on the edge of the second sheet. This length, PS, is the distance between the point P and $\overleftrightarrow{VW}$.

2. To construct a pair of parallel lines using patty paper:
   a) Move the corner to another point on $\overleftrightarrow{VW}$ and label it T. The original line may also be referred to as $\overleftrightarrow{ST}$.

   b) Use the marked edge of the second sheet to create a line segment, $\overline{TQ}$, which is perpendicular to $\overleftrightarrow{ST}$ and has a length of PS.

   c) Crease the paper to form $\overleftrightarrow{PQ}$. Now $\overleftrightarrow{PQ} \parallel \overleftrightarrow{ST}$ and the distance between them is PS.

Homework: Section 32

1. The picture at right represents a circle with center at T and diameter LA.

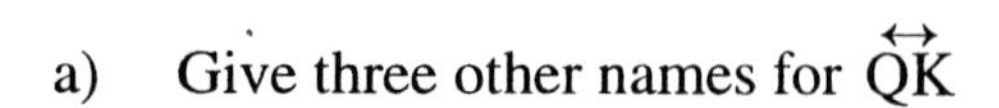

   a) Give three other names for $\overleftrightarrow{QK}$

   b) Give three other names for $\angle MTK$.

   c) Name three points of $\overrightarrow{KQ}$

   d) How many diameters of the circle contain point E?

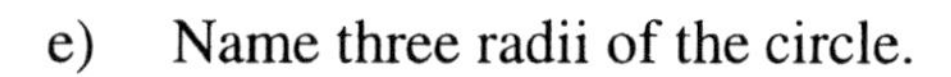

   e) Name three radii of the circle.

   f) Name two adjacent angles in the picture.

   g) Name a point of $\overrightarrow{TC}$ which is not on $\overline{TC}$.

   h) AQ + KQ = ________ .

   i) AT = QA + ________ .

   j) Name three segments which have the same length as $\overline{ET}$.

   k) If the measure of $\angle KTC$ is 40° and the measure of $\angle ETC$ is 45°, what is the measure of $\angle QTE$?

2. What kind of figure is suggested by these descriptions and pictures?

   a) an infinite flat surface

   b) two points and all the points between them

   c) the union of two rays with a common end point

   d) the union of $\overrightarrow{AB}$ and $\overrightarrow{BA}$

   e) the set of all points which are 6 units from point K

   f) the end of a pin

   g) two angles which have a common side between their non-common sides

   h)

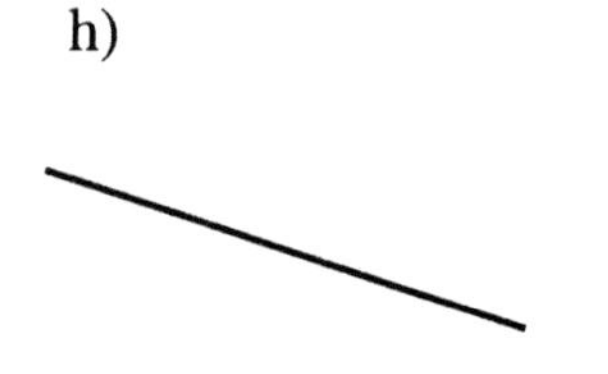

   i)

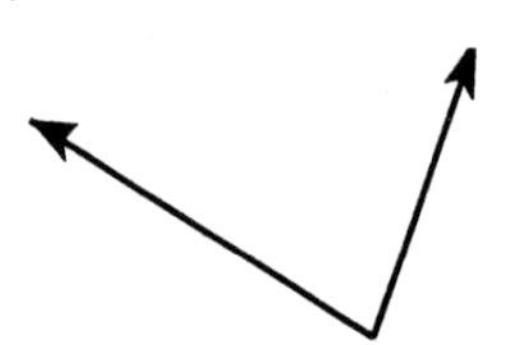

   j)

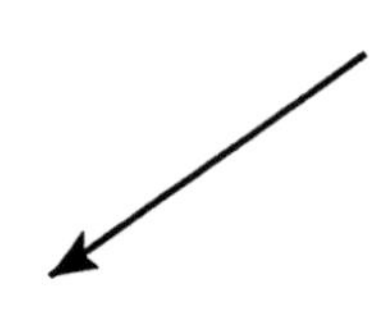

   k)

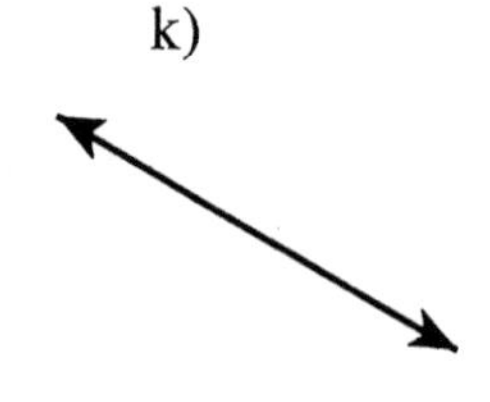

3. In the figure below:

   a) Measure the lengths of $\overline{PA}$, $\overline{PB}$, $\overline{PC}$, $\overline{PD}$, and $\overline{PE}$.

   b) Find the measures of ∠PAC, ∠PBC, ∠PCE, ∠PDE, and ∠PEB.

   c) What is the distance from P to $\overleftrightarrow{AC}$?

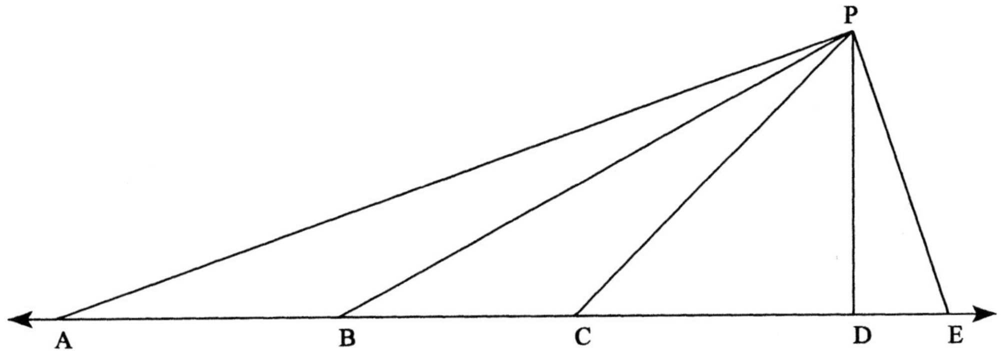

4. Use your ruler and protractor to draw the following figures:

   a) A rectangle with sides of length 8 cm and 5 cm.

   b) A triangle with two perpendicular sides.

   c) A pair of parallel lines.

5. The sides of a square have their lengths reduced by 25% and the result has an area of 9 sq cm. How many centimeters were in the perimeter of the original square?

6. a) Draw a line through P which is parallel to the line $\overleftrightarrow{AB}$.

. P

   b) How many different lines can be drawn through P which are parallel to $\overleftrightarrow{AB}$?

   c) Our common sense tells us that if m is a line and P is a point not on the line, then there is exactly one line which contains P and is parallel to m. However, unlike our other discoveries, we cannot <u>prove</u> this. We accept it as an "assumption" or a "postulate". The geometry that best describes our world as we see it is known as Euclidean geometry. It is named after the Greek mathematician Euclid, who was the first writer to organize geometry into a logical system of definitions, assumptions, and theorems. One of his assumptions is that given a line and a point P not on it, there is <u>exactly one</u> line which contains P and is parallel to the given line. This assumption is called "The Parallel Postulate" or "Euclid's Fifth Postulate". (There are some non-Euclidean geometries which do not have the parallel postulate. With these geometries the assumption is made that such parallel lines are not unique or that there no parallel lines.)

7. a) Do this Patty Paper Activity.

Step 1. Draw a pair of non-perpendicular intersecting lines on a sheet of patty paper..

Step 2. You have created four angles. In a clockwise fashion, label them $\angle 1$, $\angle 2$, $\angle 3$, $\angle 4$, respectively.

Step 3. On a second sheet trace a copy $\angle 1$. Now compare $\angle 1$ and $\angle 3$ by placing the copy of $\angle 1$ over $\angle 3$. What can you say about these angles?

Step 4. Repeat the previous step with $\angle 2$ and $\angle 4$.

Step 5. What conclusion can you draw from this activity?

b) Measure the numbered angles in each picture:

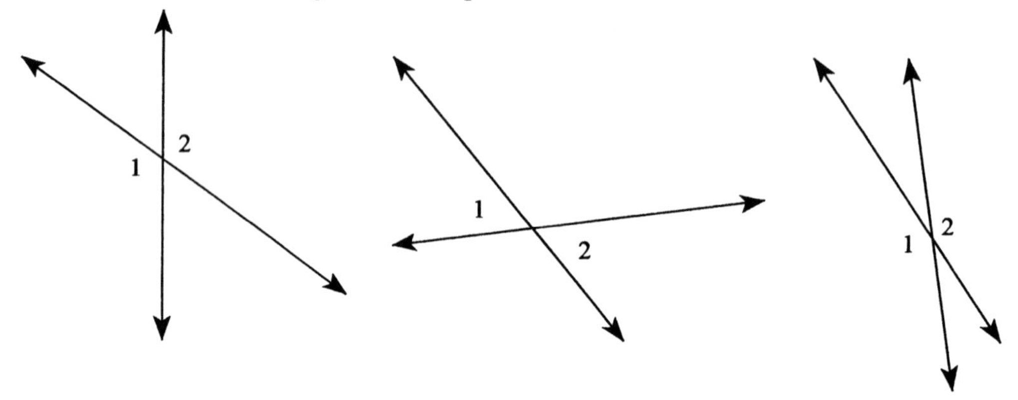

What do you notice about the measures of $\angle 1$ and $\angle 2$ in each figure?

c) These pairs of non-adjacent angles formed by intersecting lines are called <u>vertical angles</u>. The activities in parts a) and b) <u>suggest</u> that vertical angles have equal measures. In other words, vertical angles are congruent. For convenience, we will agree that in situations involving two different angles, if we say that the angles are equal (which is impossible), we really mean that they have equal measures.

It is easy to explain why any two vertical angles are equal. First, consider any pair of intersecting lines with vertical angles labeled $\angle 1$ and $\angle 2$.

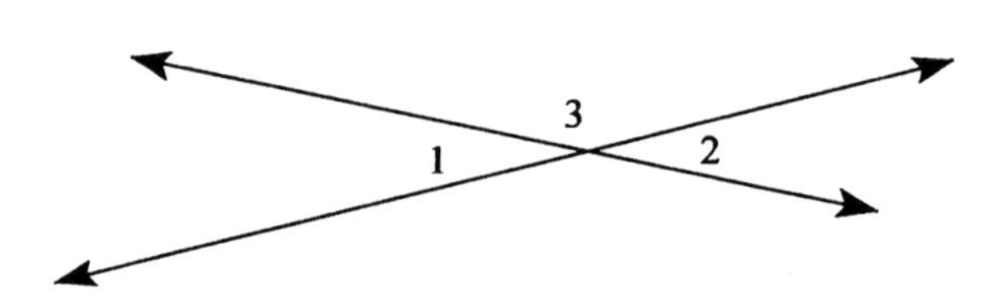

Since $\angle 1$ and $\angle 3$ form a line, we know that $m\angle 1 + m\angle 3 = 180°$. Similarly, $m\angle 2 + m\angle 3 = 180°$. By simple algebra, we get: $m\angle 1 = m\angle 2$.

8. In each of the figures below, two lines are cut by a third line(which is called a transversal). This forms two pairs of angles which make a Z - shape. The pairs are called alternate interior angles. In each picture, one pair of alternate interior angles is numbered.

a) Measure all the numbered angles.

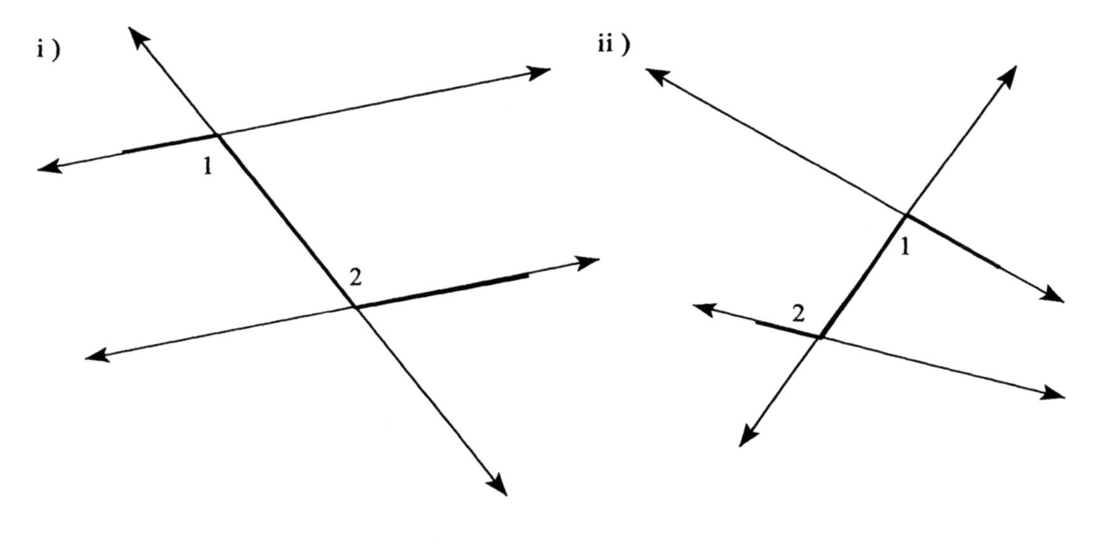

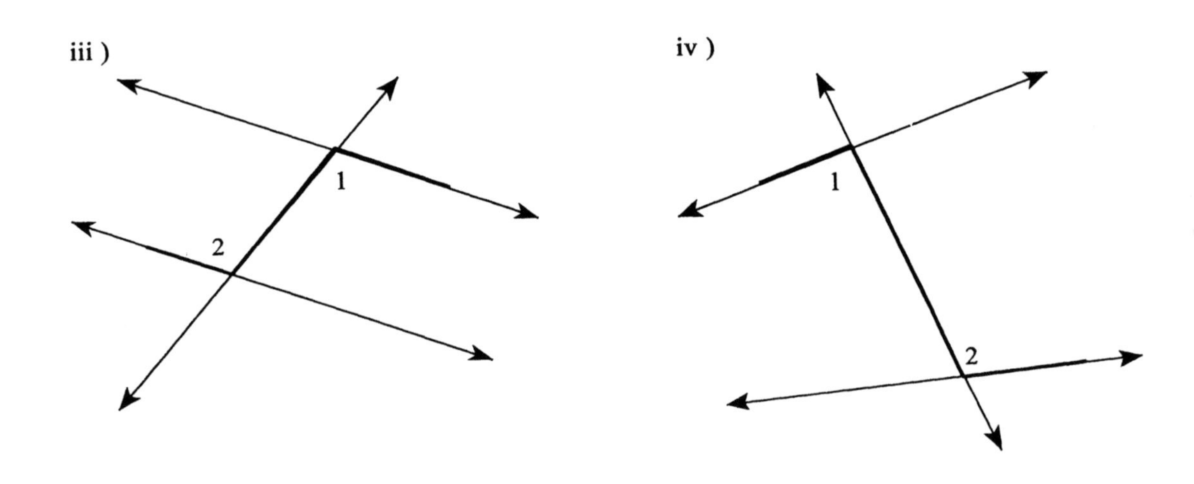

b) For the figures on the left, in which the two lines are obviously parallel, what do you notice about the measures of the alternate interior angles?

c) For the figures on the right, in which the two lines are obviously not parallel, what do you notice about the measures of the alternate interior angles?

Section 33: Sum of Measures of Angles of a Triangle

The last homework exercise in Section 32 suggests a very important relationship between parallel lines and alternate interior angles: If two lines are parallel, their alternate interior angles are equal; and if two lines are not parallel, their alternate interior angles are not equal. (This is abbreviated as "Two lines are parallel if and only if (iff) their alternate interior angles are equal.")

This relation provides a very useful definition of parallel lines. Recall that children learn in the early grades that parallel lines are lines in a plane that never intersect. This is a nice description, but because it is a negative statement, it is not at all useful. (It's impossible to verify that two lines never intersect, because lines are infinite.) The second description of parallel lines - as two lines that are everywhere the same distance apart - requires construction of perpendicular transversals. This definition, in terms of alternate interior angles, gives us a criterion for judging whether two lines are parallel by using any transversal.

Example 1:

Suppose we know that $\angle AKV$ and $\angle CVK$ are equal.

Then we know that $\overleftrightarrow{BA}$ is parallel to $\overleftrightarrow{CD}$.
This is written as $\overleftrightarrow{BA} \parallel \overleftrightarrow{CD}$.

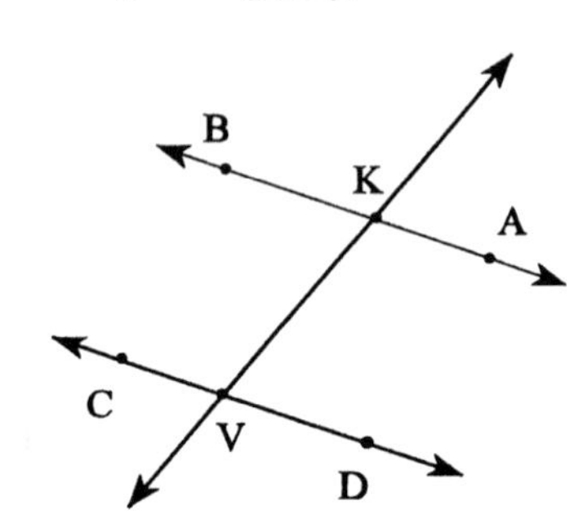

Example 2

Assume that $m\angle 1 = 100°$ and $m\angle 2 = 115°$.

Then we know that line m is not parallel to line n.

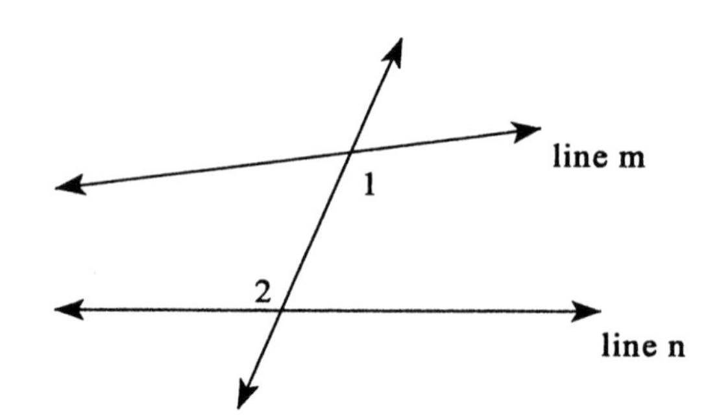

Activity 1

You have been given a triangle.

1. Label the angles 1, 2, 3.
2. Measure each angle and record your results.
3. What do you notice about the sum of these measures? (Remember that measurements are always approximate.)
4. Now:
   a) Draw a line on a sheet of paper.
   b) Tear your triangle into three pieces - each containing an angle.
   c) Arrange the angles adjacently on the line.
   d) Since the three angles together make up an angle with measure 180°, what does this suggest about the sum of the measures of the angles of a triangle?

This activity provides a concrete foundation for understanding one of the most important relationships of geometry: **The sum of the angles of a triangle is 180°.** (We will agree that this means the sum of the measures of the angles is 180°.)

To prove that this angle relationship is true for all triangles, we draw "any old triangle" and name it $\triangle ABC$. Now we label its angles, $\angle 1$, $\angle 2$, and $\angle 3$. Then we draw the line through B which is parallel to $\overleftrightarrow{AC}$.

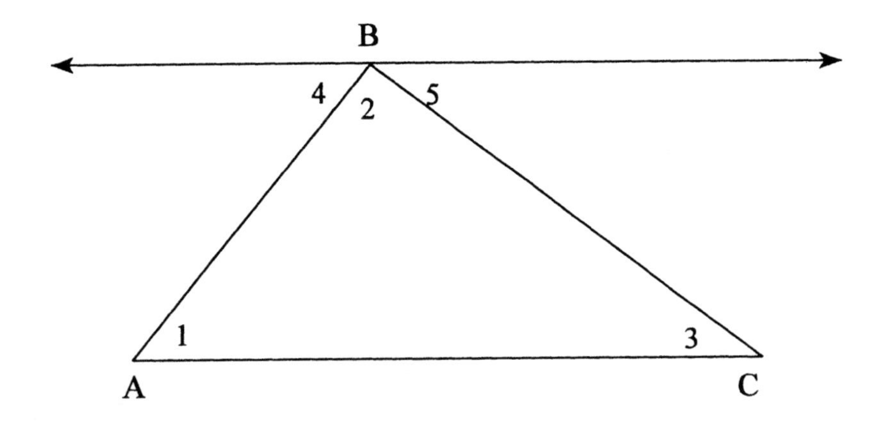

Since $\angle 4$ and $\angle 1$ are alternate interior angles of parallel lines, $m\angle 4 = m\angle 1$. Similarly, $m\angle 5 = m\angle 3$. Now notice that $m\angle 4 + m\angle 2 + m\angle 5 = 180°$. (Why?) So, by simple algebra, we get: $m\angle 1 + m\angle 2 + m\angle 3 = 180°$.

Activity 2

1. To bisect an angle:

   a) Trace $\angle ABC$ on patty paper.

   b) Fold the paper so that ray $\overline{BA}$ lies exactly on ray $\overline{BC}$ and crease the patty paper.

   c) Open the paper and draw ray $\overline{BD}$ in the crease.

   d) We now have two adjacent angles, $\angle ABD$ and $\angle CBD$. Fold the paper again along $\overline{BD}$ and notice that $\angle ABD \cong \angle CBD$. So $m\angle ABD = m\angle CBD = \frac{1}{2} m\angle ABC$. Since ray $\overline{BD}$ cuts $\angle ABC$ into two congruent angles, we say that $\overline{BD}$ bisects $\angle ABC$ or that $\overline{BD}$ is the angle bisector of $\angle ABC$.

C

B

A

2. To bisect a segment:

   • Q   S •

   a) Trace $\overline{AB}$ and Q and S on a sheet of patty paper.

   b) Find the **midpoint** of $\overline{AB}$ and label it M.

   A   B

   c) Draw the line containing Q and M. Note $\overleftrightarrow{QM}$ bisects the segment $\overline{AB}$. Draw another line that bisects $\overline{AB}$. How many other lines bisect this segment?

   d) Construct a line through S that is perpendicular to segment $\overline{AB}$.

   e) Construct a line perpendicular to $\overline{AB}$ that passes through M. How many other lines can you construct that are perpendicular to segment $\overline{AB}$?

   f) The last line you drew through M has two properties: It is perpendicular to $\overline{AB}$ and it bisects $\overline{AB}$. How many lines other lines do both of these things?

3. In part 2 above, you discovered that for each segment there is only one line that is both perpendicular to the segment and is a bisector of the segment. Such a line is called **the perpendicular bisector** of the segment. Here is a quick way to construct the perpendicular of a segment:

   a) Trace $\overline{AB}$ on a sheet of patty paper and label its midpoint M.

   A   B

   b) Fold the paper again so that $\overline{MA}$ lies on top of $\overline{MB}$. Crease the paper along this fold then open the paper. Draw the line created by the crease. On this line mark the point C somewhere above segment $\overline{AB}$ and the point D somewhere below the segment.

   c) Is $\overleftrightarrow{CD}$ perpendicular to $\overline{AB}$? How do you know?

   d) Does $\overleftrightarrow{CD}$ bisect $\overline{AB}$? How do you know?

   e) What is the relationship between $\overleftrightarrow{CD}$ and $\overline{AB}$?

4. Continue this exploration on the patty paper that you used in part 3.

   a) Using the edge of a sheet of paper to mark lengths, compare DA and DB; CA and CB; and MA and MB.

   b) What seems to be true about the distance of <u>any</u> point on the perpendicular bisector from A and B?

   c) Let P be a point on the perpendicular bisector of $\overline{AB}$. What do you know about AP and PB?

What has this activity suggested about the distances from the endpoints of a segment to any specified point on the perpendicular bisector of the segment?

Homework: Section 33

1. Is it possible for $\overleftrightarrow{AC}$ and $\overleftrightarrow{TA}$ to be parallel lines? Explain your answer.

2. In the picture below $m_1 \parallel m2$; $m_1 \perp m_3$; and $\angle 1 = 120°$.

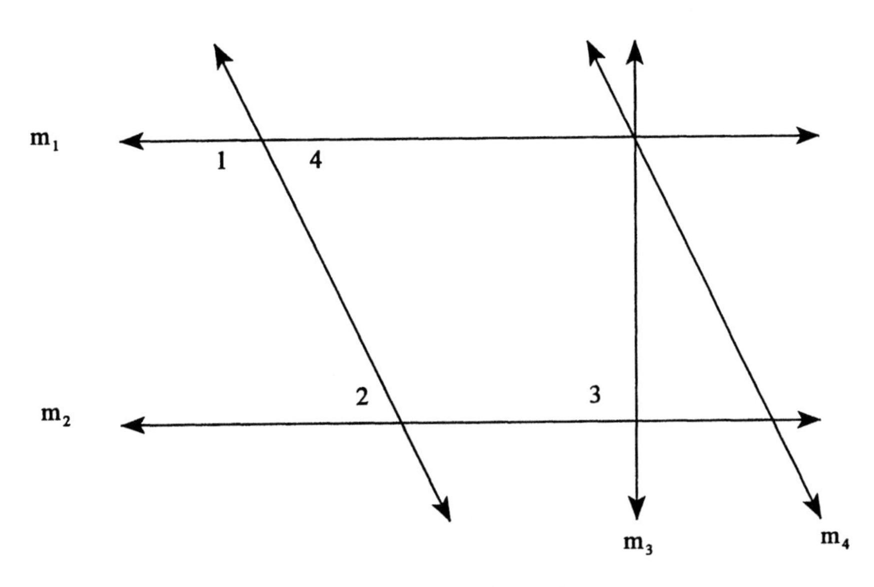

a) What is the measure of $\angle 4$? How do you know?
b) What is the measure of $\angle 2$? How do you know?
c) What is the measure of $\angle 3$? How do you know?
d) Is $m_4 \perp m_3$? How do you know?

3. Complete these statements:

a) If $m_1 \parallel m_2$ and $m_2 \perp m_3$, then ...

b) If $m_1 \parallel m_2$ and $m_3 \parallel m_2$, then ...

c) If $m_1 \perp m_2$ and $m_2 \perp m_3$, then ...

4. In the figure at right, $\overline{AB} \parallel \overline{CD}$ and $\angle CAB$ is 90° and $\angle CDB$ is 40°.
   a) What is $m\angle DCA$? How do you know?
   b) What is the distance from $\overline{CD}$ to $\overline{AB}$?
   c) What is the measure of $\angle ABD$? How do you know?

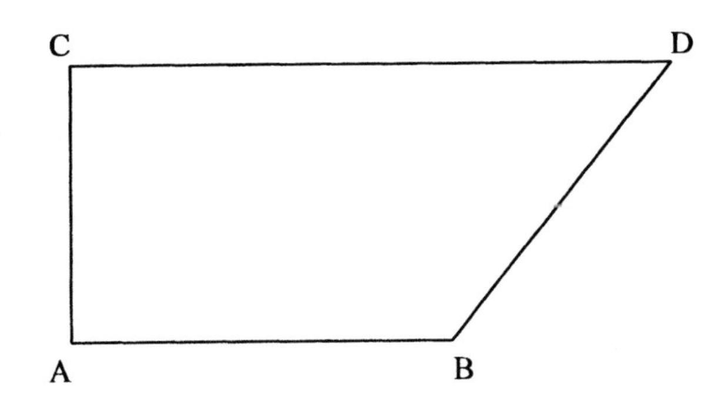

5. In the figure at right, $\overrightarrow{CB}$ bisects $\angle ACD$.
(This means that $m\angle ACB = m\angle BCD$.) Find x.

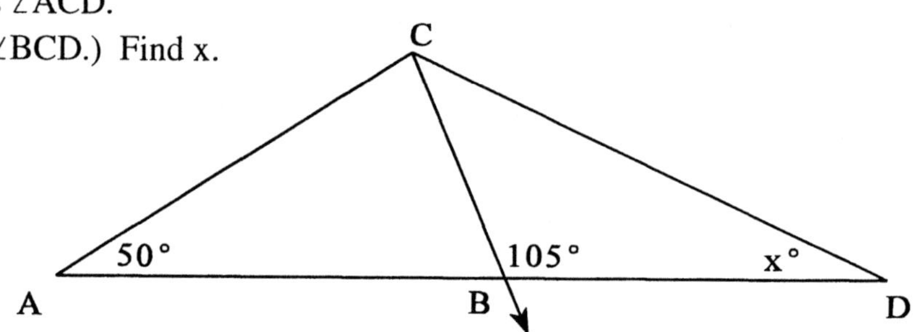

6. In the figure $\overline{AB} \perp \overline{BC}$.
Find x and y.

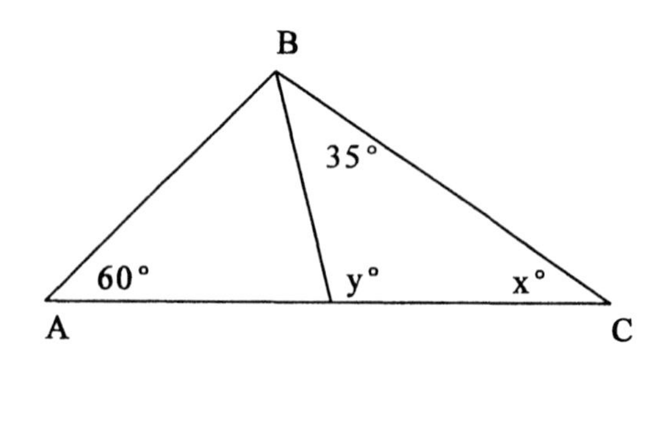

7. $\overline{FD} \parallel \overline{GE}$ and $\overline{CG}$ bisects $\angle FCE$.
Find x.

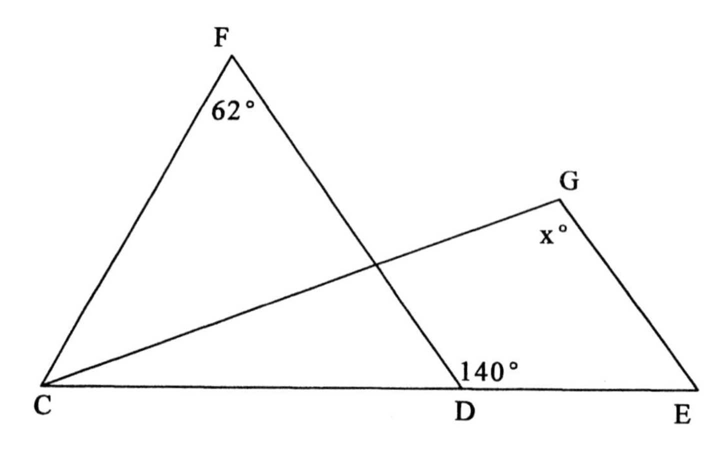

8. The area of the bottom of a rectangular fish tank is 2400 sq cm, and the tank holds 120 liters of water. How deep is the tank?

9. The volume of a box is 76 cubic inches. If the length of the edges are whole numbers, what could they be?

10. A satellite sends out test signals at set intervals: Signal 1 is sent every 2 hours 15 minutes. Signal 2 is sent every 1 hours 45 minutes. Both signals were sent at precisely 10:00 AM today. When will the signal next go out at the same time?

11. Find answers **mentally**:
   a) The ratio of adult tickets to child tickets sold at Super Screen Cinema last week was 7 to 13. What percent of the tickets were for adults?

   b) Jack bought a bicycle at a "⅓ off sale". He paid $84 for the bike. How much did he save by buying it on sale?

   c) An oil well pumps 8 barrels in 15 minutes. How long will it take to pump 72 barrels?

   d) What is $\frac{9}{7}$ of 350 ft?

   e) Last year the average price for a gallon of milk was $2.80. The price this year is 125% of that. What is the price now?

Section 34: Polygons

We have been talking about triangles, rectangles, and squares since the beginning of this course. These familiar figures are examples of polygons. A polygon is a simple, closed figure which consists entirely of segments. (A simple, closed figure is one which: lies in a plane, begins and ends at the same point, and can be drawn without lifting the pen and without passing through any points more than once.)

A polygon separates the plane which contains it into three distinct parts - the polygon itself, the points inside the figure, and the points outside of it.

Example 1 These figures are not polygons:

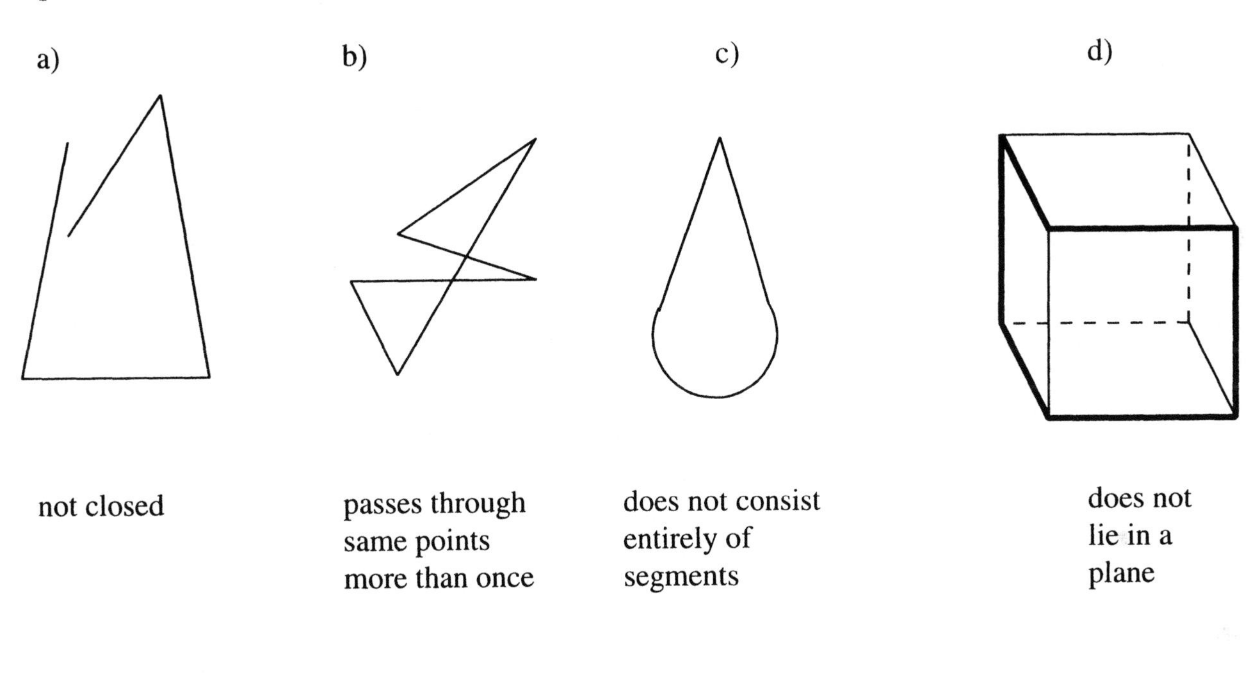

Example 2 These figures are polygons:

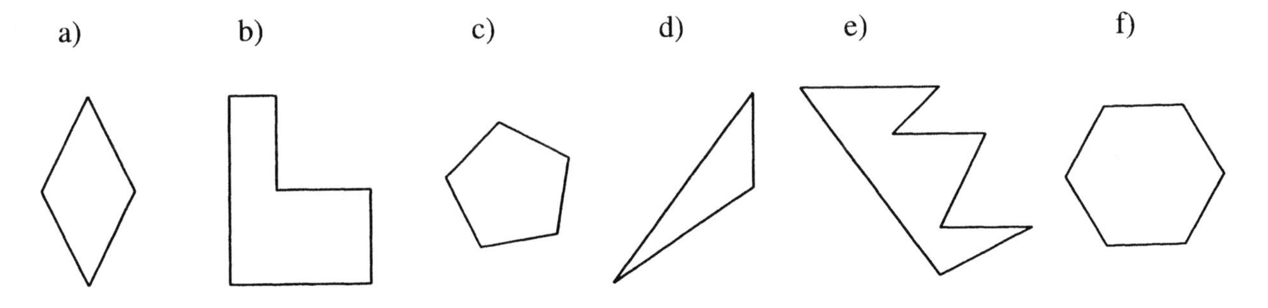

We will focus on convex polygons. A figure is convex if, for every two points A and B inside the figure, the entire segment $\overline{AB}$ is inside. Notice that figures b) and e) above are not convex - they have "dents" in them. For convenience, we will agree that polygon means convex polygon.

The segments which form a polygon are called its sides, and the endpoints of these segments are vertices. (Each one is a vertex.) Notice that a polygon has the same number of sides, angles, and vertices.

The names of common polygons are based on how many sides they have:

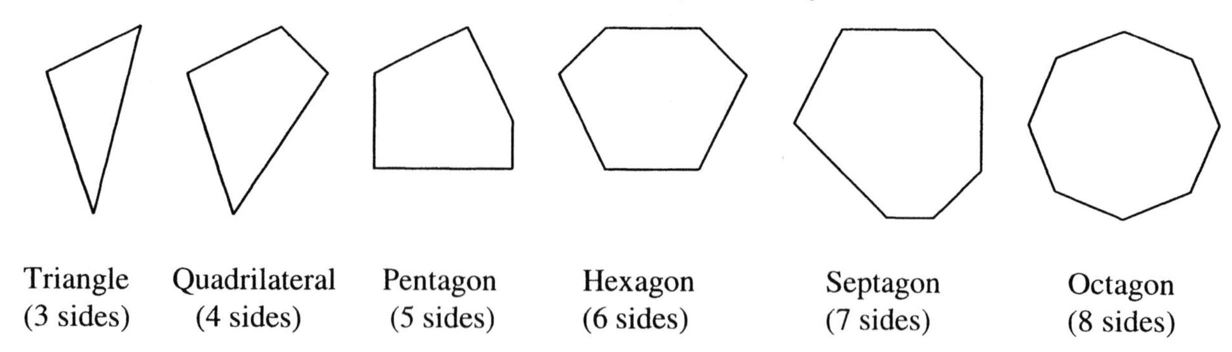

Other polygons are designated numerically according to their number of sides: 15-gon, 37-gon, 187-gon, or n-gon (for an arbitrary number of sides).

A particular polygon is named by listing its vertices in clockwise or counterclockwise order, beginning with any vertex.

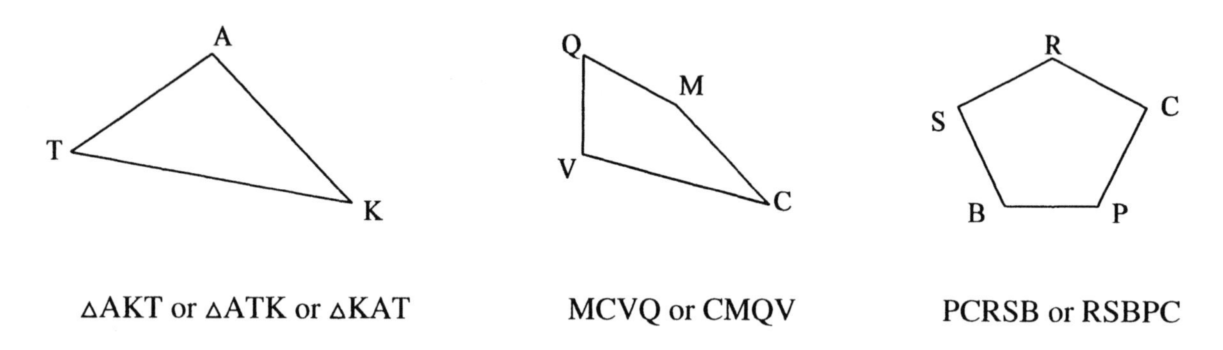

A segment which connects two non-adjacent vertices of a polygon is called a diagonal.

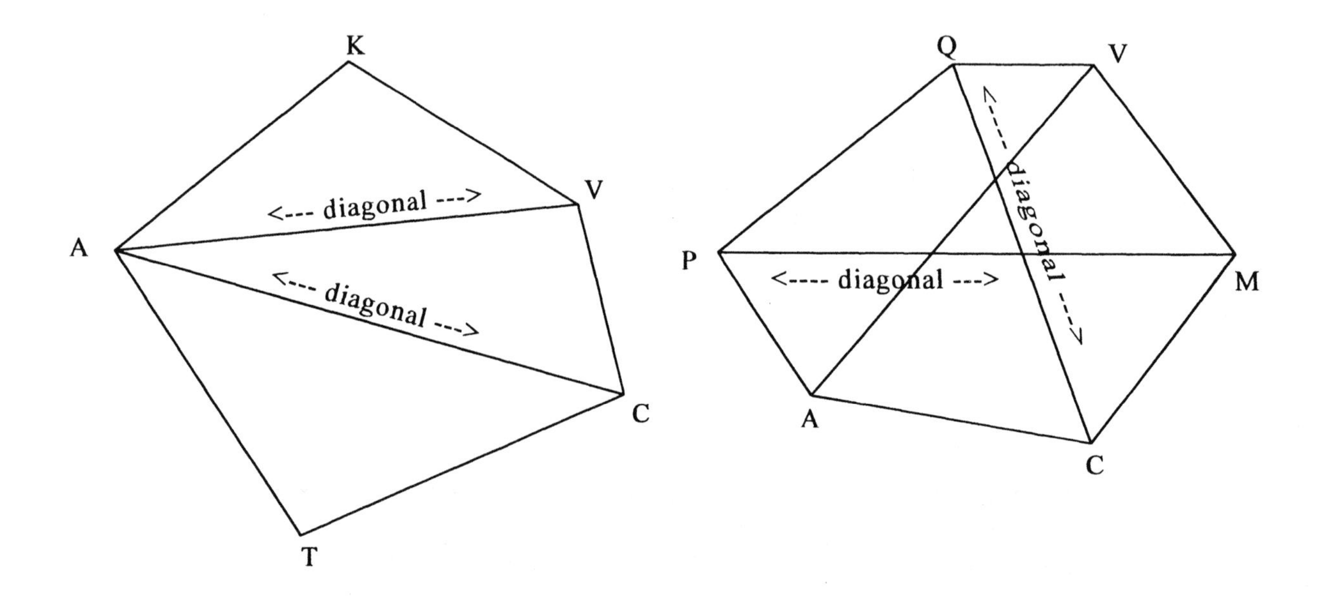

Activity

1. For each of the polygons shown below, draw all the diagonals from vertex K.

a) b)

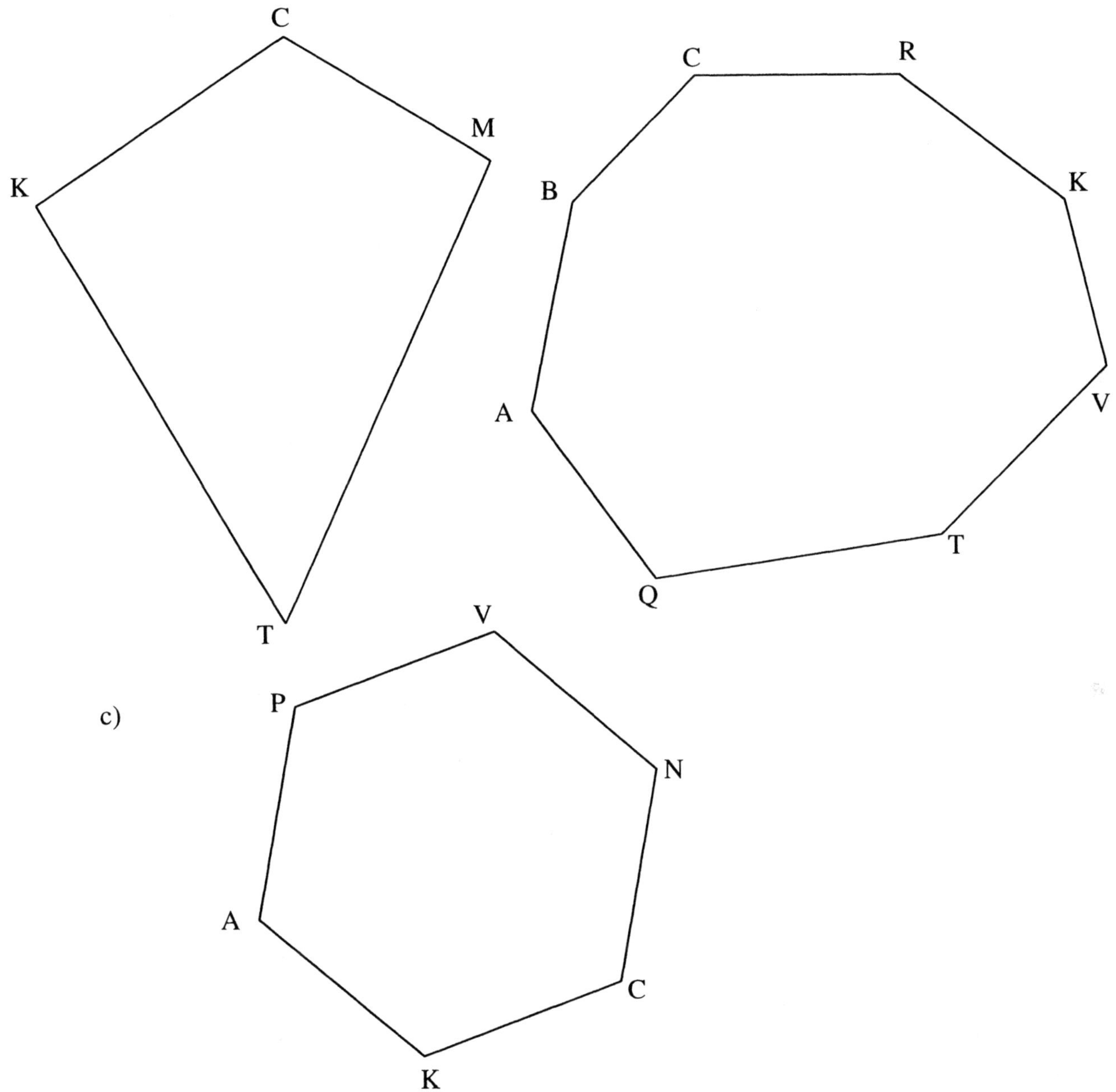

2. a) These diagonals separate each polygon into triangles. What relationship do you notice between the number of sides of the polygon and the number of triangles formed by the diagonals?

   b) For polygon ATMCBRKQVSLD, if all the diagonals were drawn from vertex M, how many triangles would be formed?

3. For the polygon at right, the sum of its angles can be found by adding the measures of angles 1 though 15. Notice that if we rearrange the order of these angles, we can write this sum as:

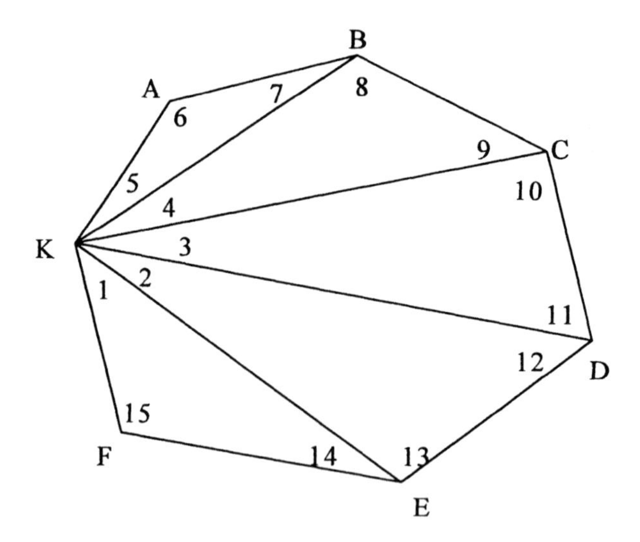

$$(\angle 5+\angle 6+\angle 7) + (\angle 4+\angle 8+\angle 9) + (\angle 3+\angle 10+\angle 11) + (\angle 2+\angle 12+\angle 13) + (\angle 1+\angle 14+\angle 15)$$

a) What is the sum of the angles within each set of parentheses? How do you know?

b) What is the sum of the angles of the polygon?

4. What is the sum of the angles of an 18-gon? a 15-gon? a 91-gon? an n-gon?

If all the sides of a polygon have the same length, the polygon is equilateral. If all of its angles have the same measure, the polygon is equiangular. A polygon which is both equilateral and equiangular is said to be regular.

Homework: Section 34

I. . Which of these are convex polygons? For those that are not, explain why.

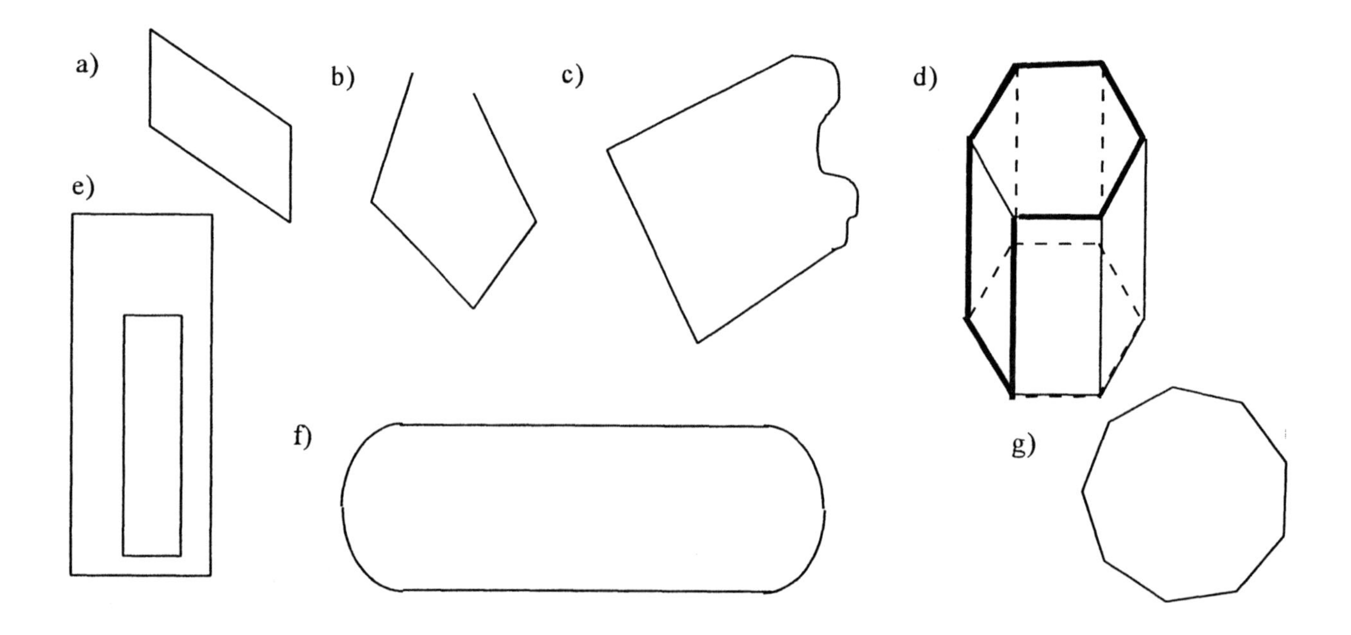

2. What is the common name of a regular quadrilateral?

3. What is the sum of the angles of an octagon? a 27-gon?

4. a) What is the sum of the angles of a regular 15-gon?
   b) What is the measure of each angle?

5. What is the measure of each angle of a regular 36-gon? a regular n-gon?

6. How many diagonals does a triangle have? a quadrilateral? pentagon? a hexagon? a septagon? an octagon? a 14-gon? a 39-gon? an n-gon?

7. a) Can a triangle be equilateral and not regular? How do you know?

   b) Can a quadrilateral be equilateral and not regular? How do you know?

II. Construct each of these figures, **cut them out and bring them to class.**

1. An equiangular pentagon which is not regular.
2. An equilateral quadrilateral which is not regular.
3. A quadrilateral with one pair of parallel sides.
4. A quadrilateral with two pairs of parallel sides.
5. a) A single triangle which has these properties:
      i) perimeter is 12 inches
      ii) length of each side is a whole number of inches.
      (Make your measurements as close to these specifications as possible.)
   b) Do you expect all the triangles constructed by members of the class to be alike (disregarding slight errors in measurement)? Explain.

III. Which of these figures have a center? For those that do, located the center. Explain what you mean by center.

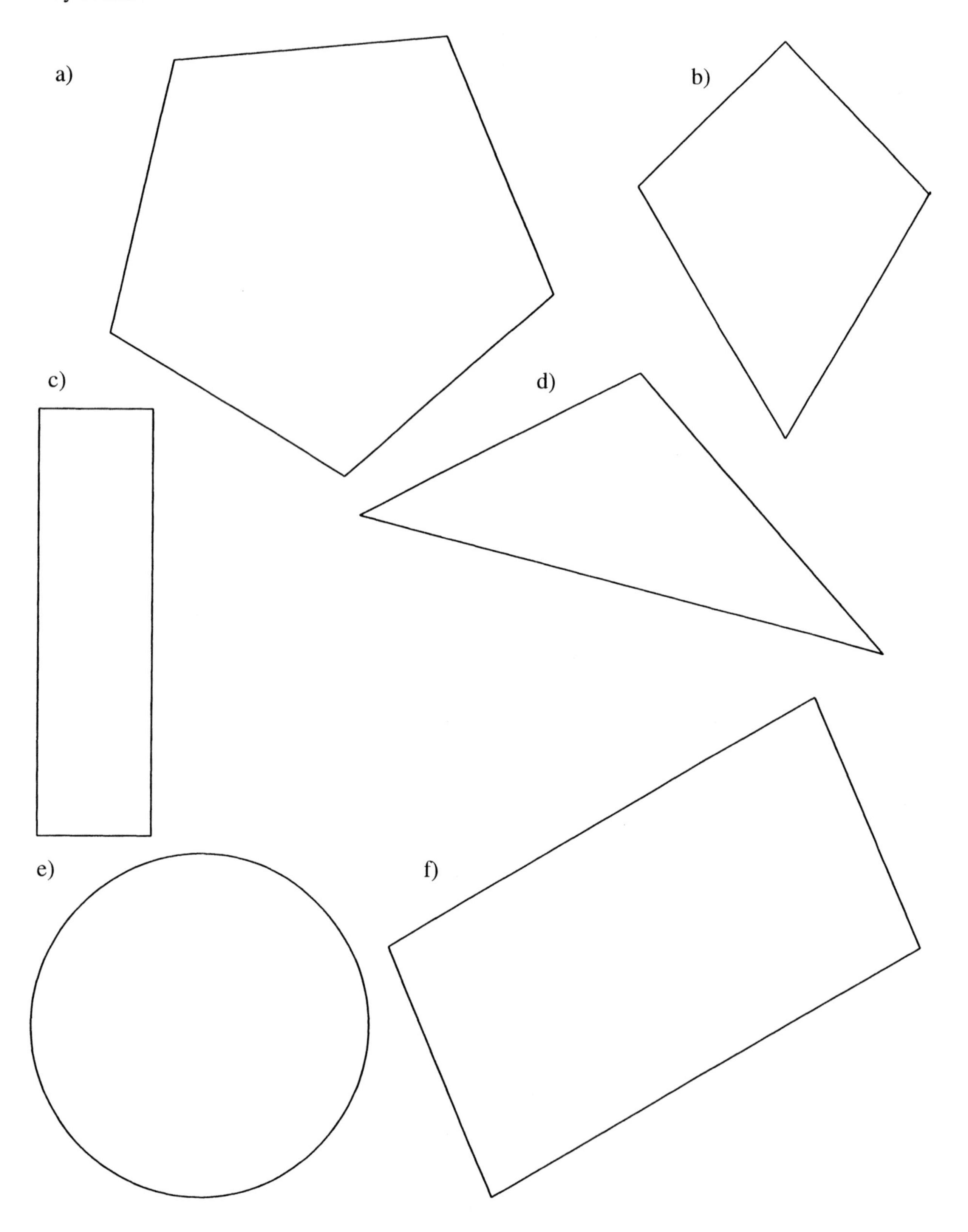

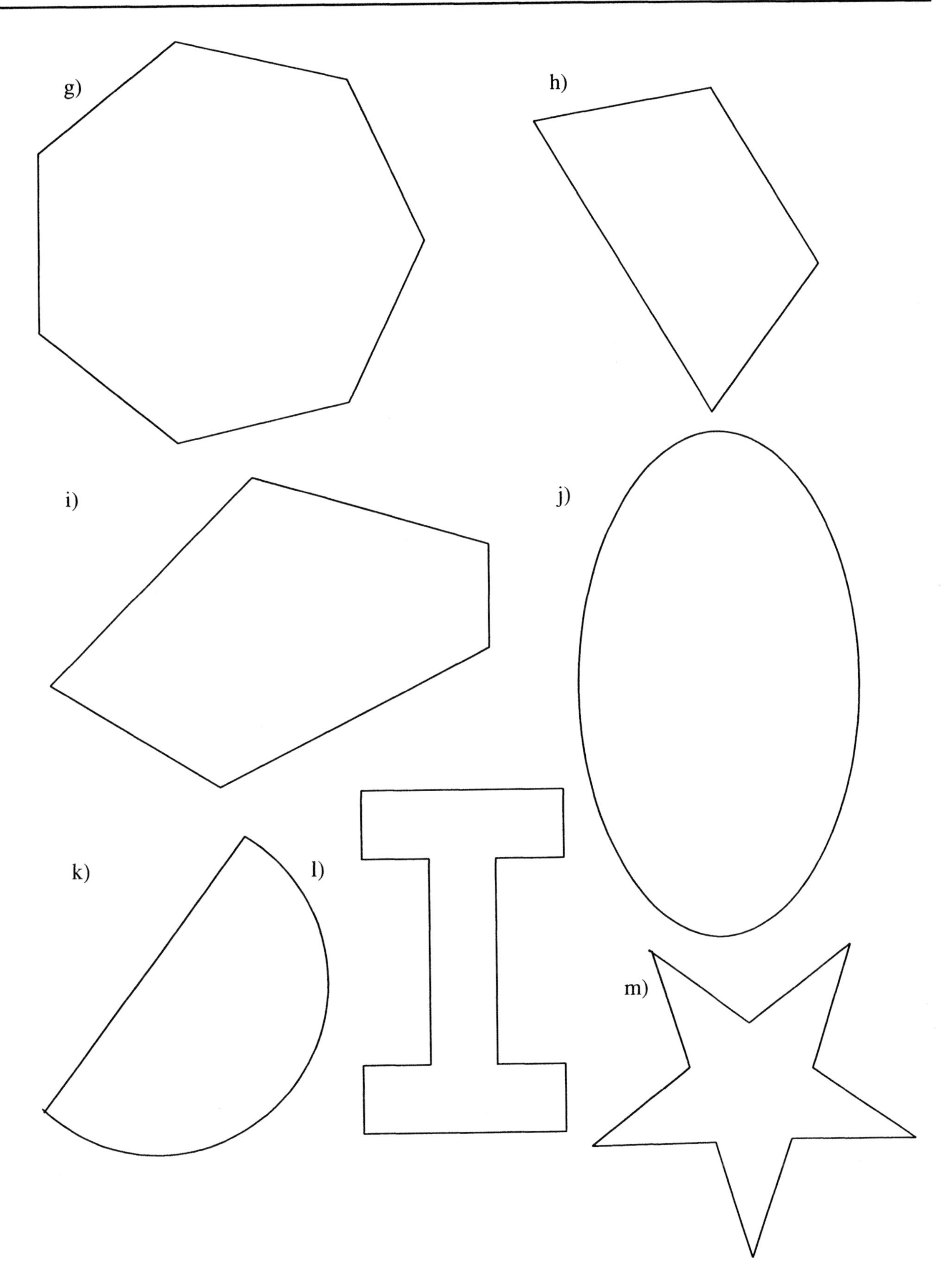
g)
h)
i)
j)
k)
l)
m)

Section 35: Centers and Lines of Symmetry

The homework for Section 34 included work with several particular kinds of triangles and quadrilaterals. In Exercise II. 5 you considered three triangles with special properties:

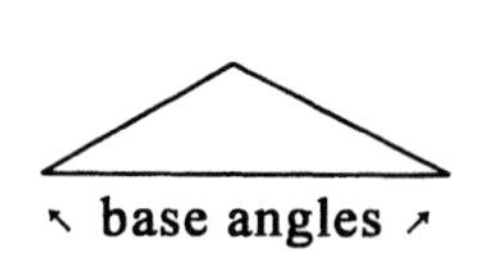

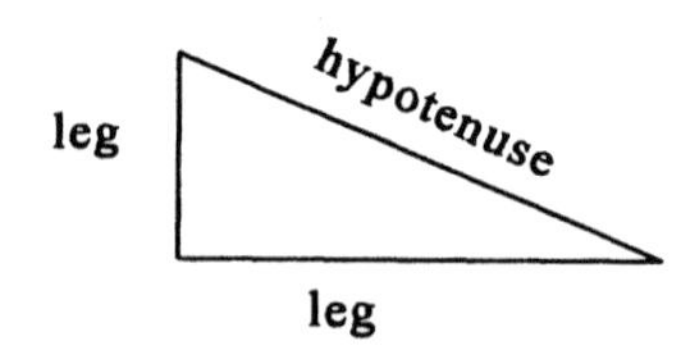

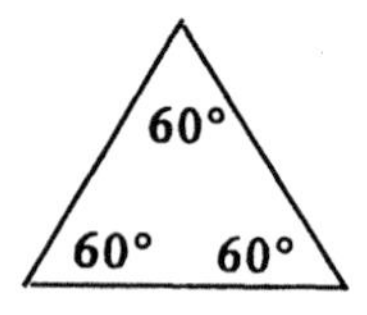

An isosceles triangle has two sides of equal length. The angles opposite those sides are called base angles.

A right triangle has one angle which measures 90°. The sides sides which form the right angle are called legs, and the longest side is called the hypotenuse.

An equilateral triangle has all sides equal and all angles equal. (It is regular.)

You are already familiar with some special properties of squares (all sides of equal length, and all angles measure 90°) and rectangles (all angles measure 90°, and both pairs of opposite sides have equal length). In Exercises II. 2, 3, and 4, you constructed three other interesting quadrilaterals.

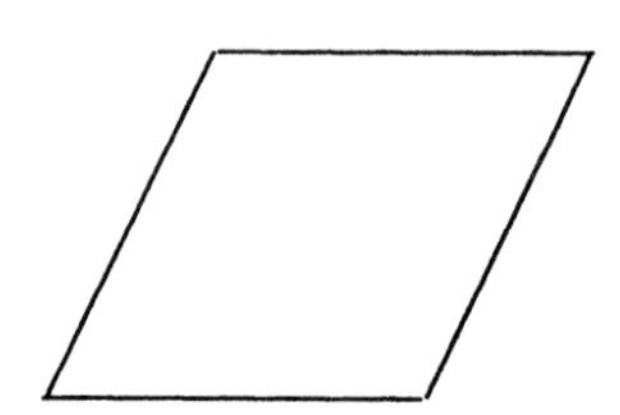

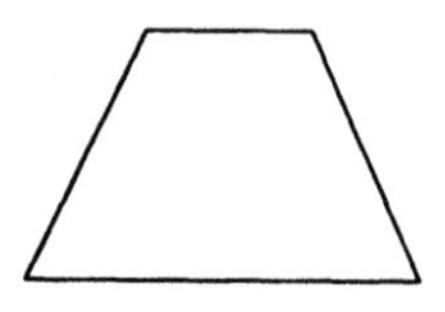

A rhombus has all sides of equal length.

A parallelogram has two pairs of parallel sides.

A trapezoid has one pair parallel sides.

These triangles and quadrilaterals, as well as, regular polygons with any number of sides, have several important properties which merit further attention.

Activity

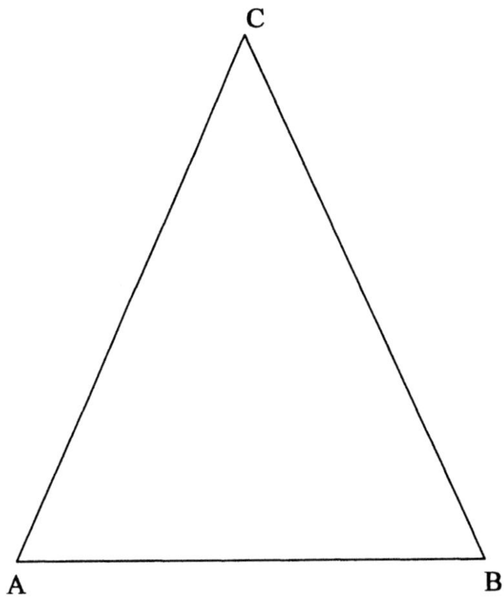

1. a) Use patty paper to compare AC and BC. What kind of triangle is △ABC?

   b) Trace △ABC on a sheet of patty paper.

   c) Find the perpendicular bisector of $\overline{AB}$. (Recall this is done by folding the paper until A lies on top of B and one half of the segment lies on top of the other half and creasing the paper.) Let M be the midpoint of $\overline{AB}$. Now MA = ______.

   d) What are the measures of ∠CMA and ∠CMB? How do you know?

   e) Fold the paper again along $\overline{CM}$ and notice that △CMA fits exactly on △CMB. (When a figure can be folded along a line so that one half fits exactly on the other half, the line is called a line of symmetry. So $\overleftrightarrow{CM}$ is a line of symmetry for △ABC. Does △ABC have any other lines of symmetry?)

   f) Since ∠ACM fits on ∠BCM, what can you say about the relationship between $\overrightarrow{CM}$ and ∠C?

Also notice that ∠A fits exactly on ∠B. (Remember that ∠A and ∠B are called the base angles of △ABC.) So we have demonstrated that base angles of an isosceles triangle are congruent.

This activity certainly doesn't prove that any of these relationships regarding an isosceles triangle will always be true. In fact, they are based on a fundamental assumption about triangles which we accept without proof: If each side of a triangle matches in length with some side of a second triangle, then the two triangles are congruent. This is usually referred to as the side-side-side axiom, or simply SSS.

Regarding △CMA and △CMB:

i) You found that $\overline{AC}$ and $\overline{BC}$ have the same length (within the limitations of measurement).
ii) MA = MB because M is the midpoint of $\overline{AB}$.
iii) $\overline{CM}$ is the third side of each triangle (and it's certainly equal to itself!)

Therefore, since we know that the three sides of △CMA each match one of the sides of △CMB, we accept by SSS that the two triangles are congruent. This is written symbolically as △CMA ≅ △CMB. The order of the vertices in the name of each triangle is very important; it specifies how one figure fits on top of the other.

2. a) Copy this hexagon and the names of its vertices on a sheet of patty paper.

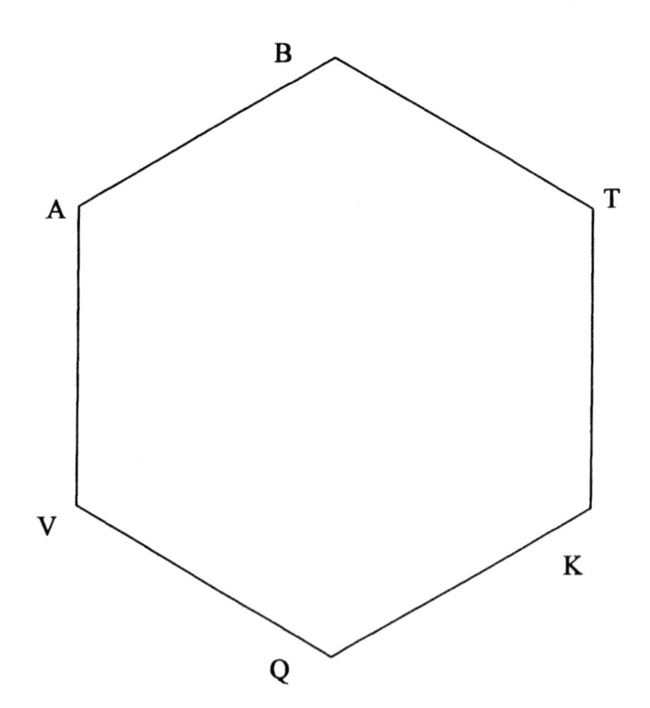

b) Find all lines of symmetry for the figure by folding the paper in various ways.

Notice that the hexagon has two different kinds of lines of symmetry--diagonals and perpendicular bisectors. Each diagonal line of symmetry bisects a pair of opposite angles; and each perpendicular bisector intersects a pair of opposites sides.

Also notice that all lines of symmetry intersect at a point which is clearly the center of the figure. Label it C.

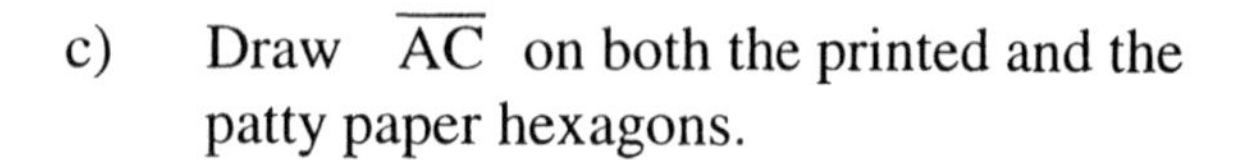

c) Draw $\overline{AC}$ on both the printed and the patty paper hexagons.

d) Fit the patty paper figure on top of the printed figure so that all vertices correspond by name. Put the point of your compass on C and <u>rotate</u> the patty paper clockwise until vertex A of the patty paper figure lies on vertex B of the printed figure.

Notice that the patty paper hexagon again fits exactly on the printed figure. When a copy of a figure can be rotated around a point, so that the copy fits exactly on the original one or more times <u>before</u> returning to its starting position, the point is called a <u>center of rotation</u>.

e) Measure the angle formed by the two segments named $\overline{AC}$ (one on each hexagon). This is called the <u>angle of rotation</u>. As the patty paper is rotated a complete cycle (until all vertices have returned to their starting positions), how many times does the copy fit exactly on the original hexagon? What is the relationship between the number of sides of a regular polygon and its angle of rotation?

3. a) Copy the figure at right on a piece of patty paper. Label the vertices A, K, M, V to correspond with those printed here. Does this figure have any lines of symmetry? If so, what are they?

b) Fold the patty paper so that the crease forms diagonal $\overline{AM}$. Fold the paper again to form diagonal $\overline{KV}$. Name the point of intersection C.

c) Draw diagonals $\overline{AM}$ and $\overline{KV}$ on the printed figure.

d) Fit your patty paper figure on top of the parallelogram, and put the point of your compass on C.

e) Rotate the patty paper clockwise until each vertex is again on top of a printed vertex. What is the angle of rotation for the parallelogram? (This is the angle formed by $\overline{AC}$ on the printed figure and $\overline{AC}$ on the patty paper.) In one complete cycle, how many times will the patty paper figure fit exactly on the stationary figure?

It is interesting that although a parallelogram has no lines of symmetry, it does have a <u>center of rotation</u>.

4. a) Trace this figure on a sheet of patty paper and label the vertices as shown. How many lines of symmetry does this rectangle have?

b) Fold the patty paper to show the lines of symmetry and the diagonals. Notice that all these lines intersect at a common point. Label it C. This is the center of rotation of the rectangle. What is the angle of rotation?

5. a) Copy this pentagon on a sheet of patty paper.

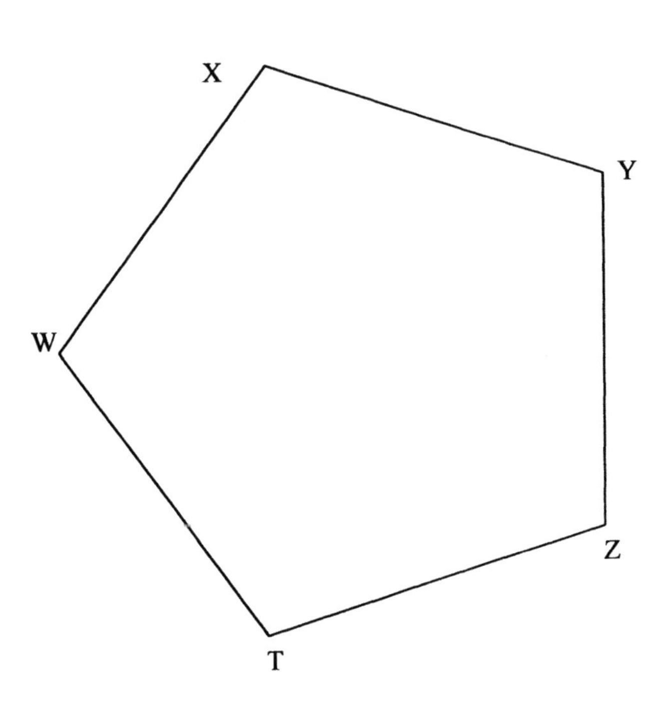

b) Fold the paper to form all lines of symmetry.

Notice that this figure has only perpendicular bisector lines of symmetry; and their common point of intersection is the center of rotation of the pentagon.

6. a) Trace this triangle on a sheet of patty paper. Notice that all three sides have different lengths, and hence angles have different measures. (This type of triangle is called a scalene triangle.) Beginning with the smallest angle, and continuing in order of size, label the vertices of the angles D, F, and E on both triangles. Does this figure have any lines of symmetry?

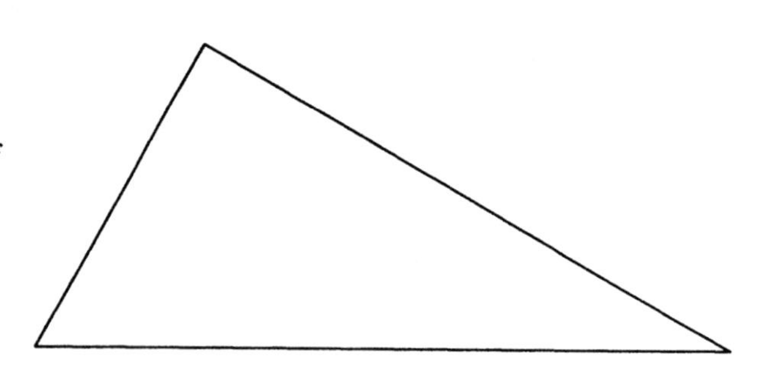

   b) Does this triangle have a center?

Obviously, we cannot not find a point where lines of symmetry meet. Moreover, there is no center of rotation. In the next section we will investigate other ways to find the center of a figure.

7. a) Trace this circle on a sheet of patty paper or use your compass to draw one.

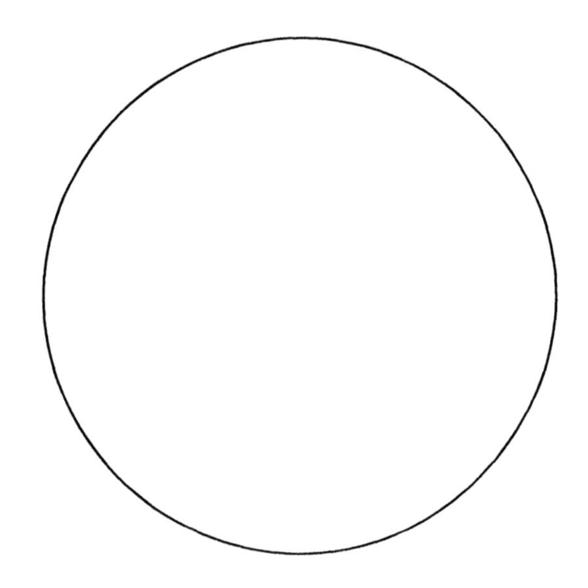

   b) Find several lines of symmetry by folding the paper. How many lines of symmetry does the circle have?

   c) Label the center of the circle C. What is the angle of rotation for the circle?

Homework: Section 35

1. a) Draw all lines of symmetry for each figure on pages 146 and 147.

   b) Which figures have a center of rotation? For those that do, what is the angle of rotation?

2. For each figure, state a congruence for the two figures which would fit on each other if the figure were folded along the dotted line of symmetry:

a)

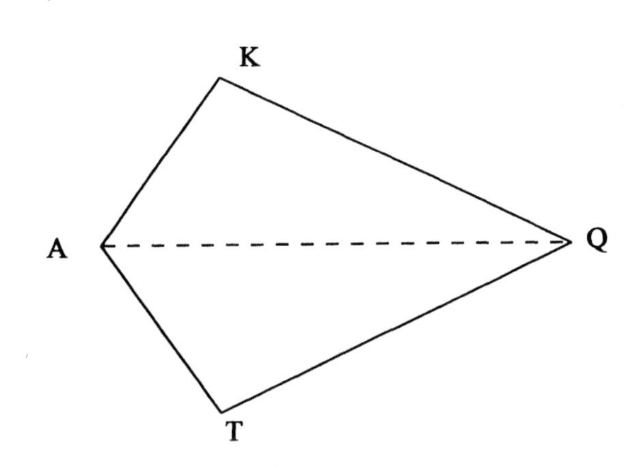

b)

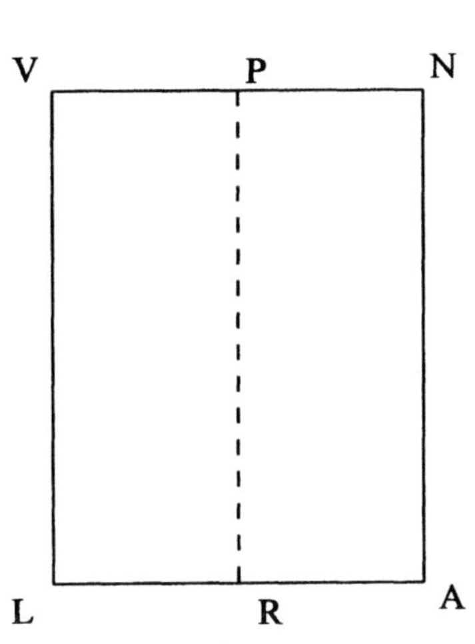

c)

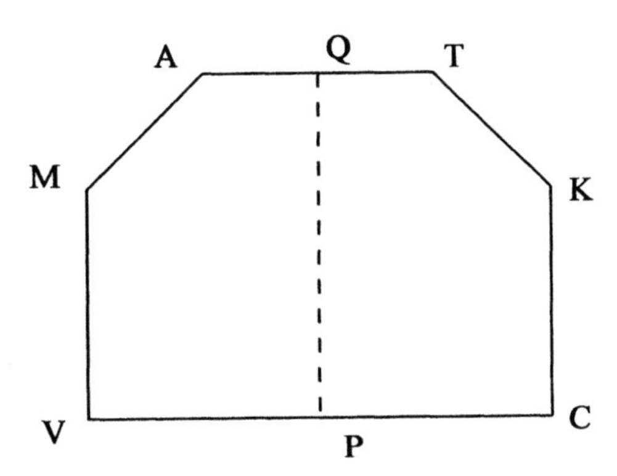

4. a) Pick any three points on the circle and mark them A, T, K.

   b) Draw △ATK.

   c) Construct the perpendicular bisectors of any two sides of△ATK.
   <u>Note</u>: The intersection of these perpendicular bisectors is the center of the circle!

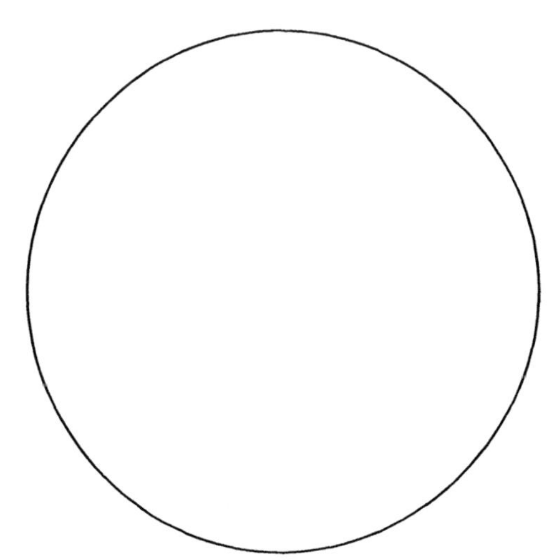

Section 36: Discovery Activities with Patty Paper

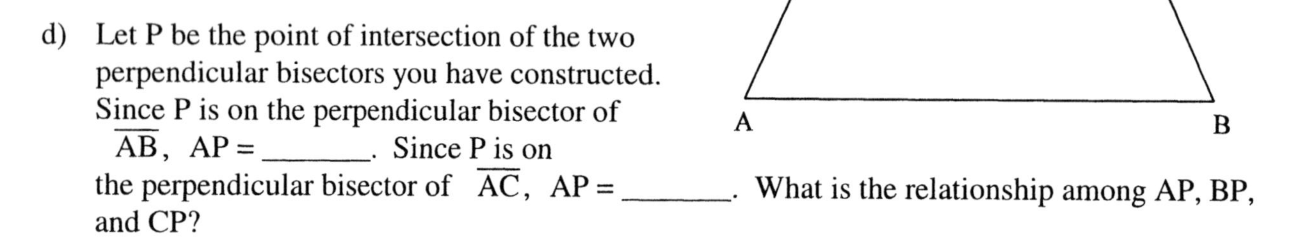

1. a) Trace $\triangle ABC$ on a sheet of patty paper and find the perpendicular bisector of $\overline{AB}$.

   b) Notice that the third vertex of the triangle, C, lies on this perpendicular bisector. What is the relationship between AC and BC? What kind of triangle is $\triangle ABC$?

   c) Now construct the perpendicular bisector of $\overline{AC}$. Does vertex, B, lie on this perpendicular bisector?

   d) Let P be the point of intersection of the two perpendicular bisectors you have constructed. Since P is on the perpendicular bisector of $\overline{AB}$, AP = _______. Since P is on the perpendicular bisector of $\overline{AC}$, AP = _______. What is the relationship among AP, BP, and CP?

   e) What do you think will happen when you draw the perpendicular bisector of $\overline{BC}$? Construct it and see if you are correct.

   f) Since P is on the perpendicular bisector of $\overline{BC}$, BP = _______. Does this give you any more information than you already had?

   g) You have discovered that the point of intersection of the perpendicular bisectors of the sides of a triangle is the same distance from each vertex. If you want to locate this point of intersection for a particular triangle, how many of the perpendicular bisectors do you have to draw?

   j) Use your compass to draw a circle centered at P with radius AP. The circle should contain all vertices of the triangle. Why?

   k) A circle that contains all of the vertices of a polygon is called the **circumscribed circle** of the polygon. Its center is the point of intersection of the perpendicular bisectors of each side.

2. a) Copy the triangle at the top of the page again.

   b) Now find the angle bisector of each angle.

   c) Notice that the three angle bisectors meet at a point. Label it Q.

   d) Find the distance from Q to one of the sides of the triangle, say $\overline{BC}$. (Remember that you need to draw a segment that contains Q and is perpendicular to the side.)

e) Open your compass to this distance and draw a circle centered at Q. Notice that this circle just "touches" each side of the triangle, it is said to be tangent to each side. Such a circle is called an **inscribed circle**. (Because of our limitations in using construction tools, the circle may not "touch" each side of the triangle.) This tells us that if we are looking for a point that is equidistant from all sides of a figure, the first step is to bisect all the angles of the figure.

3. a) Copy this hexagon on a sheet of patty paper.

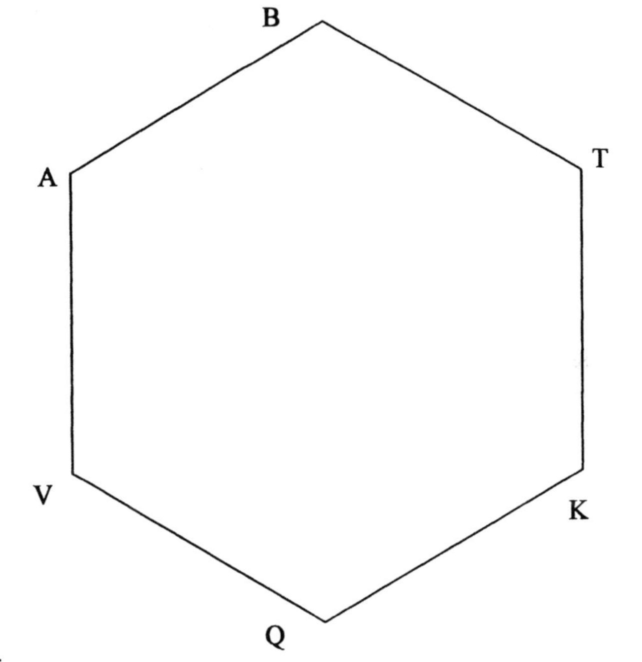

b) Recall from your previous work that all lines of symmetry intersect at a point. Draw two lines of symmetry for this figure--one diagonal and one perpendicular bisector. Label their point of intersection C.

c) Put the point of your compass at C (the center of the hexagon) and put the pencil point at one of the vertices. With this opening as radius, draw the circumscribed circle of the hexagon.

d) Now open your compass to represent the distance between C and the sides of the hexagon. (Remember that distance is measured perpendicularly: put the point of your compass on the center of the hexagon, and put the pencil point on the midpoint of one of the sides which is intersected by the perpendicular bisector line of symmetry.) With this opening as radius, draw the inscribed circle.

4. a) Copy the figure at right on a sheet of patty paper.

b) Find the perpendicular bisectors of the sides of this figure. Can you construct a circumscribed circle? If so, do it. If not, explain why not.

c) Find the angle bisectors of this figure. Can you construct an inscribed circle? If so, do it. If not, explain why not.

Recall: Although a parallelogram has no lines of symmetry, it does have a center of rotation--but this point is not the same distance from all vertices or from all sides, so it is not the center of either an inscribed or a circumscribed circle.

5. Does the rectangle have an inscribed or circumscribed circle? If so, draw it/them with your compass.

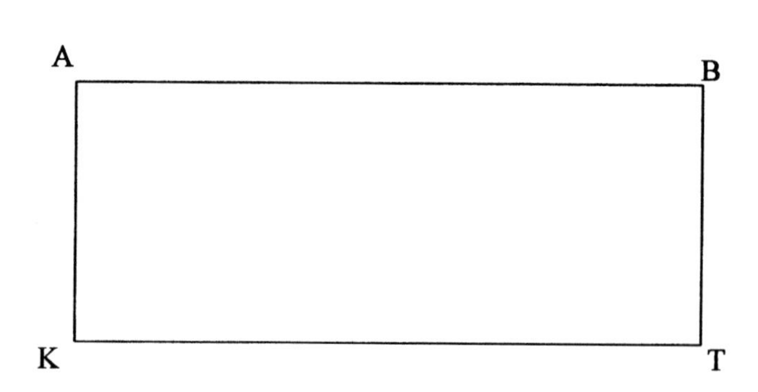

6. a) Trace this triangle on a sheet of patty paper.

   b) On your patty paper triangle, locate the midpoint of the side $\overline{DF}$. Label it M. Next, fold the paper so that the crease forms $\overline{EM}$. $\overline{EM}$ is a median of the triangle. (A median is a segment which connects a vertex of a triangle with the midpoint of the opposite side.)

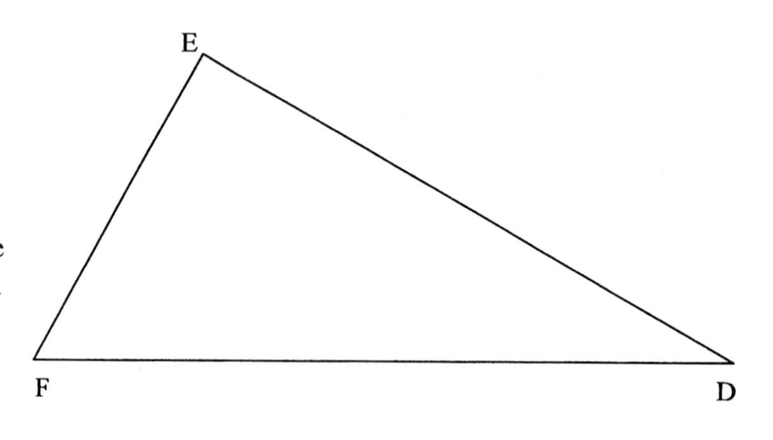

   c) Now fold the paper to form the perpendicular bisector of $\overline{FD}$. Notice that it does not contain vertex E.

   d) Finally, fold the paper to bisect $\angle E$. For this scalene triangle, unlike the isosceles triangle at the beginning of this activity, all of these segments are different:

      i) the perpendicular bisector of $\overline{FD}$
      ii) the bisector of $\angle E$
      iii) the median from E to $\overline{FD}$

   e) Draw all these segments on the printed triangle above (using whatever tools you find convenient).

Homework: Section 36

1. a) Draw any three non-collinear points. Name them A, B, and C.

   b) Draw $\overline{AB}$ and $\overline{AC}$, and construct their perpendicular bisectors. Name the point of intersection of the perpendicular bisectors, K.

   c) Draw a circle with radius AK and center K. Within the limitations of construction, it should contain all vertices of the triangle.

   Note: This exercise demonstrates that any three non-collinear points lie on a circle!!

2. Suppose Joe and Suzy each construct a triangle having sides of 5, 8, and 11 centimeters. Will the triangles necessarily be congruent? How do you know?

3. Find x.

   AP = PC

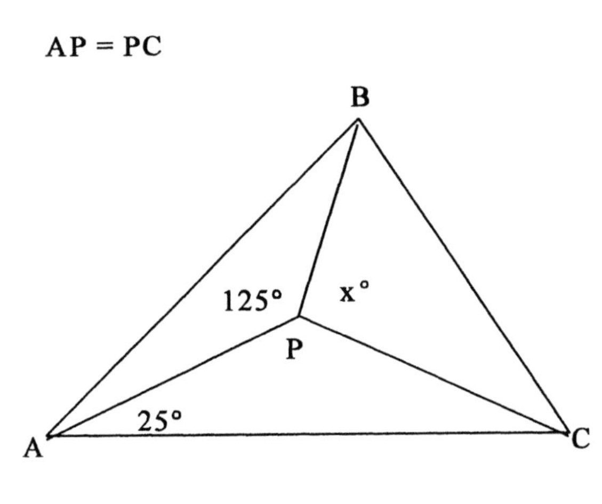

4. Find x.

   BD = AD

   EF = CF

   $\overline{AB} \parallel \overline{CE}$

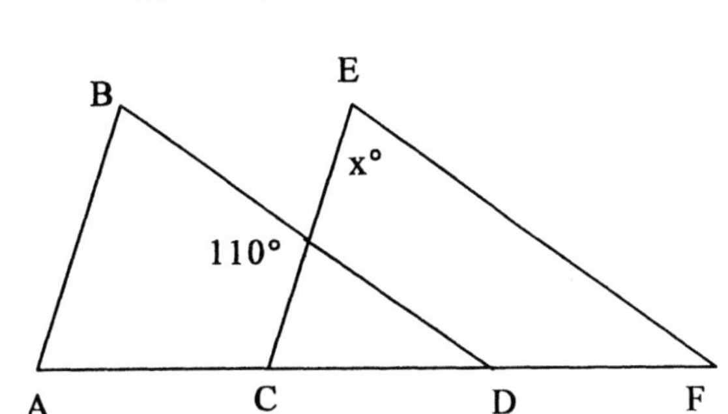

5. Suzy bought some Puffo yeast which causes dough to double in size every 6 minutes. Last week, it took two recipes of bread dough an hour to half fill her big mixing bowl. How long will it take one recipe of bread dough to fill the bowl?

6. For each of the following figures construct the circumscribed circle and the inscribed circle, if possible.

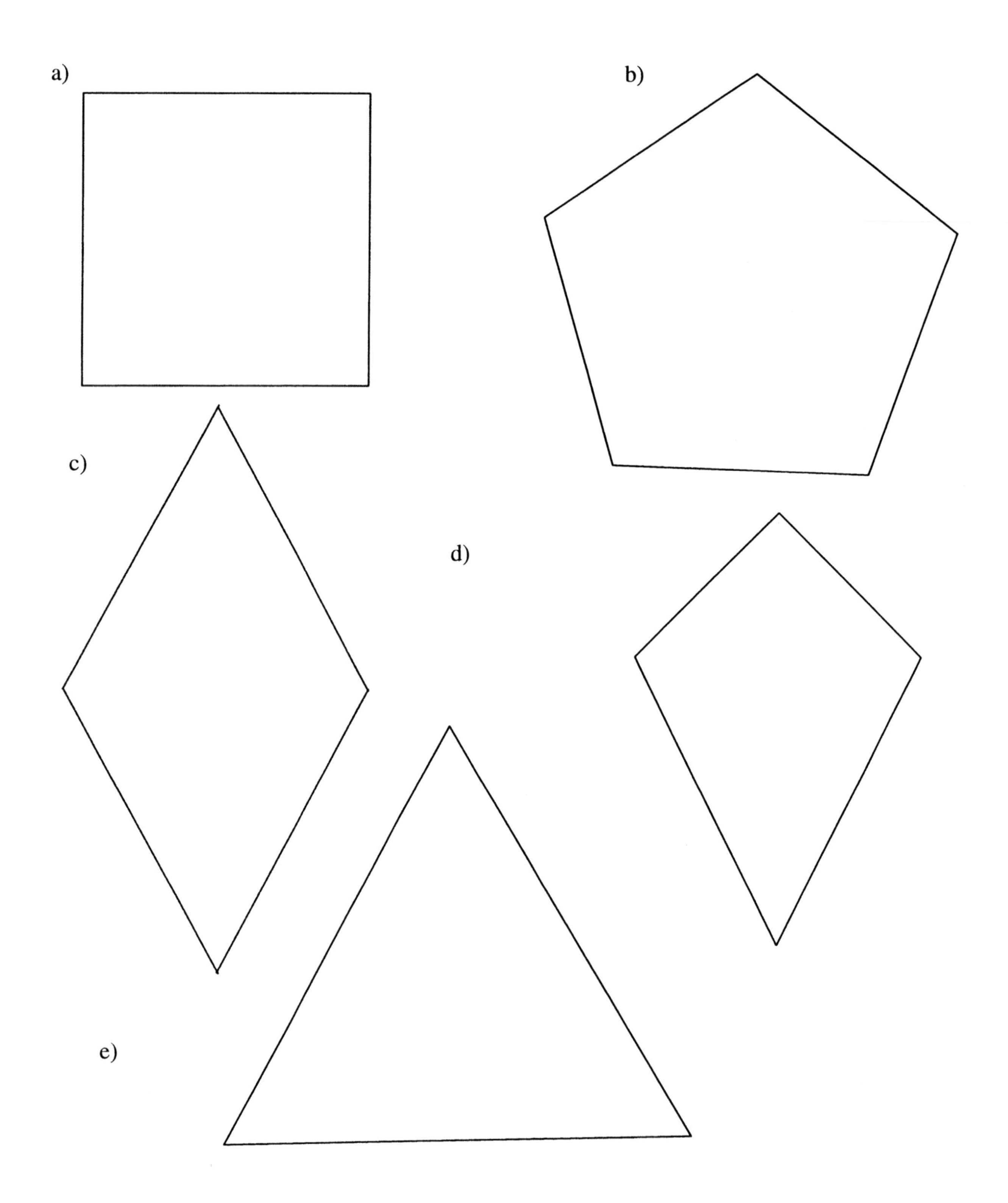

Section 37: Recap

Several very important concepts and relationships were introduced in the classroom and homework activities of Section 35 and 36. We are listing them here for emphasis and clarity:

- Two figures are <u>congruent</u> if one is an exact copy of the other. One could be placed on the other in such a way that they would "fit".
  If the figures are polygons, the congruence can be described symbolically by listing the vertices for each polygon in the order in which they "fit".
  For example, $\triangle KMQ \cong \triangle TVS$ implies that $\triangle TVS$ would fit exactly on $\triangle KMQ$ if T lies on K, V lies on M, and S lies on Q. This clearly implies that KM = TV, MQ = VS, KQ = TS, $\angle K = \angle T$, $\angle M = \angle V$, and $\angle Q = \angle S$.

- The perpendicular bisector of a segment is the line which contains the midpoint of the segment <u>and</u> is perpendicular to the segment. It consists of all the points which are the same distance from each endpoint of the segment.

  If $\overleftrightarrow{TV}$ is the perpendicular bisector of $\overline{AB}$, then $\overline{TV} \perp \overline{AB}$, the midpoint of $\overline{AB}$ is on $\overleftrightarrow{TV}$, and if K is on $\overleftrightarrow{TV}$, then AK = BK.

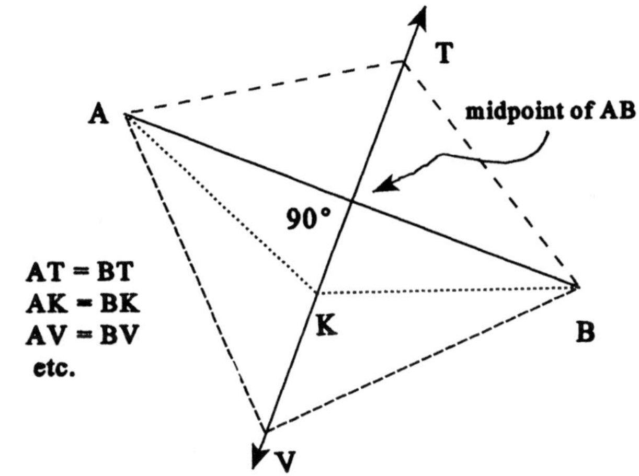

- A line of symmetry separates a figure into two parts which fit exactly when the figure is folded along that line.

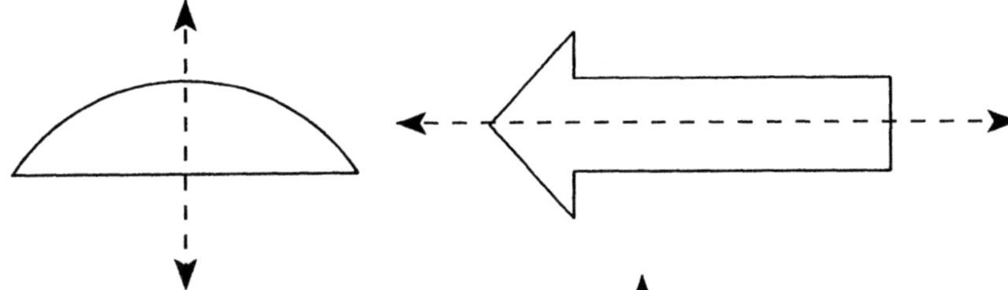

- If a figure has two or more lines of symmetry, they intersect at a center of rotation.

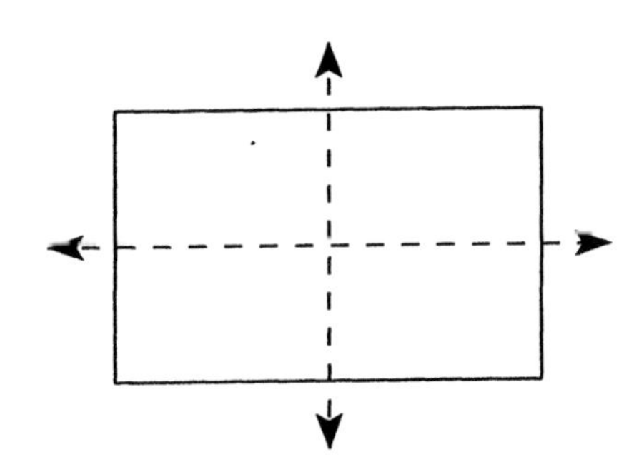

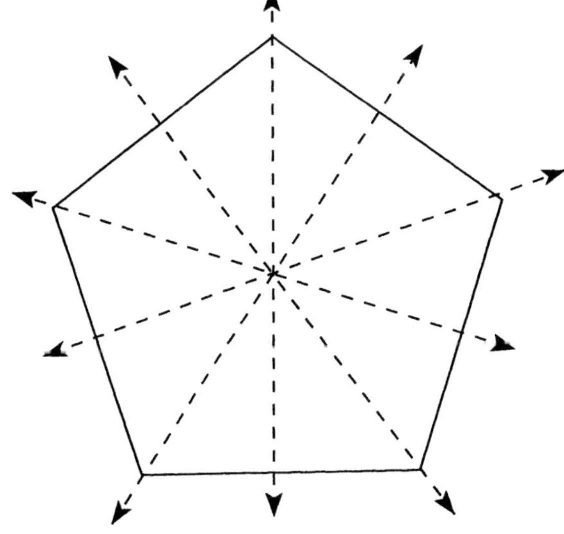

* When a copy of a figure can be rotated around a point so that the copy fits exactly on the original, one or more times before returning to its starting position, the point is called a center of rotation.

* As the copy of a figure rotates from its starting position to the next position where it fits exactly on the original, it moves through a certain fraction of a cycle: $\frac{x}{360}$ degrees. x is the measure of the angle of rotation for the figure.

* A parallelogram has a center of rotation even though it has no lines of symmetry.

* A regular polygon has a center of rotation which is also the center of an inscribed and circumscribed circle.
The segments drawn from the center to each vertex separate the regular n-gon into n congruent triangles.
If the regular polygon has an even number of sides, the center can easily be located by drawing two diagonal lines of symmetry.
If the regular polygon has an odd number of sides, the center can be located by bisecting two angles or by constructing the perpendicular bisectors of two sides.

Example 1:

Angle of rotation: 45°

Example 2:

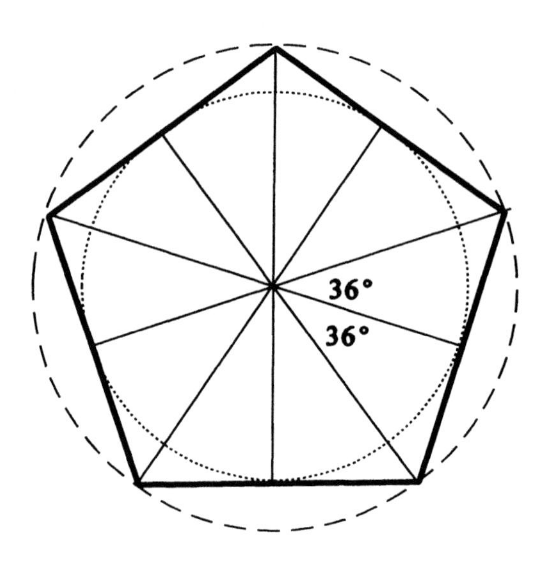

Angle of rotation: 72°

* The center of a circle can be located by picking three points on the circle, drawing the triangle defined by those points, and drawing the perpendicular bisector of two sides of the triangle.

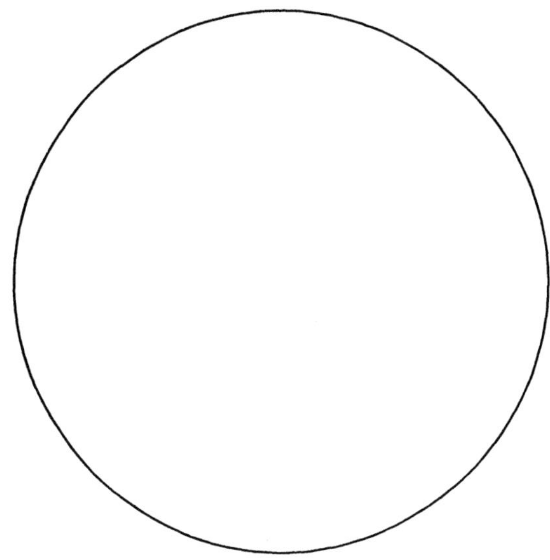

* Any three non-collinear points lie on a circle. It can be constructed by drawing the triangle defined by the three points, and locating the center of the circle as the point of intersection of the perpendicular bisectors of two sides of the triangle.

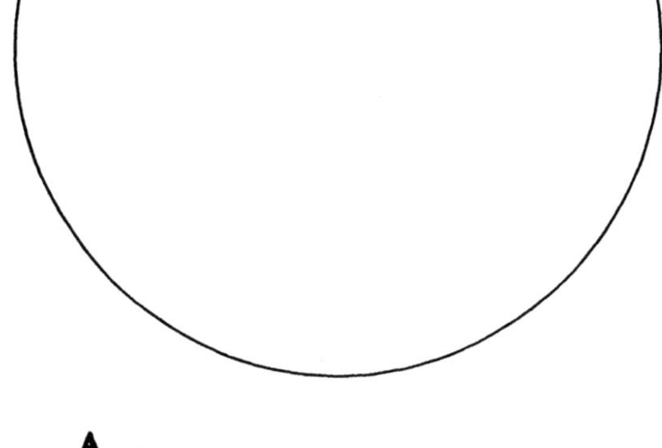

* If the three sides of one triangle have the same respective lengths as three sides of another triangle, we assume that the two triangles are congruent. (SSS)

* A median of a triangle is a segment connecting a vertex and the midpoint of the opposite side.

* In an isosceles triangle the line of symmetry is also the perpendicular bisector of the base, the bisector of the vertex angle, and the median to the base.

* The base angles of an isosceles triangle are congruent.

* A scalene triangle has no line of symmetry and no center of rotation.

* For a scalene triangle: the perpendicular bisector of a side, the median to that side, and the bisector of the angle opposite that side are all different segments.
  However, in any triangle all three perpendicular bisectors meet at a common point (which is the center of a circumscribed circle); all three angle bisectors meet at a common point (which is the center of an inscribed circle); and all three medians meet at a common (which is the center of gravity of an actual scalene triangular object).

Homework: Section 37

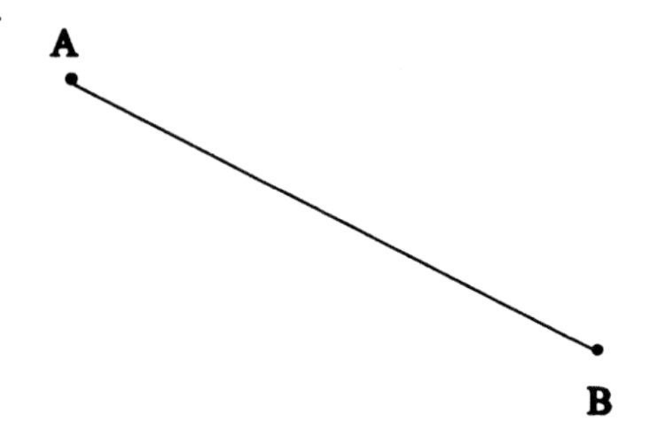

1. a) Open your compass to a radius which is more than half of AB.
   b) Construct two circles having this radius, one with center at A and one with center at B.
   c) Label the points of intersection of the two circles, T and V.
   d) Draw $\overleftrightarrow{TV}$
   Note: $\overleftrightarrow{TV}$ is the perpendicular bisector of $\overline{AB}$!!

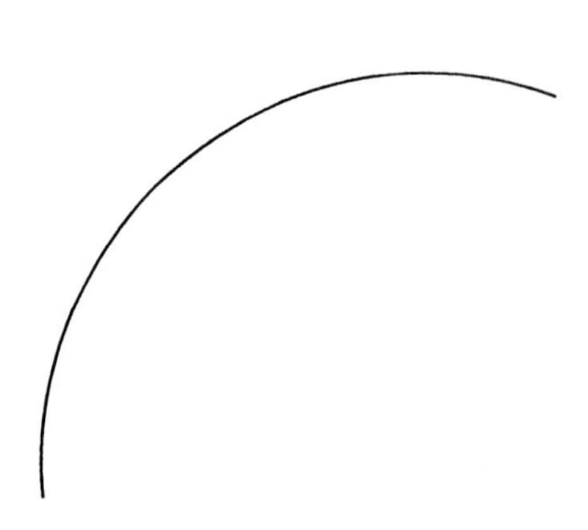

2. a) The arc at right is a portion of a circle. Select three points on the arc and label them K, M, and T.
   b) Draw $\triangle$KMT.
   c) Construct the perpendicular bisectors of two sides of the triangle.
   d) The intersection of these lines is the center of the circle which contains this arc. Complete the circle.

3. Locate the center of rotation for each figure. What is the angle of rotation for each?

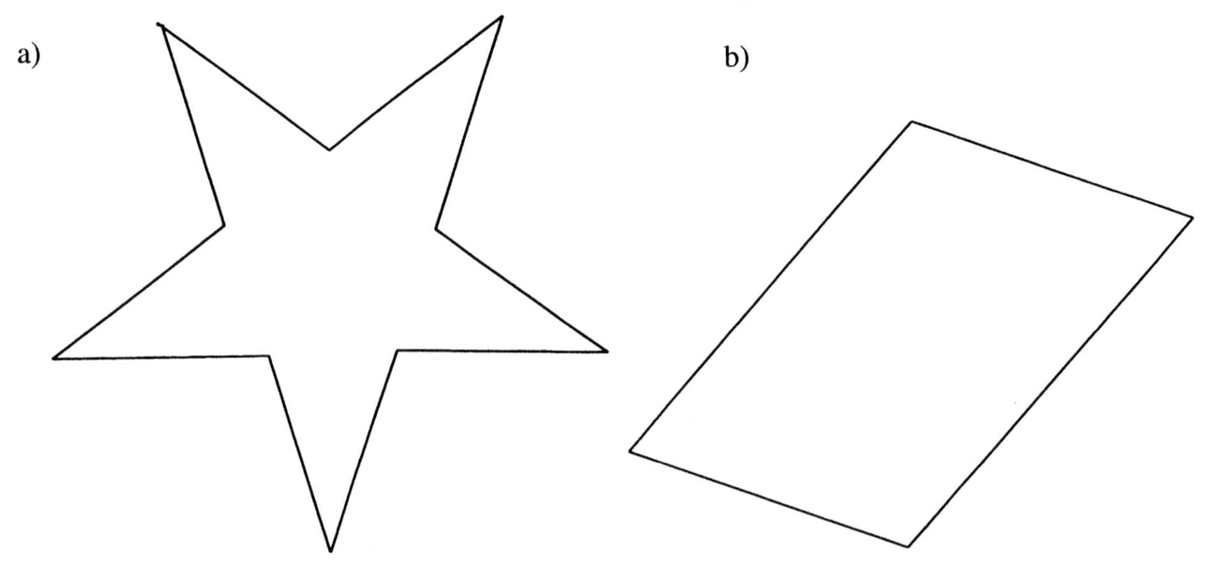

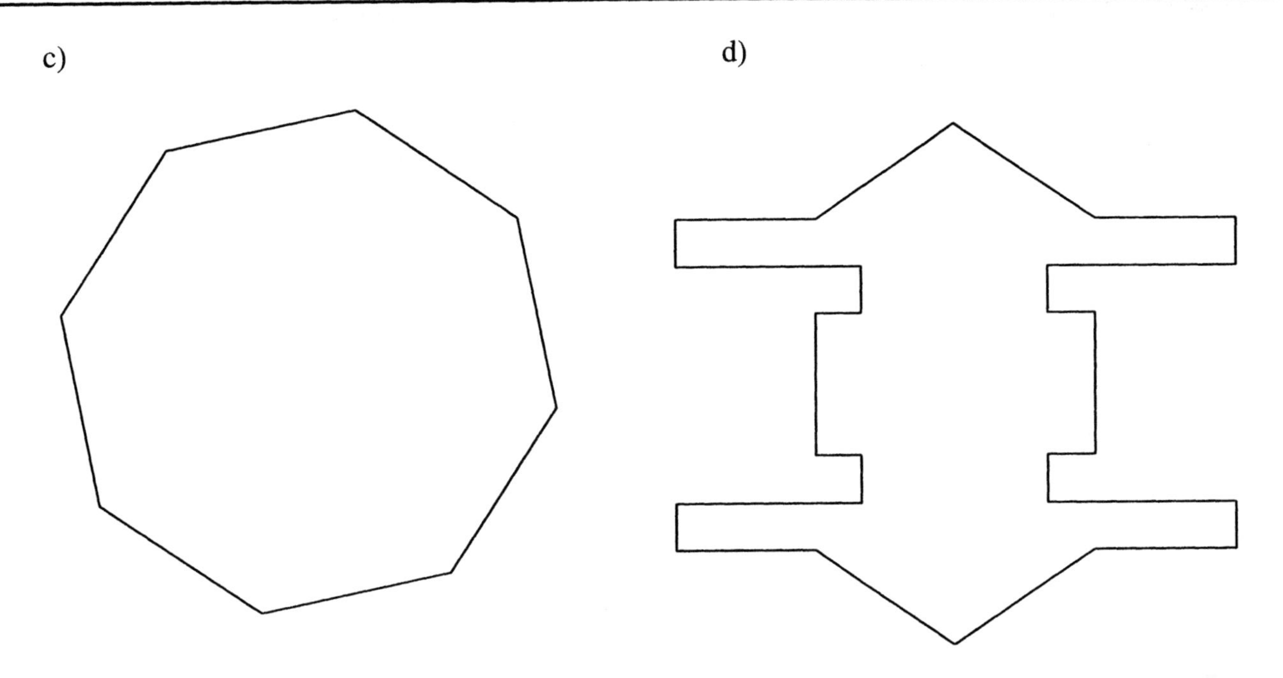

4. State a congruence for each pairs of figures.

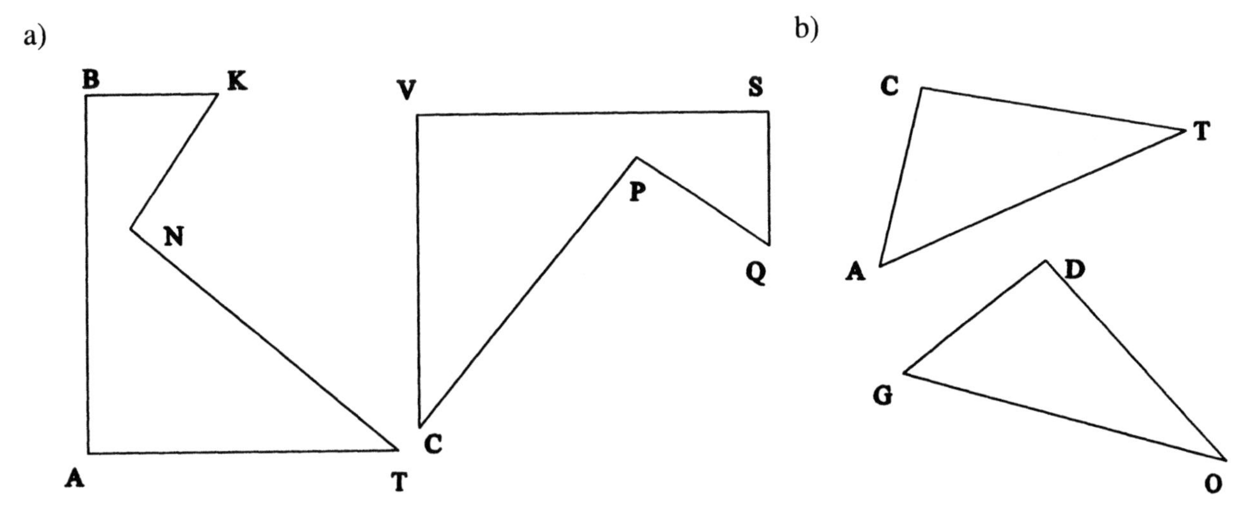

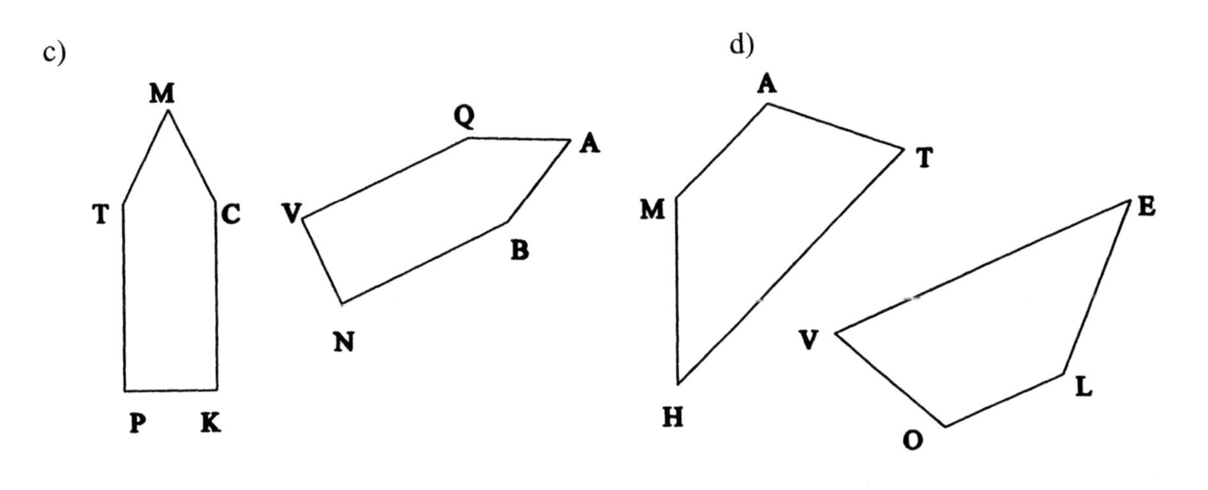

5. Construct the inscribed and circumscribed circles, if possible, for each figure.

a)

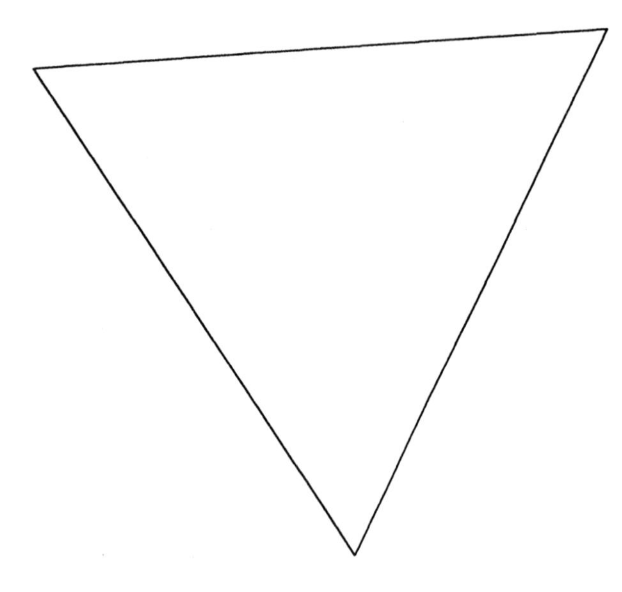

b)

c)

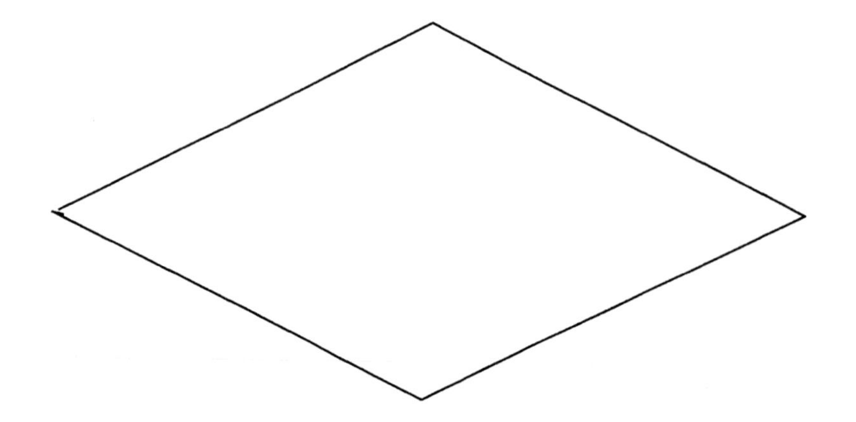

II.

1. On sturdy paper construct these figures using a ruler and protractor. Label them (include your name), cut them out and bring them to class.

   a) a triangle with two 50° angles and a side of 14 cm between those two angles

   b) a triangle with two 50° angles and a side of 14 cm which is not between those two angles

   c) a triangle with sides of 10 cm and 16 cm and an angle of 30° between those two sides

   d) a triangle with sides of 10 cm and 16 cm and an angle of 30° opposite the 10 cm side

   e) a triangle with angles of 40° and 60°

   f) a regular six-sided figure

   g) a six-sided figure that is equilateral but not equiangular

   h) a triangle with sides of 7 cm, 10 cm, and 14 cm

2. **Explain** what you did to construct the figures in parts b), d), and g).

Section 38: Congruent Figures

The concept of congruence has permeated the last two sections. We have talked a lot about one figure "fitting exactly" on another, and you have made copies of figures by tracing them on patty paper.

Now we turn our attention to constructing congruent figures (duplicates of each other) without tracing. This can be done by specifying measurements (for example: construct a rectangle with length 8 cm and width 5 cm) or by duplicating sides and angles of an existing figure using ruler, protractor, and/or compass. For example: construct an angle, with vertex at C, which is congruent to this angle:

We begin with the construction of congruent figures using specified measurements; and the first question to be considered is: In order to construct congruent figures of a particular type, what measurements must be specified?

For segments, angles, circles, and regular polygons, only one measurement is required - segments are congruent if they have the same length; angles are congruent if they have the same opening between rays (degrees); circles are congruent if they have the same radius; regular n-gons are congruent if the have the same length of one side.

Two measurements must by specified to guarantee congruence of rectangles (length and width) and rhombuses (length of side and measure of one angle).

Example 1: Construct a rectangle of length 8 cm and width 5 cm.

Since we know that all angles must be 90°, all rectangles constructed with these specifications will be exactly like this one:

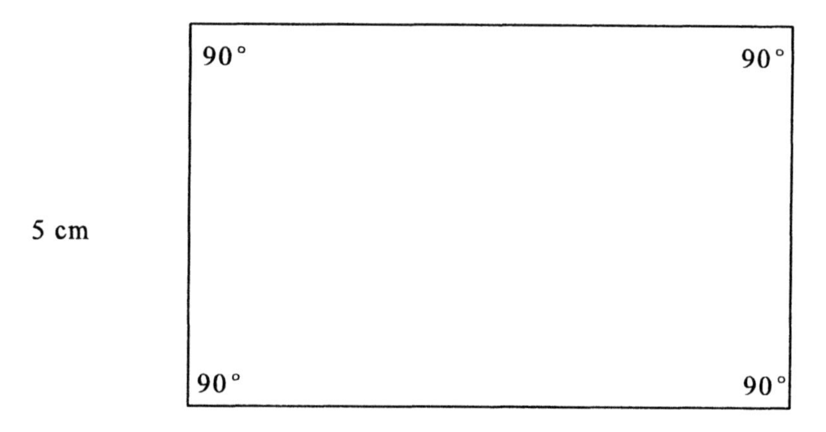

Example 2: Construct a rhombus having a side of 5 cm and one angle of 30°.

Since all sides are equal and opposite sides are parallel, we know that any rhombus with these specifications must be congruent to this one:

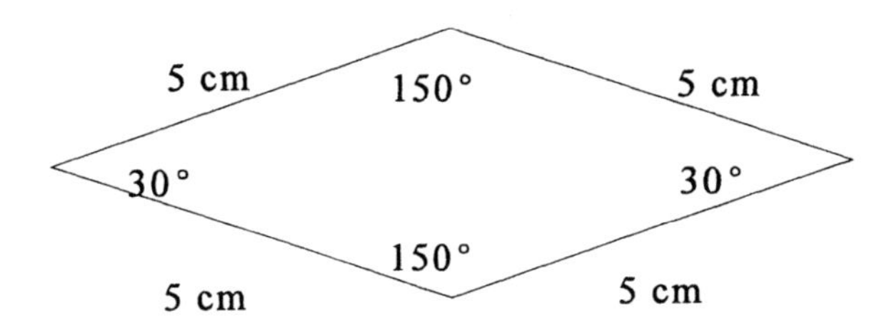

For non-equilateral triangles, three measurements can sometimes be sufficient to guarantee congruence as the following examples indicate.

Example 3: Construct a triangle with sides of 2, 3, and 4 inches.

First draw a 4 inch segment and name it $\overline{AB}$. Then draw a circle of radius 2 inches with center at A, and a circle of radius 3 inches with center at B. Name one of the points of intersection of the circles K.

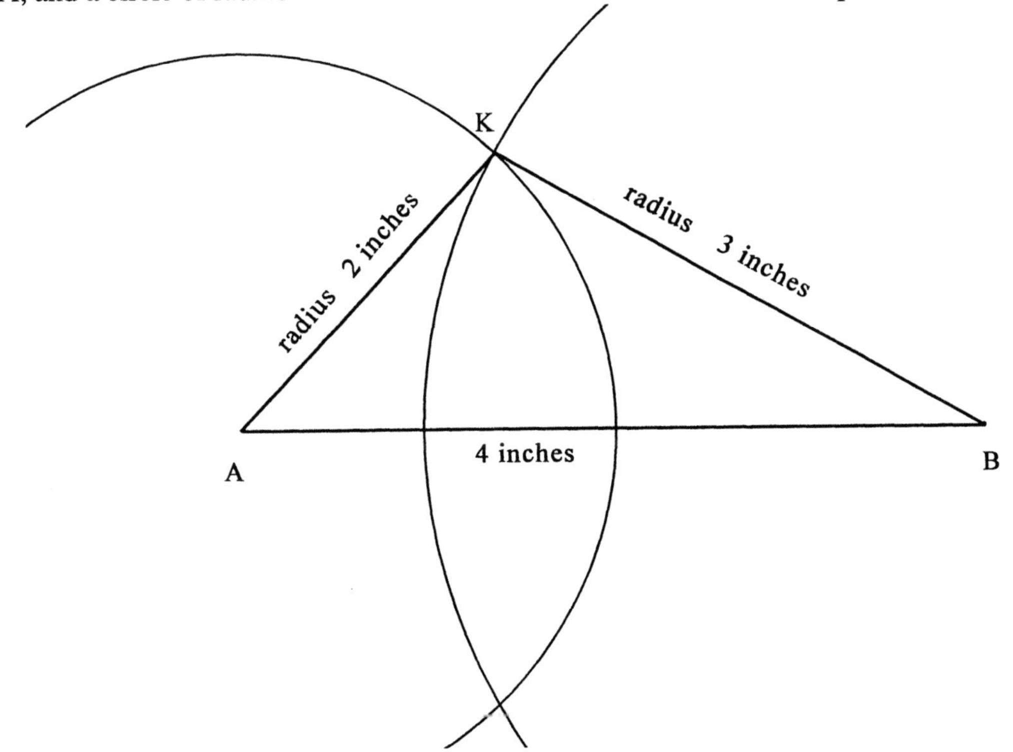

It is obvious that $\triangle ABK$ has the required specifications, and that any other triangle which is constructed with these specifications will be congruent to $\triangle ABK$. This is the side-side-side (SSS) congruence which we discussed previously.

Example 4: Construct $\triangle AKT$ having these specifications: $AK = 5$ cm, $\angle K = 50°$, and $KT = 4$ cm.

First draw $\overline{AK}$. Then construct an angle of 50° with vertex at K. Next, locate point T by measuring a distance of 4 cm from K as shown below:

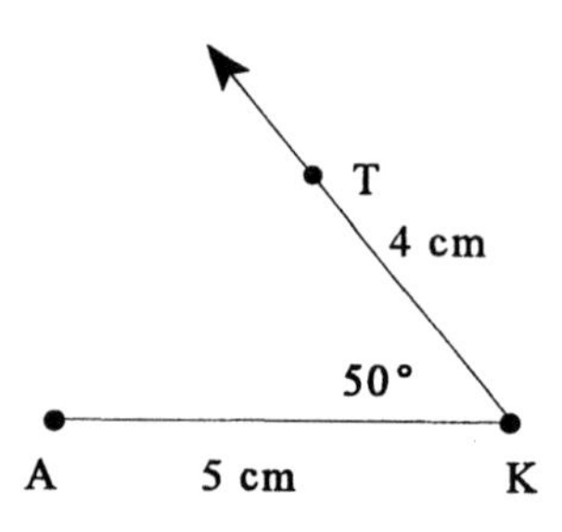

Now all three vertices of the figure are determined, so all triangles constructed with these specifications will be exactly like $\triangle AKT$. This is the side-angle-side (SAS) congruence.

Example 5: Construct a triangle which has angles of 35° and 45° and a side of 6 cm between those angles.

First draw a 6 cm segment and name it $\overline{AB}$.
Draw a 35° angle which has $\overrightarrow{AB}$ as one side.
Draw a 45° angle which has $\overrightarrow{BA}$ as one side.

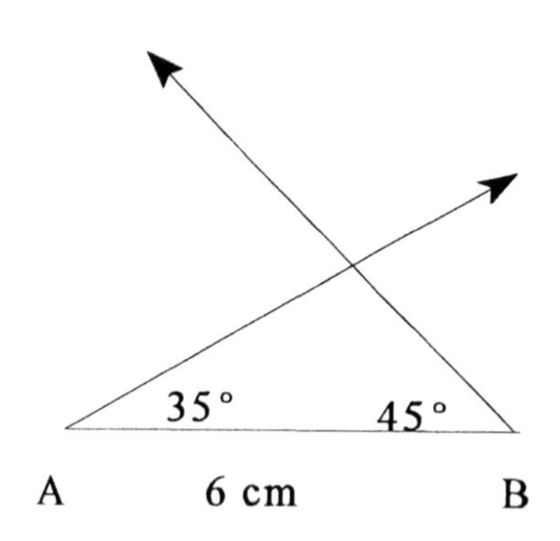

Notice that the third vertex is fixed -- all triangles constructed with these specifications must be exactly alike. This is the angle-side-angle (ASA) congruence.

Example 6: Construct a triangle which has angles of 35° and 45° and a side of 4 cm opposite the 45° angle.

Notice that if two angles of the triangle are 35° and 45°, then the third angle must be 100°.

Draw a 4 cm segment and name it $\overline{AB}$. Then draw a 35° angle with vertex at A, and a 100° angle with vertex at B as shown below.

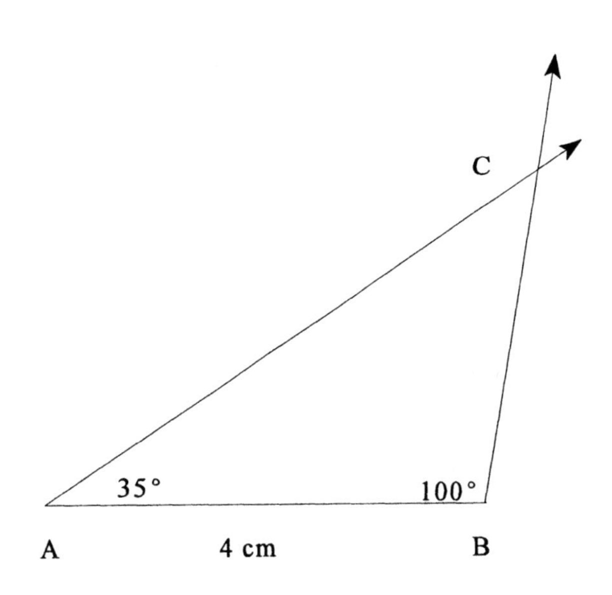

The third vertex of the triangle is now fixed.

It is obvious that this triangle has the given specifications, and that any other triangle constructed with these specifications will be congruent to △ABC.

So we see that specification of the length of one side and the measures of any two angles is enough to guarantee congruence of triangles.

But specification of any three parts of a triangle is not necessarily enough to assure congruence.

Example 7: Construct a triangle which has sides of 7 cm and 4 cm and an angle of 30° opposite the 4 cm side.

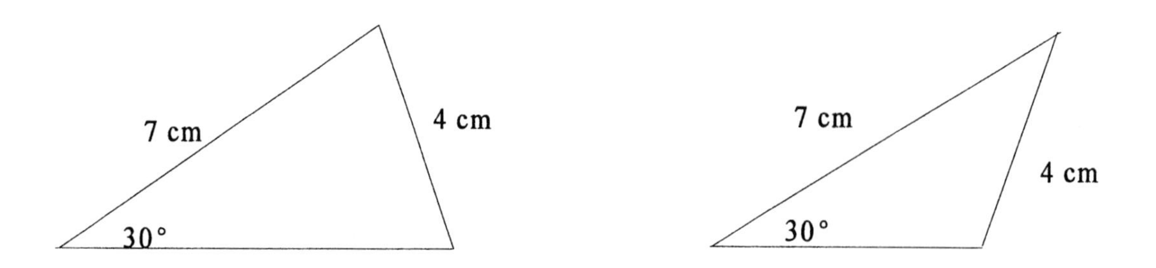

Notice that both of these triangles fit the given specifications, but the triangles certainly are not congruent.

In general, to guarantee congruence for non-regular n-gons, measurements must be specified for 2n - 3 consecutive parts (sides and angles). This will be explained and demonstrated in Example 8 below.

Now suppose that a figure is to be constructed, not according to specified measurements, but by duplicating an existing figure.

Example 8: Construct a polygon which is congruent to the one below.

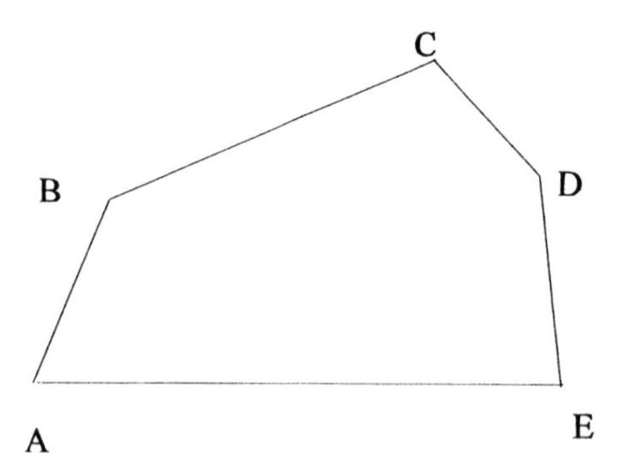

We begin by duplicating a side. (We choose $\overline{AE}$.) This can be done by measuring or by copying with a compass.

Next, duplicate an angle at one endpoint of $\overline{AE}$. (We choose $\angle A$.) This, too, can be done by measuring with a protractor or copying with a compass.

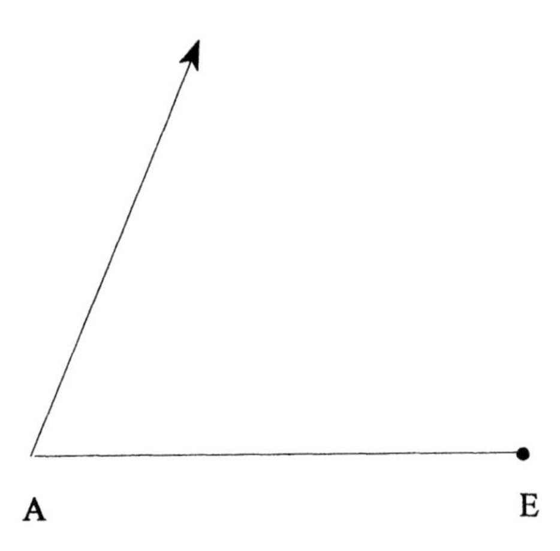

Now copy $\overline{AB}$, then $\angle B$, then $\overline{BC}$, then $\angle C$, then $\overline{CD}$.

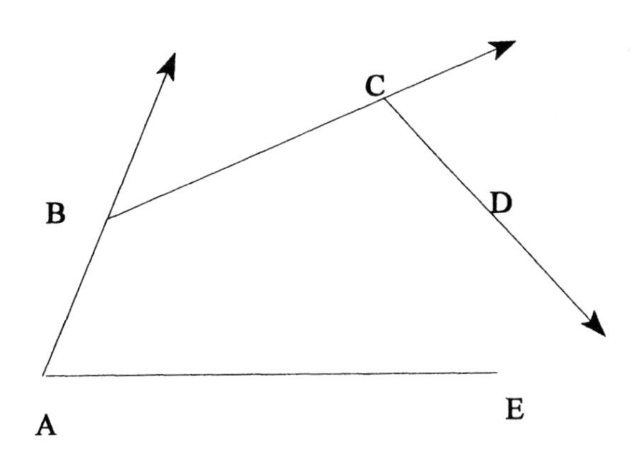

Now all the vertices are determined, so it is not necessary to copy $\angle D$ or $\overline{DE}$ or $\angle E$.

Notice that in order to make a copy of an n-gon, the last three parts need not be duplicated. And since an n-gon has 2n parts (n sides and n angles), we have demonstrated that $2n - 3$ consecutive parts must be specified or copied.

Important Note Since we copy polygons by measuring angles and segments, these constructions are subject to the possible error inherent in all measurements. So when we construct polygon TRVKS as a duplicate of polygon ABCDE, we say that the two figures are congruent - but it is understood that their corresponding parts are approximately equal (within the limitations of measurement).

Homework: Section 38

1. Copy this triangle by using the SAS congruence, then copy it again using the ASA congruence. Finally, copy it a third time using the SSS congruence. Which method is easiest?

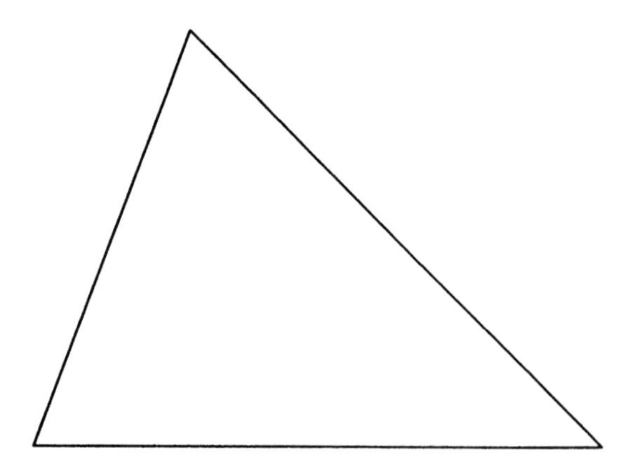

2. Construct a rectangle with length 7.0 cm and width 4.6 cm. Construct the two diagonals of the rectangle.

3. Construct a parallelogram with one side 8.2 cm, one side 4.6 cm, and one angle of 50°.

4. a) Construct a rhombus with sides of 4.0 cm.
   b) Construct another rhombus which is not congruent to the first one.
   c) Measure the diagonals of each rhombus.
   d) For each rhombus, measure the angles formed by the diagonals.

5. Copy these figures with ruler, protractor, and/or compass..

   a)

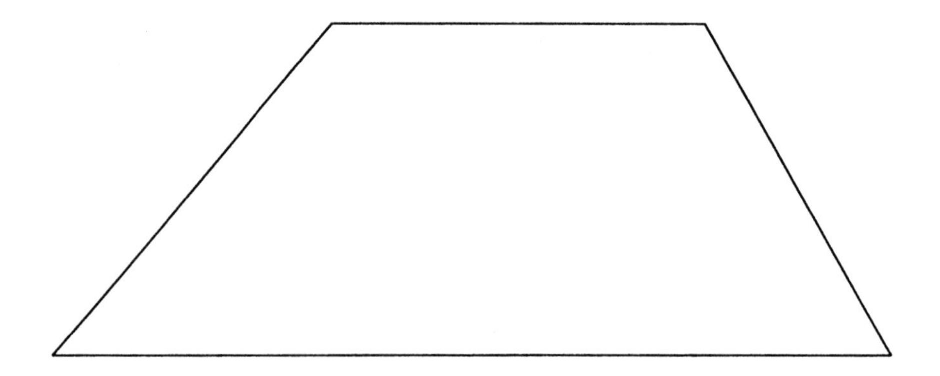

b)

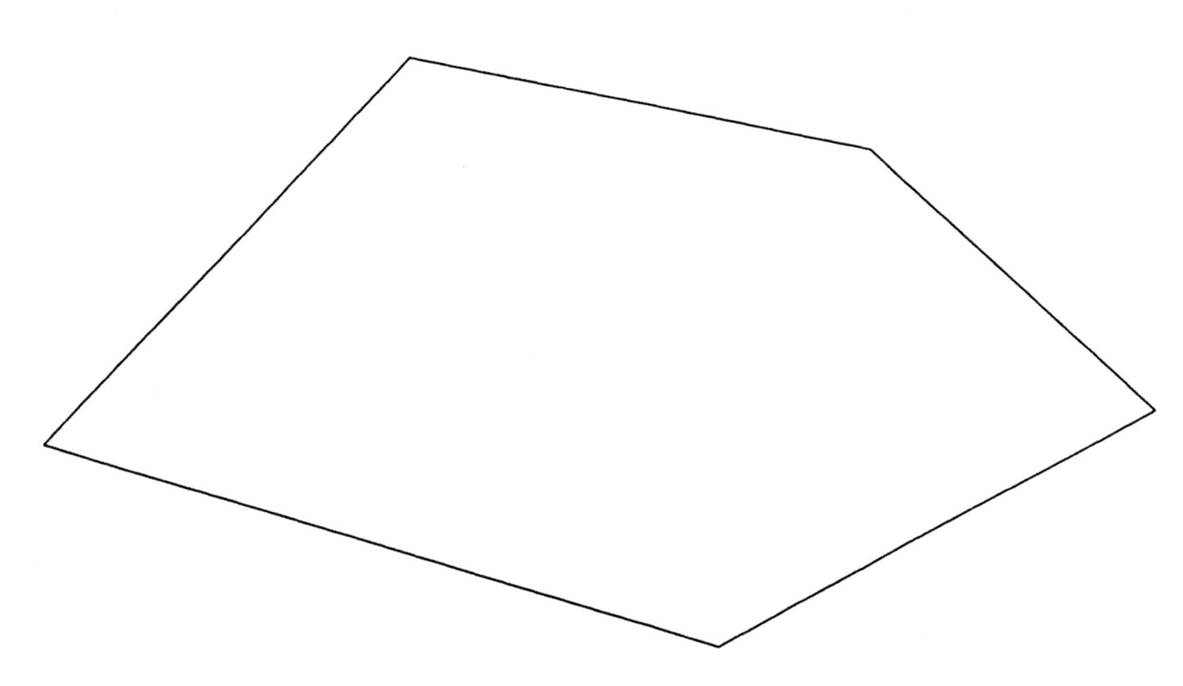

6. List all the information which can be deduced from each statement.
   a) KRVT is a rectangle.
   b) $\triangle$RST is isosceles and $\angle S = 120°$.
   c) $\triangle ATQ \cong \triangle KRM$
   d) V is on the perpendicular bisector of $\overline{AK}$.
   e) PTQAC is a regular polygon.
   f) ABCD is a rhombus

7. Use the information given in the picture to find x.

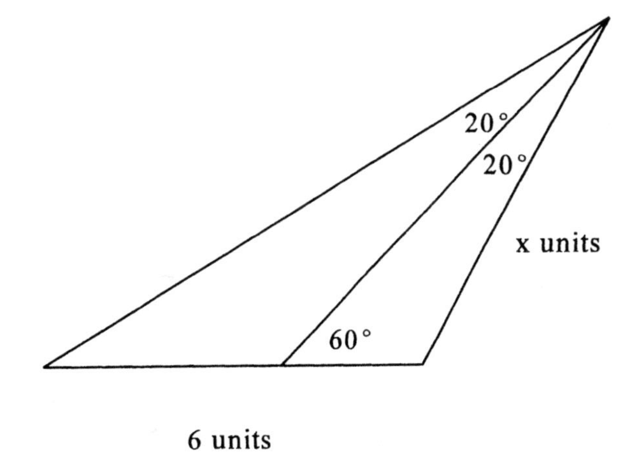

8. Let AMKT be a parallelogram.
   a) Explain why $\triangle AMK \cong \triangle KTA$.
   b) Explain why AM = KT and MK = AT.

9. a) Copy this set of figures on one piece of patty paper:

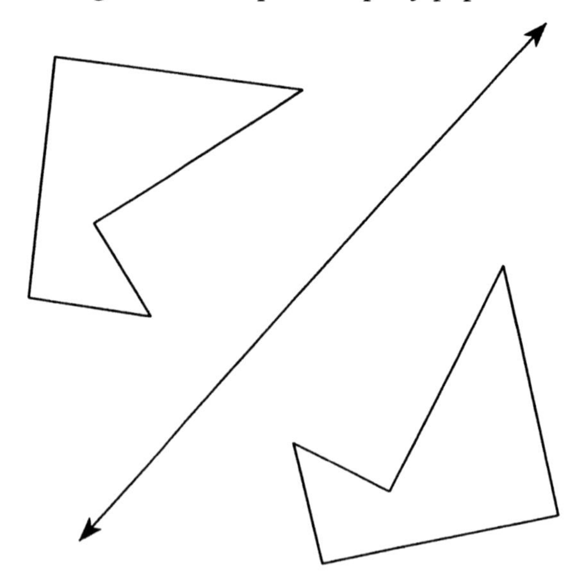

b) What happens when the paper is folded along the line?

These two polygons are said to be reflections of each other, and the line is called a line of reflection.

c) If the line in each picture below is a line of reflection, draw the reflected figure.

i) ii) iii)

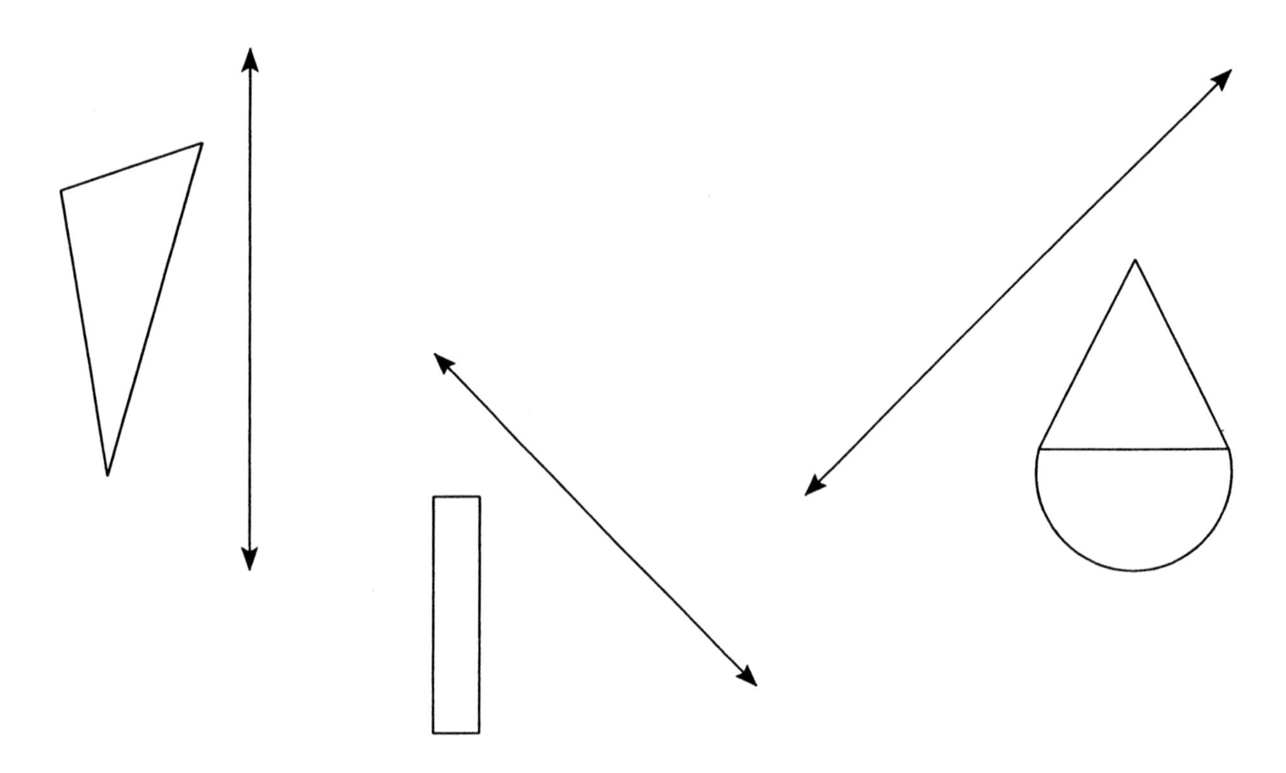

Section 39: Similarity

Figures which are copies of each other are called congruent -- all of their corresponding parts have equal measures. There are figures which are not the same size (and hence are not congruent) but they have the same shape. In other words, one figure is an "enlargement" of the other. Such figures are called similar. These are some examples of similar figures:

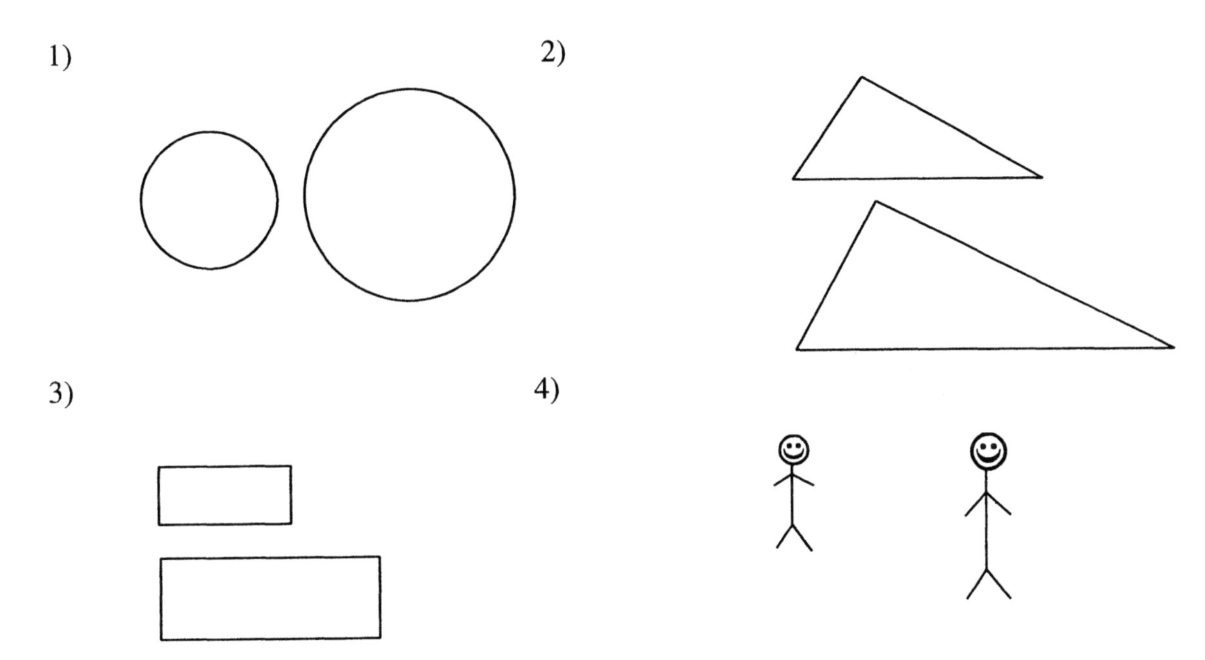

If two polygons are similar, then their corresponding angles are equal and their corresponding sides are proportional. This means that if one pair of corresponding sides has the ratio 2 to 3, then all pairs of corresponding sides have the same ratio.

If Δ ABC is similar to ΔRKV we write $\Delta ABC \sim \Delta RKV$. This means $m\angle A = m\angle R$, $m\angle B = m\angle K$, $m\angle C = m\angle V$, and $\frac{AB}{RK} = \frac{BC}{KV} = \frac{AC}{RV}$.

Example 1: Assume $\Delta KQM \sim \Delta TSP$. If KQ = 12 units, TS = 8 units, and TP = 6 units, what is KM?

Since the triangles are similar, we know that $\frac{KQ}{TS} = \frac{KM}{TP}$.

So $\frac{12}{8} = \frac{x}{6}$

$8x = (12)(6)$
$x = 9$

Therefore KM = 9 units.

In order to be sure that two triangles are similar, it is sufficient to know that they have two pairs of angles which are respectively equal.

Example 2:

Suppose $\triangle$ ABC and $\triangle$RST have m$\angle$A = m$\angle$S and m$\angle$C = m$\angle$R. Then we can say that $\triangle$ ABC ~ $\triangle$STR.

Suppose $\triangle$KQC and $\triangle$TAM have m$\angle$Q = 40°, m$\angle$C = 60°, m$\angle$M = 80°, and m$\angle$T = 40°. Since the sum of the angles of a triangle is 180°, we know that m$\angle$K = 80° and m$\angle$A = 60°. So $\triangle$KQC ~ $\triangle$MTA.

Example 3: In the figure below AB$||$RS. Find KS.

Since AB$||$RS, m$\angle$A = m$\angle$S and m$\angle$B = m$\angle$R. (Why?) So $\triangle$ABK ~ $\triangle$SRK. Hence:

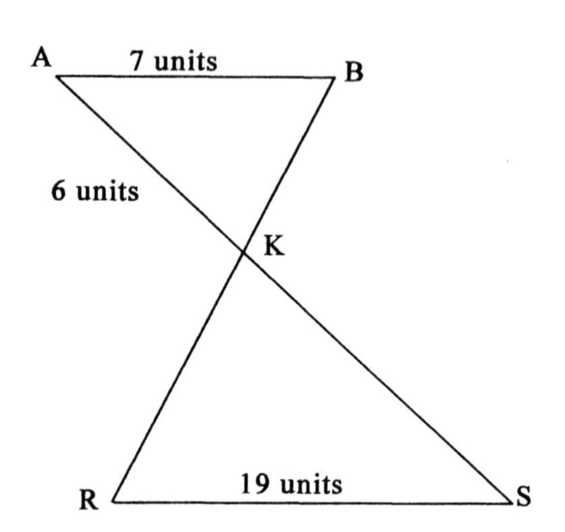

$$\frac{AB}{RS} = \frac{AK}{KS}$$

$$\frac{7}{19} = \frac{6}{x}$$

$$7x = (6)(19)$$

$$x = 16\frac{2}{7}$$

Example 4: A 6-foot man, standing next to a tree, casts a 2-foot shadow. If the shadow of the tree is 10 ft., how tall is the tree?

The picture at right emphasizes that this problem involves two similar triangles. Since m$\angle$B = m$\angle$R (both are right angles) and m$\angle$A = m$\angle$S (because the sun hits both objects from the same direction), we know that $\triangle$ABC ~ $\triangle$SRT. Thus:

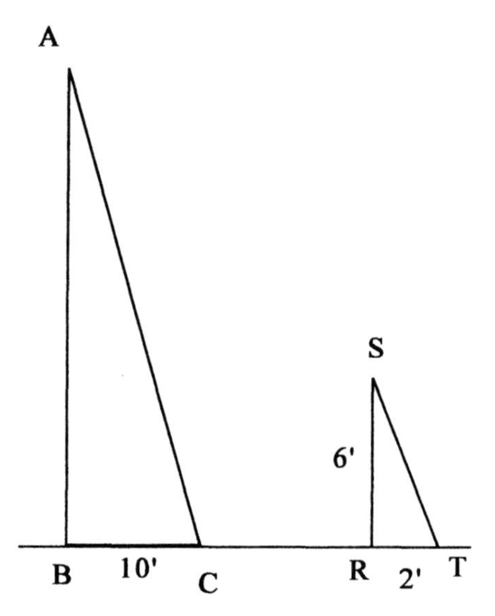

$$\frac{AB}{SR} = \frac{BC}{RT}$$

$$\frac{x}{6} = \frac{10}{2}$$

$$x = 30.$$

Therefore the tree is 30 ft. tall.

Similarity of figures is used when we make scale drawing such as blueprints for a house.

Example 5: What is the distance across the lake represented in the picture at right?

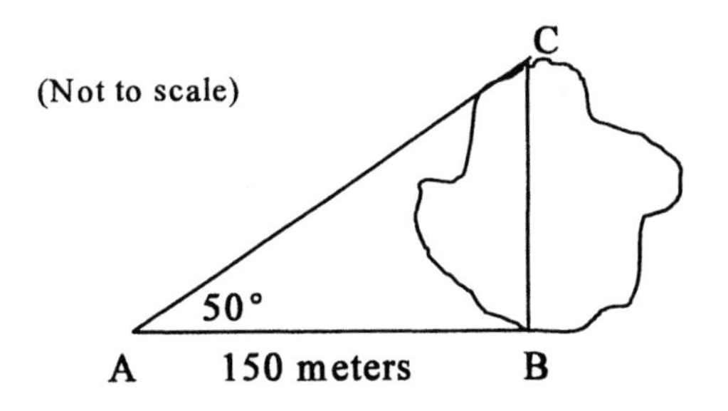

A scale drawing of this figure can be used to solve the problem. We will draw a similar triangle A'B'C'.

First, we choose a reasonable scale. For example, let 1 cm represent 30 meters. Since AB = 150 meters, then A'B' = 5 cm. m∠B'A'C' = 50° and m∠A'B'C' = 90°. We draw the figure and measure B'C' to be about 5.9 cm. This means that BC is about 5.9(30) meters which is 177 meters. So it is approximately 180 meters across the lake.

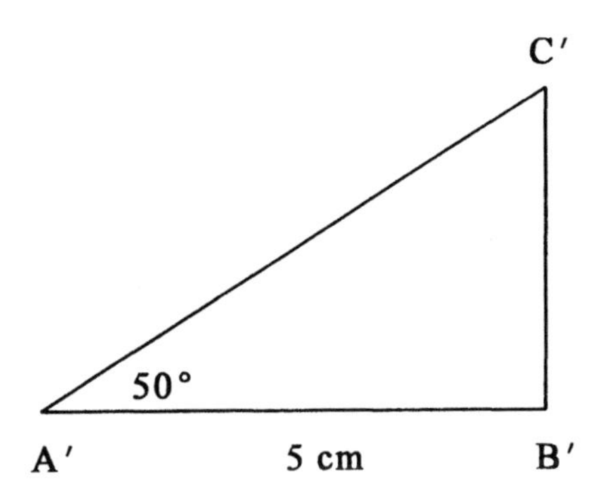

Homework: Section 39

1. In the figure below, if m∠B = m∠E, then △ABC ~ △________. x = ________ and y = ________.

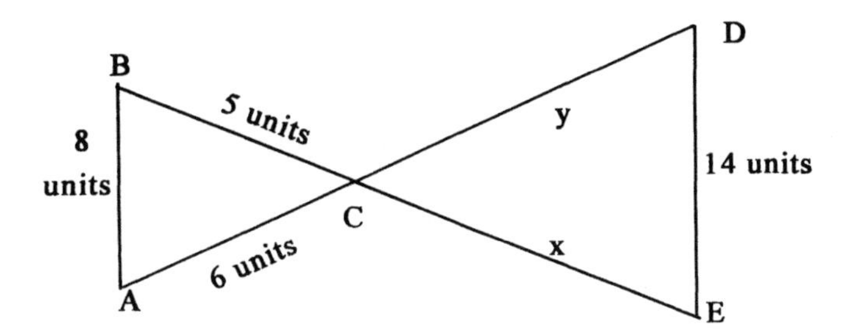

2. Assume △MNC ~ △VTP, MN = 7, NC = 8, MC = 10, and TP = 6. VP = ________.

3. In the figure at right, $\overline{BE}$ bisects m∠ABC.

   a) △ABE ~ △________,

   b) ∠A ≅ ________.

   c) If BD = 8, then BE ________.

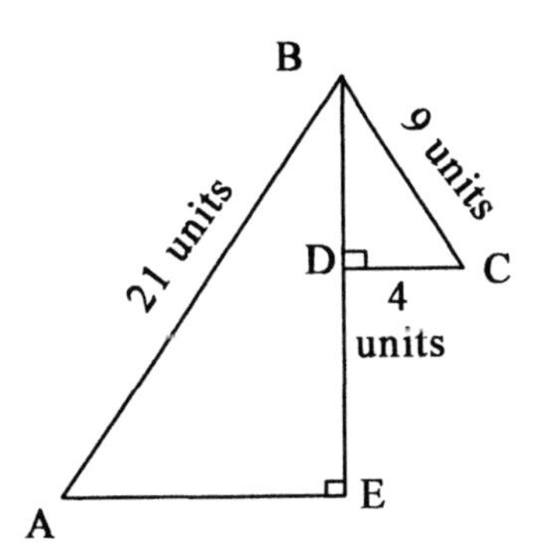

4. A water tower casts a 30 ft shadow while a 5 ft 6 in tall boy standing next to the tower casts a 4 ft shadow. How tall is the tower?

5. In the figure at right, if SQ || RV,

   then $\triangle$RTV ~ $\triangle$_______ and QV = _______.

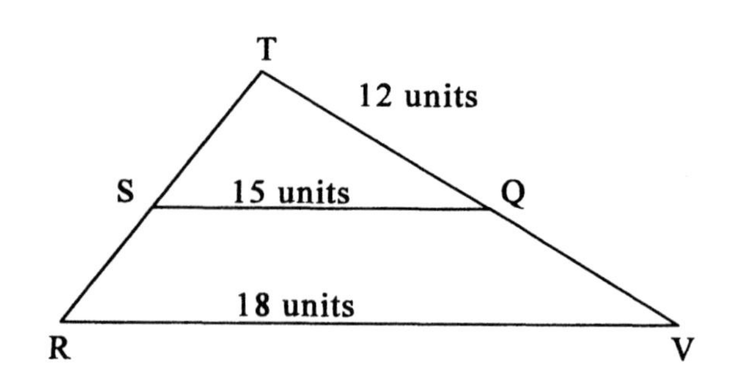

6.

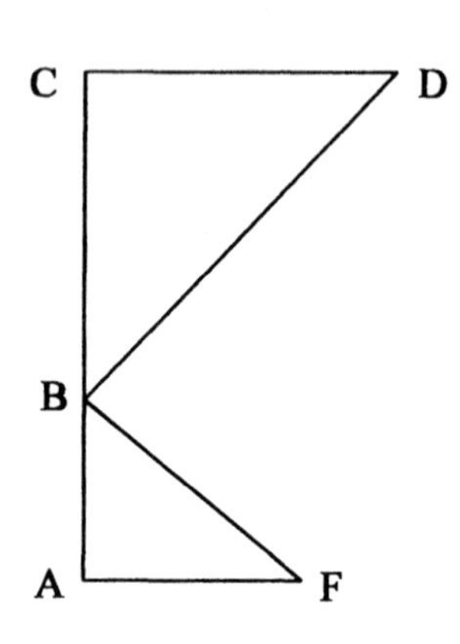

$\overline{AC} \perp \overline{CD}$
$\overline{AC} \perp \overline{AF}$
$\overline{DB} \perp \overline{BF}$

a) m$\angle$CDB = m$\angle$ _______

b) m$\angle$CBD = m$\angle$ _______

c) $\triangle$DCB ~ $\triangle$ _______

7. An army engineer wants to build a bridge across a gorge. He needs to determine how far it is across the gorge from point A to point B. He used his instruments to measure the angle at point C which is 125 meters from point B. Make a scale drawing and determine the width of the gorge.

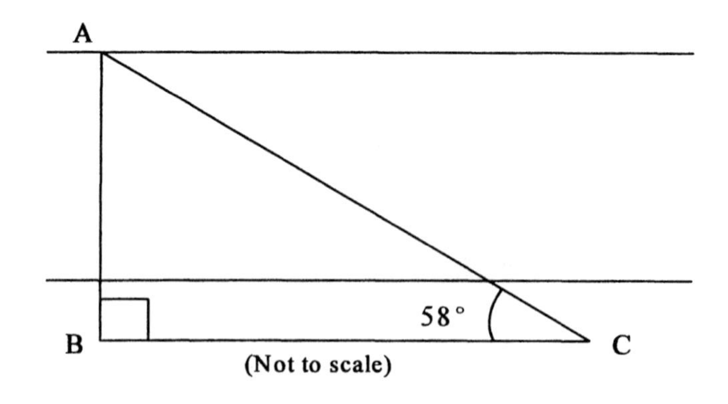

(Not to scale)

8. Make a scale drawing of a triangle which has the measurements shown below, and use it to estimate the perimeter of the actual triangle.

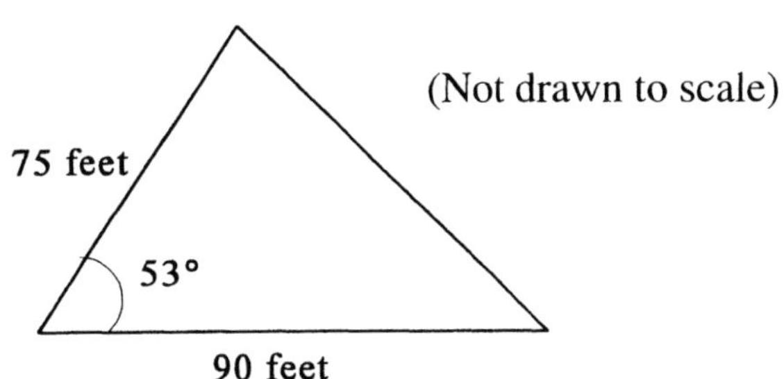

9. a) The length of a side of Square A pictured below is 4 units long. What is its area?
   b) What is the length of a side of Square B pictured below? What is its area?
   c) It is easy to see that the length of a side of Square B is three times the length of Square A. Is the area of Square B three times the area of Square A? What is the relationship between their areas?
   d) Is the perimeter of Square A one-third the perimeter of Square B? Explain.

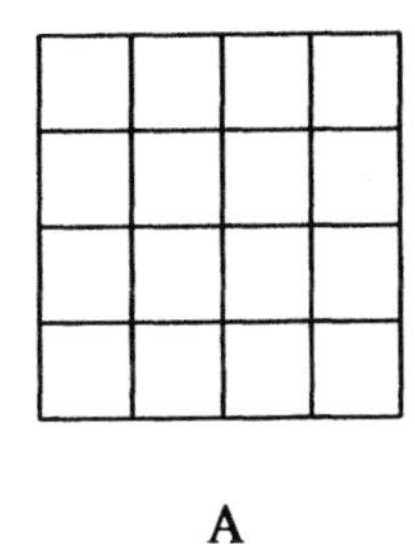

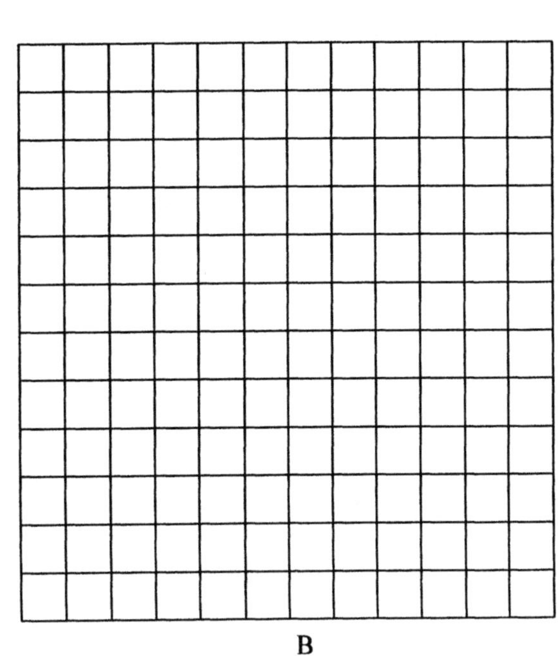

10. The radius of a gold fish pond is 10 feet. If a second pond is created and its radius is half of the radius of the first circle, is the area of the second circle half the area of the first circle? Why or why not?

11. Last year Pete's garden was 2 feet by 4 feet and was surrounded by a low fence. This year he has doubled the length and doubled the width.
    a) Does he need twice as much fencing as he used last year? Why or why not?
    b) Last year he used 6 bags of fertilizer to cover the entire garden. This year he plans to spread the fertilizer in the same way that he did last year. How many bags should he buy?

12. If two polygons are similar, what must they have in common?

13. If two polygons are similar, what can be different?

14. If two hexagons are regular, must they be similar? If yes, why? If no, give and example.

15. If two hexagons are similar, must they be regular? If yes, why? If no, give an example.

16. If two pentagons are similar, what additional information is needed in order to determine if they are congruent?

17. A copier may reproduce exact replicas of the original or it may enlarge or reduce the original by a certain percent. In any case, a similar figure is produced.

    a) The figure on the right is the result of setting the copier at 124%. Does this mean that the sides were increased by 24% or that entire area of the rectangle was increased by 24%? Make the appropriate measurements and answer the question.

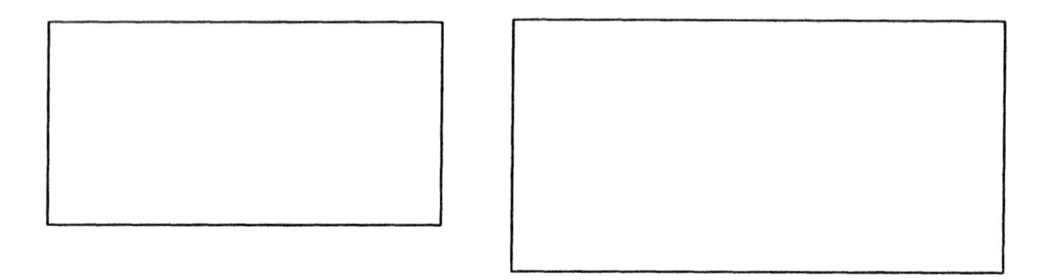

    b) The figure on the right was produced from the one on the left. What was the setting on the copier?

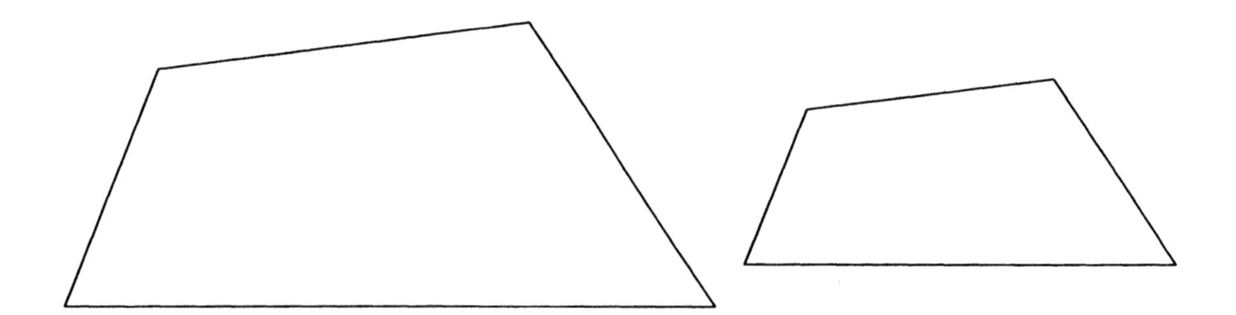

Chapter IV Review

I.

1. Name each figure:

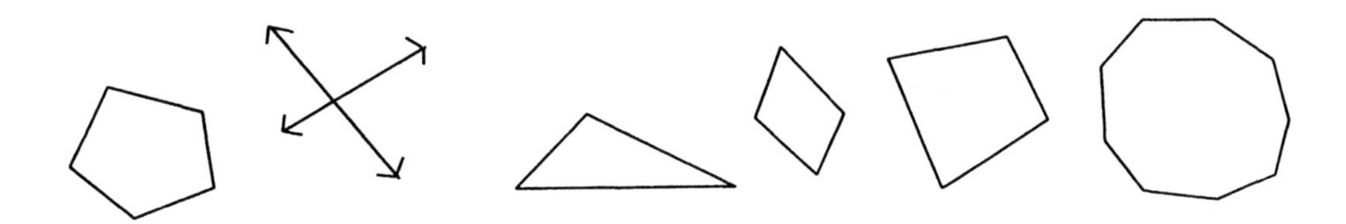

2. Which of these are polygons? If a figure is not a polygon, say why.

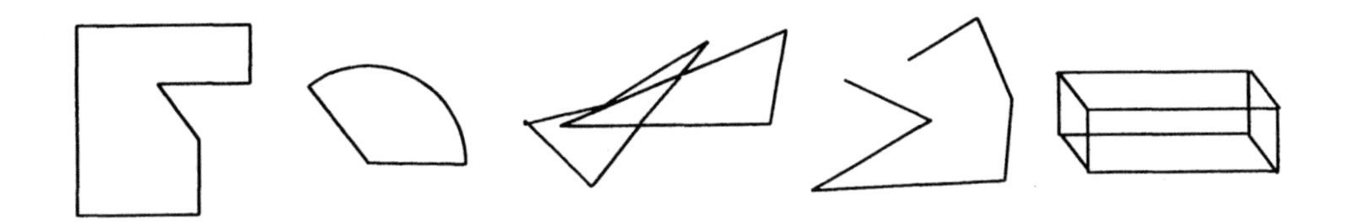

3. For each figure, draw all lines of symmetry. If there is a center of rotation, label it C and give the angle of rotation.

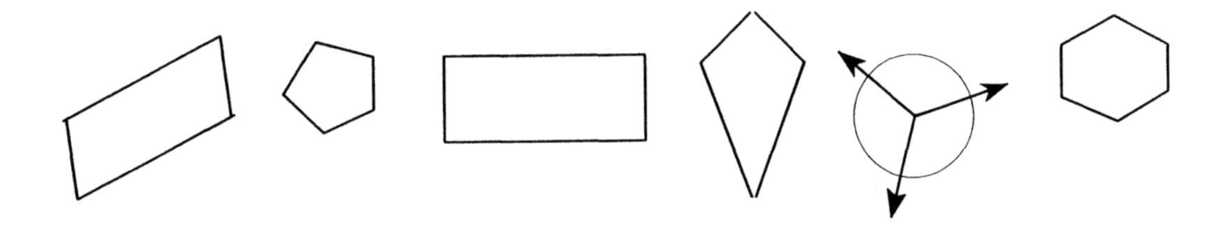

4. For each kind of figure, list the specifications which must be given to assure that figures drawn with those specifications are congruent:

   a) triangle   b) parallelogram   c) regular 11-gon   d) rectangle

   e) circle   f) non-regular n-gon

5. In the figure at right, assume that $\overline{AB} \parallel \overline{CD}$. Find x and y.

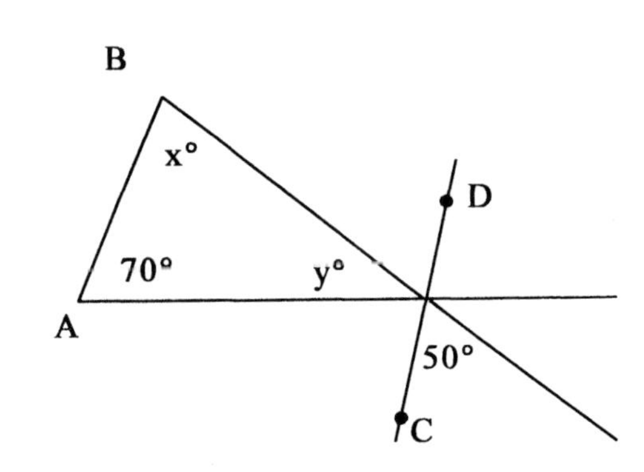

6. Let $\triangle KTA$ and $\triangle MCQ$ have these relationships: $\angle K = \angle C$, $\angle T = \angle Q$, and $AK = MC$

Can I be sure that the two triangles are congruent?

If so, fill in: $\triangle$ KTA ≅ $\triangle$______. If not, explain why not.

7. If $\overline{MT}$ is the perpendicular bisector of $\overline{PK}$, can I be sure that $\triangle MKT$ is isosceles? Can I be sure that $\triangle MPK$ is isosceles? Explain.

8. If $\triangle QCV$ is isosceles and m $\angle Q = 102°$, then m $\angle V =$ ______° and $\overline{QC}$ has the same length as segment ________.

9. a) How many lines contain points V and K? ________
   b) How many circles contain non-collinear points A, T, and Q? _______
   c) If A, B, and C are non-collinear,
      i) how many lines contain B and are parallel to $\overline{AC}$?
      ii) How many lines contain B and are perpendicular to $\overline{AC}$?
      iii) How many planes contain A, B, and C?
      iv) How many planes contain $\overline{AC}$?

10. The diagonals of a rectangle are ________ and the diagonals of a rhombus are ________.

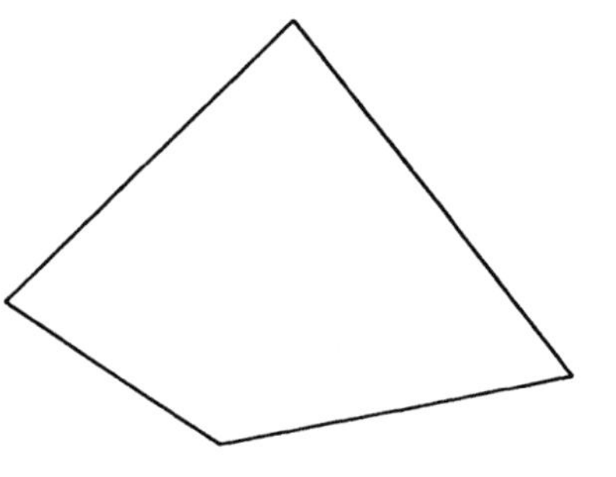

11. If this figure were reduced on a copy machine by pushing the 75% button, what would be the perimeter of the reduced figure?

12. Suppose everyone in the class constructed a triangle with sides of 6 cm and 8 cm and an angle of 45° opposite the 6 cm side. Would all the triangles have to be congruent? Explain.

13. In the figure, TRAP is a trapezoid and $\overline{TA}$ bisects $\angle RTP$. Explain why $RA = RT$ and $\overline{AT} \perp \overline{AP}$.

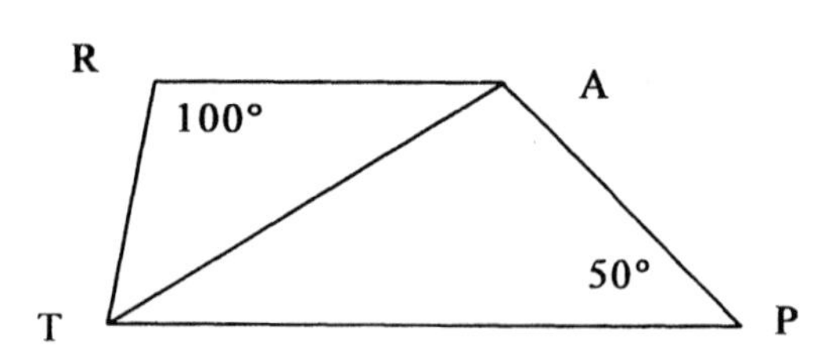

14. In the figure, KB = MB,
$\overline{QA} \perp \overline{KM}$, $\overline{AT} \perp \overline{MB}$
KQ = 10 units, TM = 9 units, and
MA = 15 units.

$\triangle$KQA ~ $\triangle$________

KM = ________ units

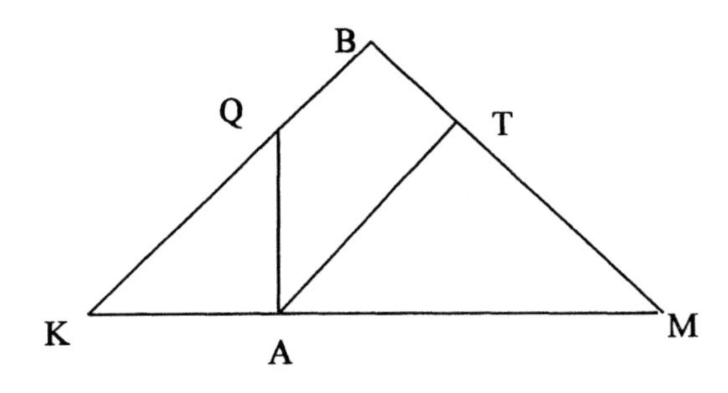

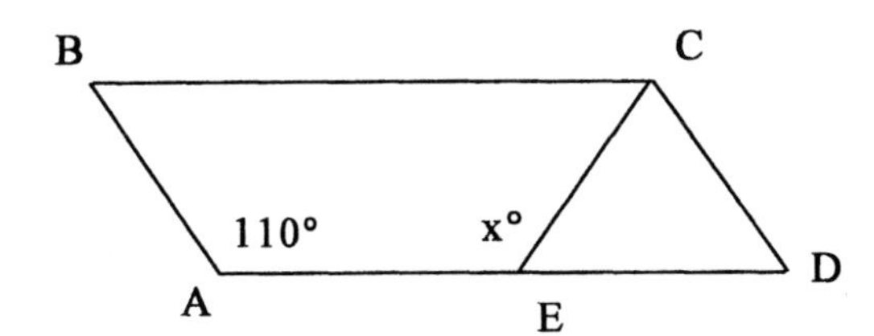

15. In the figure, ABCD is a parallelogram, and AB = EC. Find x.

16. The perimeter of a regular polygon is 972 cm and each angle is 150°. What is the length of each side?

17. Suppose that a muk is a unit for measuring angles, and each angle of a rectangle is 21 muks. How many muks is each angle of a regular hexagon? How many muks is the sum of the angles of a triangle?

18. Assume that $\overline{AB} \perp \overline{AF}$, $\overline{CD} \perp \overline{DE}$, and $\overline{AB} \parallel \overline{DE}$.
Can we be certain that $\overline{CD} \parallel \overline{AF}$?
Explain your answer.

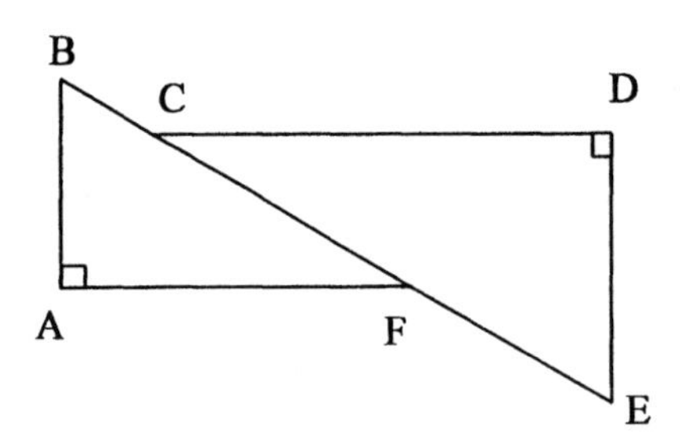

19. Explain why each of these cannot be constructed.
   a) a triangle with sides of 7, 9, and 17 inches
   b) a parallelogram with angles of 72° and 118°
   c) a triangle with angles of 92° and 88°
   d) a regular polygon with angles of 70°
   e) a circle with radius 4.3 cm and circumference 24.8 cm
   f) an equilateral triangle which is not regular
   g) a triangle with two sides of 8 cm each and angles of 100° and 42°.

20. Explain what each of these is:
    a) line of symmetry
    b) perpendicular bisector of a segment
    c) alternate interior angles
    d) regular polygon
    e) distance from a point to a line
    f) circle
    g) similar figures
    h) plane
    i) degree
    j) congruent figures
    k) vertical angles

21. Give three different descriptions of parallel lines.

22. What does "two-dimensional" mean? What does "three-dimensional" mean?

23. Suppose that a path from A to B goes through point K. What determines whether or not the path from A to B is straight?

24. What determines whether or not an infinite surface is a plane?

25. Match the corresponding vertices:

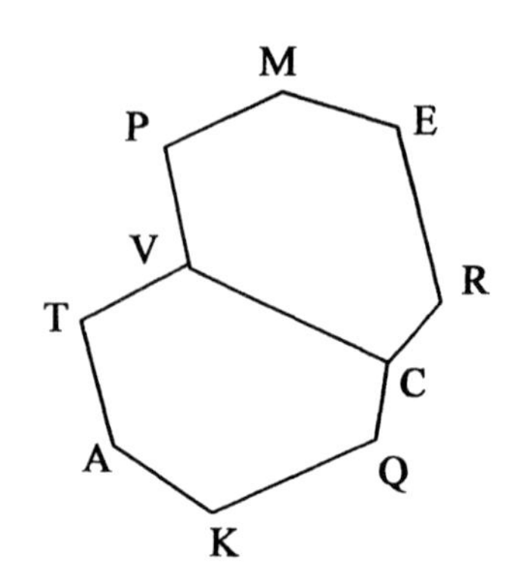

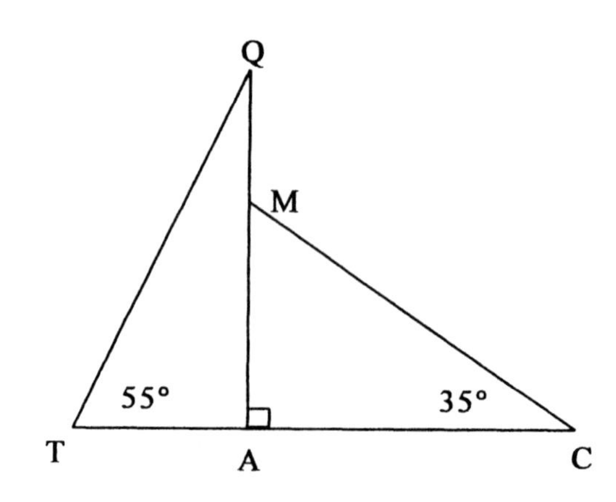

MERCVP ≅ ________

ΔTQA ~ Δ ________

II. Use any tools.

1. Construct a line which contains K and is parallel to $\overleftrightarrow{VC}$.
   Construct the perpendicular bisector of $\overleftrightarrow{VC}$.

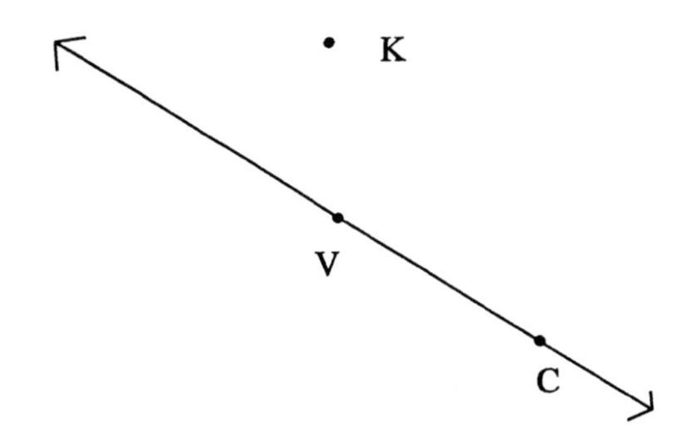

2. Draw the reflection of the figure across $\overleftrightarrow{AB}$ without folding or tracing.

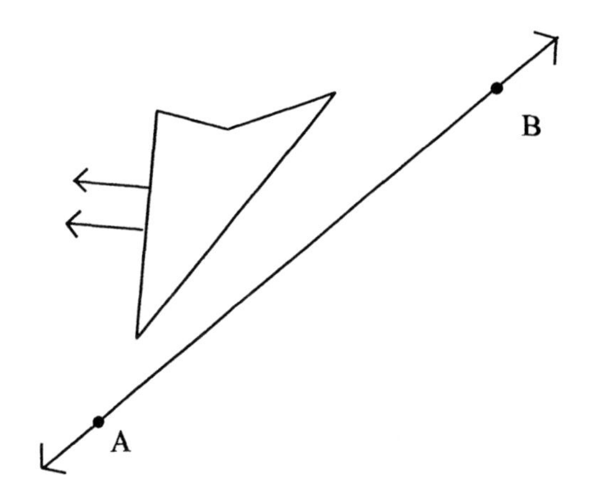

3. Construct a parallelogram which has sides of 4.0 cm and 7.0 cm, and an angle of 100°.

4. Find the distance from P to $\overleftrightarrow{AB}$.

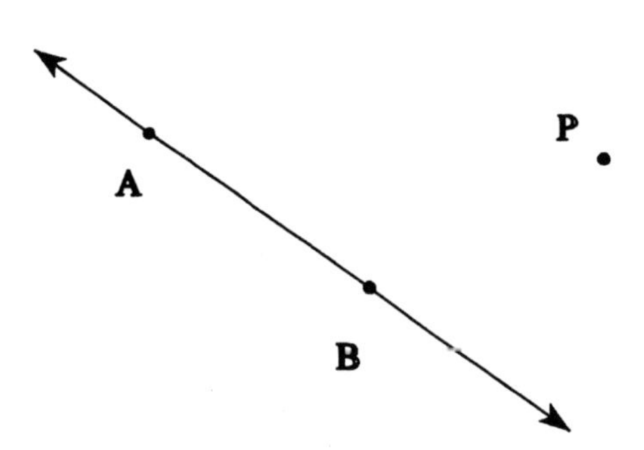

5. Complete the circle whose arc is given at right.

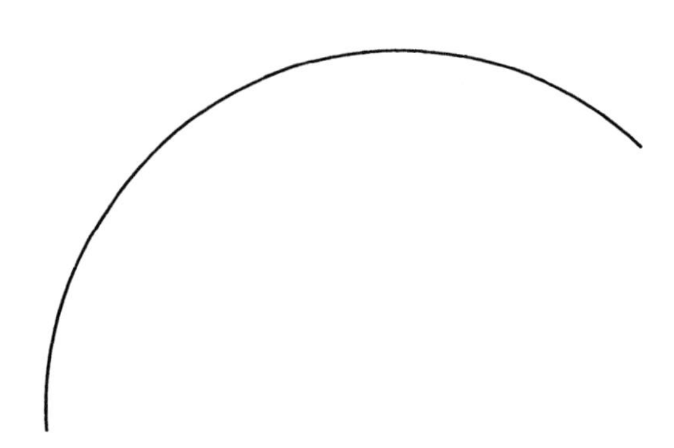

6. Construct a triangle with sides of 4.0 cm, 6.0 cm, and 7.0 cm. Construct its inscribed circle.

7. Construct a five point star. Draw its lines of symmetry. What is its angle of rotation?

8. Construct an isosceles right triangle with hypotenuse 7.0 cm.

9. Copy the figure shown at right without tracing. Label your copy ABCDE so that ABCDE ≅ KQPTV.

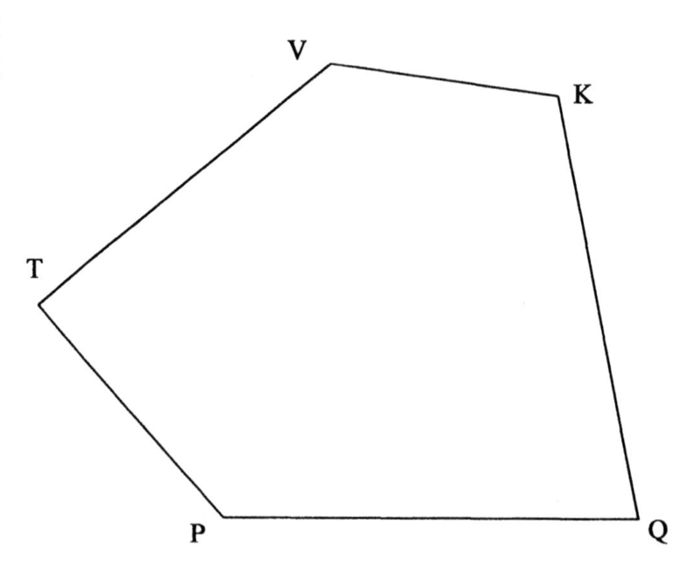

10. In Destin, Florida, Oceanside Blvd runs parallel to a straight section of beach at a distance of 200 ft from the water. First Avenue is perpendicular to Oceanside. You own a piece of property

which is bounded on three sides by Oceanside, First Ave., and the beachfront. The property is 250 ft along the water and 350 ft along Oceanside. Make a scale drawing of your property and use it to find the perimeter.

11. Construct a regular 9-gon by following these directions:

 a) Construct a circle of radius 4.0 cm. Name the center C.

 b) Choose a point on the circle. Name it A.

 c) Find the angle of rotation for a regular 9-gon.

 d) Construct an angle of this measure with vertex C and one side along segment $\overline{AC}$.

 e) Name the point, where the other side of the angle intersects the circle, B.

 f) Open your compass to the length AB.

 g) Put the point of your compass on B, and find another point on the circle (other than A) which is a distance of AB from B. Name the point K.

 h) Repeat this procedure to find points D, E, F, G, H, M (in order) on the circle such that KD = ED = FE = FG = GH = HM = AM.

 i) Draw segments AB, BK, KD, DE, EF, FG, GH, HM, MA.

12. Assume AR = CR and BR = AR. Find x.

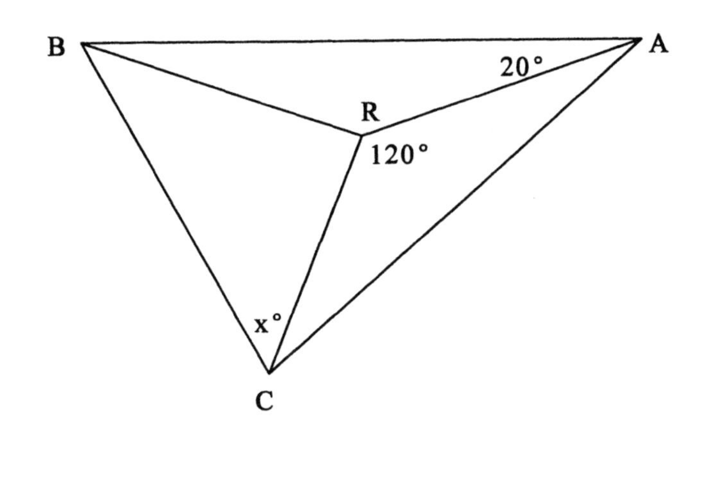

13. $\overleftrightarrow{KT}$ bisects $\angle AKM$ and AM = KM. Find x and y.

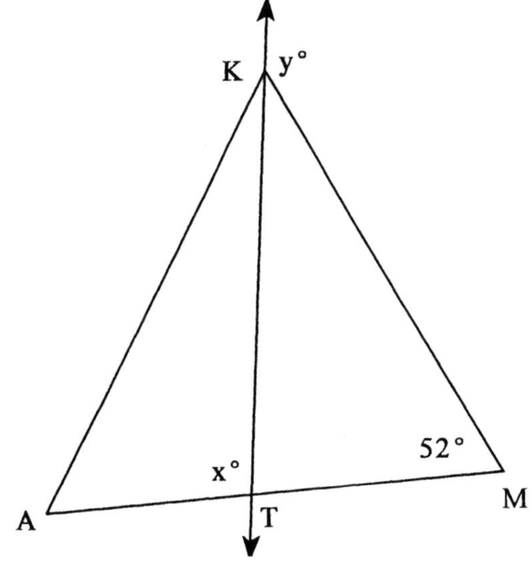

Chapter V

Section 40: Area of Triangles

In Chapter IV, we examined the units used to measure the area of regions and the volume of objects. We continue the study of these concepts, with emphasis on understanding how the dimensions of a figure can be used to compute its surface area and its volume.

As was pointed out in the first chapter, the area of a rectangle is easily determined. If its length is j units and the width is k units, then the area is jk square units.

Example: The rectangle below has length of 9 units and width 5 units. Therefore 5 rows of squares, with 9 squares in each row, fit inside of this. These 45 square units constitute the area of the figure.

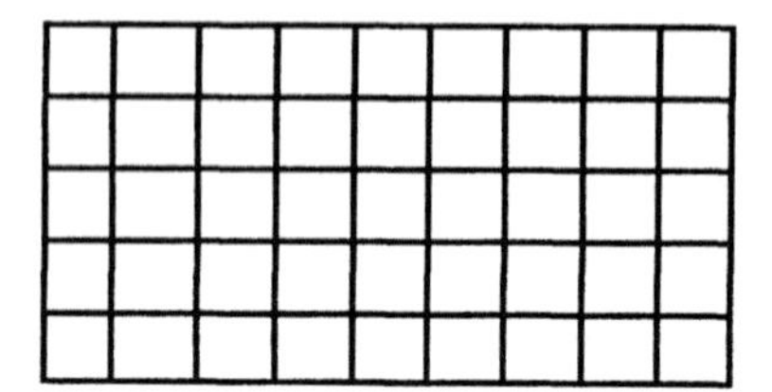

We can use this knowledge about the area of a rectangle to find the area of triangles. The technique can be most easily demonstrated with a right triangle. Consider any right triangle DEF with $\angle E = 90°$.

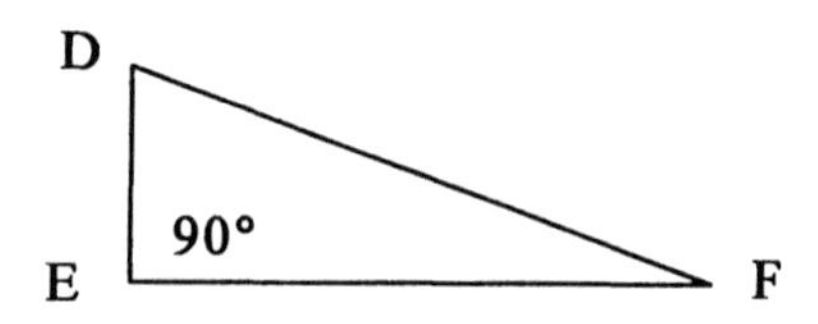

There is a line which contains D and is parallel to $\overline{EF}$, and there is a line which contains F and is parallel to $\overline{DE}$.

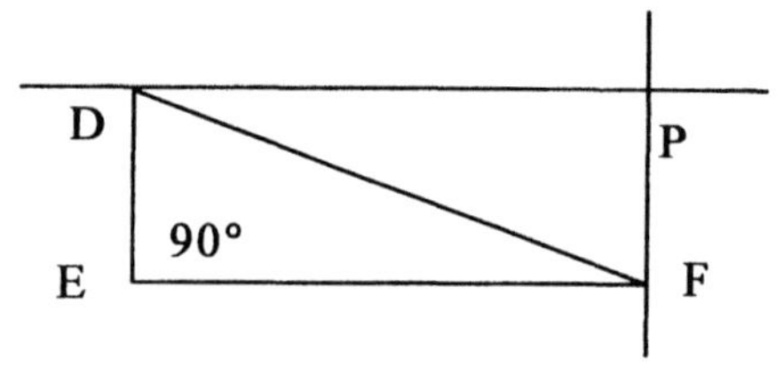

It is easy to show that DEFP is a rectangle. (How?) Therefore, $\triangle DEF \cong \triangle FPD$, and area $\triangle DEF$ = area $\triangle FPD$. This means that the area of each triangle is half the area of the rectangle. Since EF is the length of the rectangle and DE is its width, the area of $\triangle DEF = \frac{1}{2}(DE)(EF)$.

In order to use this technique on a non-right triangle, we first need a right angle to define the rectangle. In the right triangle above we already knew that $\overline{DE} \perp \overline{EF}$. In any other triangle we must create a perpendicular segment. An altitude of a triangle is a segment from a vertex perpendicular to the line containing the side opposite that vertex. Hence a triangle has three altitudes.

In the figure at right, $\overline{CT}$ is the altitude from C. It is also called the altitude to $\overline{AB}$. $\overline{AB}$ is the base for this altitude.

Similarly, $\overline{BW}$ is the altitude from B. Notice that this altitude is outside the triangle and that W is on line $\overline{AC}$ although it is not between A and C. Its base is segment $\overline{AC}$.

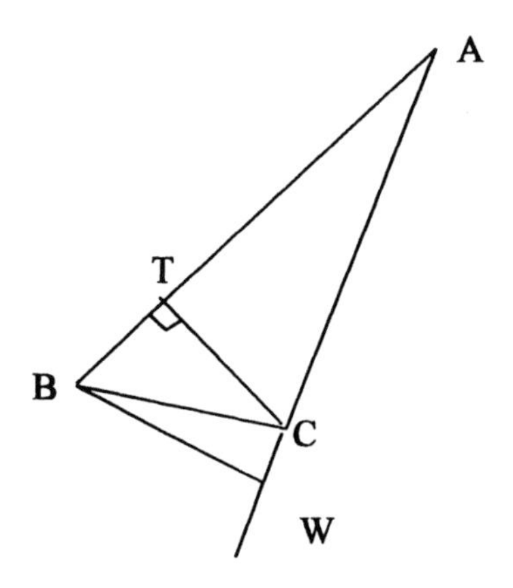

Construct the altitude to $\overline{BC}$.

Note that the words altitude and base are used to refer to the lengths of those segments as well as to the segments themselves.

Consider any triangle RST, with altitude RP drawn to base $\overline{TS}$.

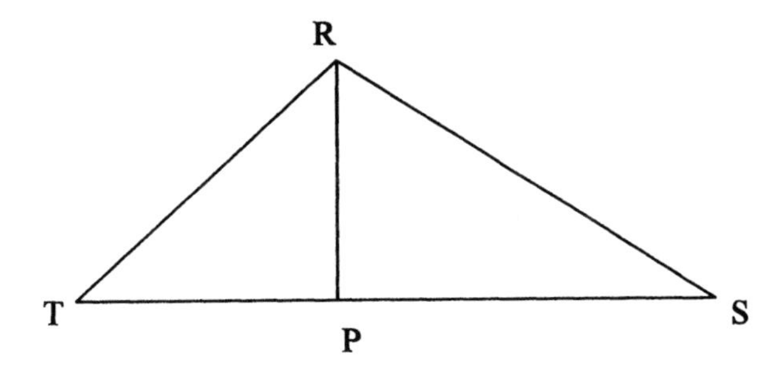

Construct a rectangle with length TS and width equal to RP.

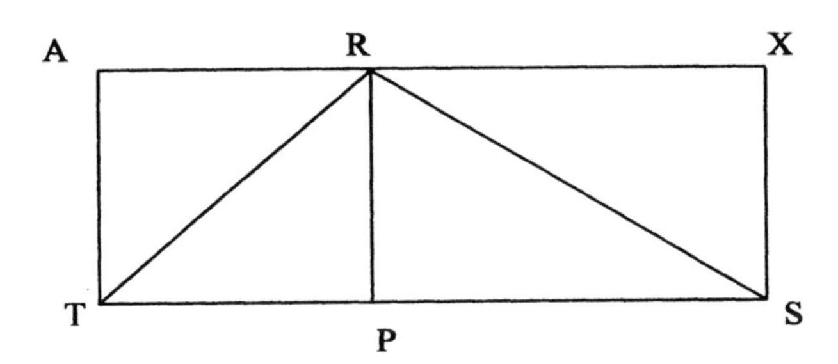

It is easy to see by SAS congruence that $\triangle TPR \cong \triangle RAT$ and that $\triangle SPR \cong \triangle RXS$. So the area of the original triangle RST is one-half the area of the rectangle created by an altitude and its base.

Thus area $\triangle RST = \frac{1}{2} \times \text{width} \times \text{length}$

$$= \frac{1}{2}(RP)(TS).$$

But $\overline{RP}$ is an altitude and $\overline{TS}$ is its base, so area $\triangle RST$ is $\frac{1}{2}ab$.

Reconsider the right triangle DEF on the previous page. $\overline{EY}$ Is the altitude to base $\overline{DF}$. Notice that the two legs are a "built-in" altitude and base: $\overline{DE}$ is the altitude to base $\overline{EF}$ and $\overline{EF}$ is the altitude to base $\overline{DE}$.

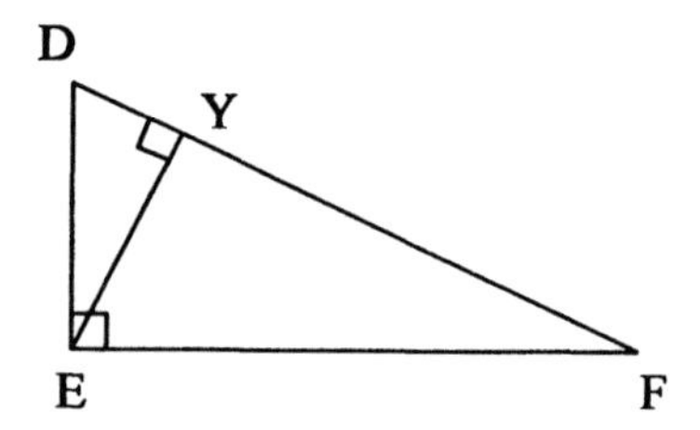

Example 1: Find the area of △ABC from the information given in the picture.

Since ∠B is a right angle, $\overline{BC}$ is an altitude of the triangle and $\overline{AB}$ is its base.

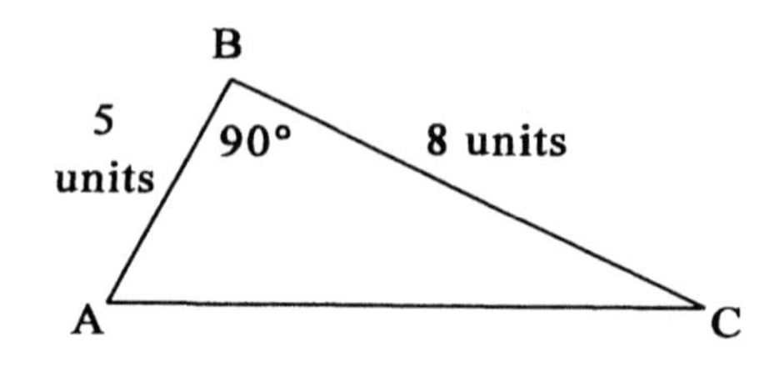

Area △ABC = $\frac{1}{2}$(AB)(BC) = $\frac{1}{2}$(5)(8) = 20 sq units.

Example 2: Find the area of △RST by making the necessary measurements:

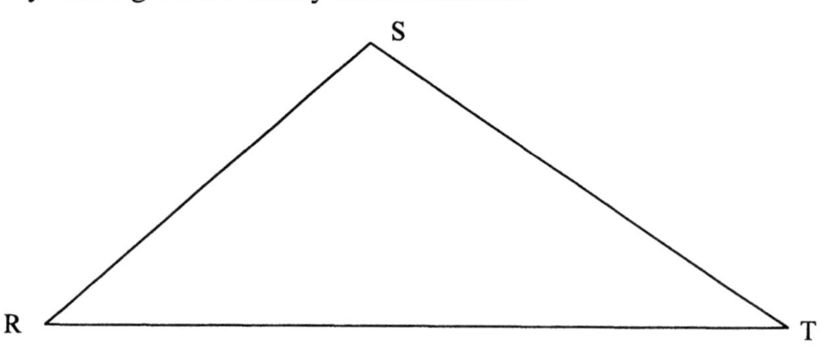

To determine the area, we must find the lengths of an altitude and its base. If we select $\overline{RT}$ as base, we construct the altitude to $\overline{RT}$.

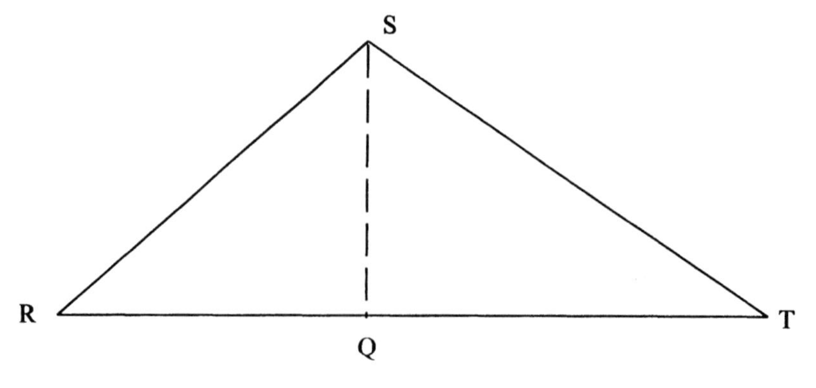

Now with a ruler find that SQ ≈ 3.4 cm and RT ≈ 9.7 cm.
Therefore Area △RST = $\frac{1}{2}$(SQ)(RT) ≈ 16.5 sq cm.

Example 3: The sides of $\triangle KTV$ have these measures: KT = 7 units; TV = 9 units; KV = 12 units. The altitude to $\overline{TV}$ is 8 units. What is the length of the altitude to $\overline{KV}$?

Recall that a triangle has three altitudes and three corresponding bases. All three pairs will yield the same result for the area of the triangle. (If the segments are measured, there may be some slight variation due to error.)

Using base TV and its altitude, we have: Area $\triangle KTV = \frac{1}{2}(8)(9) = 36$ sq units.

Using base KV and its altitude, x, we have: Area $\triangle KTV = \frac{1}{2}(x)(12) = 6x$ sq units.

Both 36 sq units and 6x sq units give the area of $\triangle KTV$. So 6x sq units = 36 sq units and x must be 6 units. Hence the length of the altitude to $\overline{KV}$ is 6 units.

Example 4: In the triangle at right AB = BC = CD. If Area $\triangle ABE$ is 20 sq units, what is the area of $\triangle ADE$?

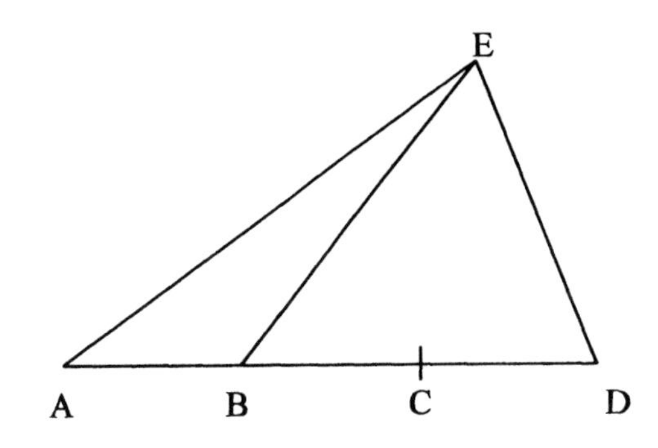

First we notice that base $\overline{AD}$ of $\triangle ADE$ is 3 times as long as base $\overline{AB}$ of $\triangle ABE$. Also, the altitude to base $\overline{AD}$ is the same as the altitude to $\overline{AB}$. (It is the perpendicular segment from E to $\overline{AD}$.) We will call the length of this altitude, a.

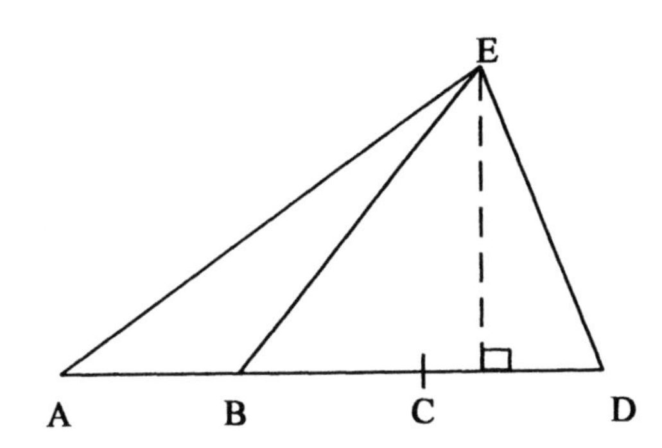

$$\begin{aligned} \text{So } \triangle AED &= \frac{1}{2}(a)(AD) = \frac{1}{2}a(3AB) \\ &= 3[\frac{1}{2}a(AB)] \\ &= 3[\text{Area } \triangle ABE] \\ &= 3(20 \text{ sq units}) \\ &= 60 \text{ sq units.} \end{aligned}$$

Hence Area $\triangle AED$ is 60 sq units.

Homework: Section 40

1. For the triangle at right, construct the altitude from P.

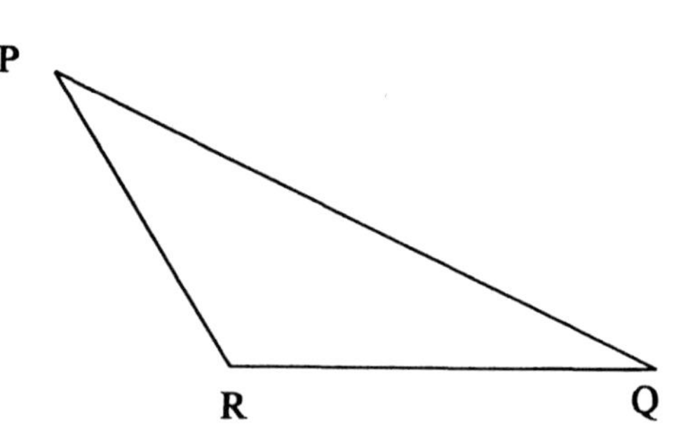

2.

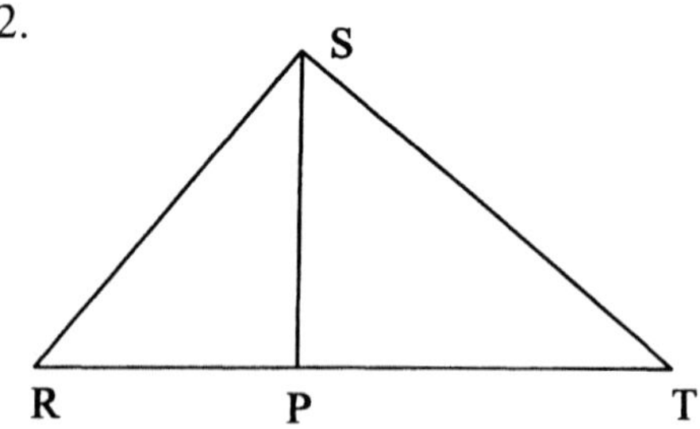

$\overline{RS} \perp \overline{ST}$; $\overline{SP} \perp \overline{RT}$
RT = 15 units
ST = 12 units
RS = 9 units
Find SP.

3. ABCDEF is a regular hexagon with center P and perimeter 78 units. The distance from P to each side is 11 units. What is the area of the hexagon?

4. Find the area of the figure.

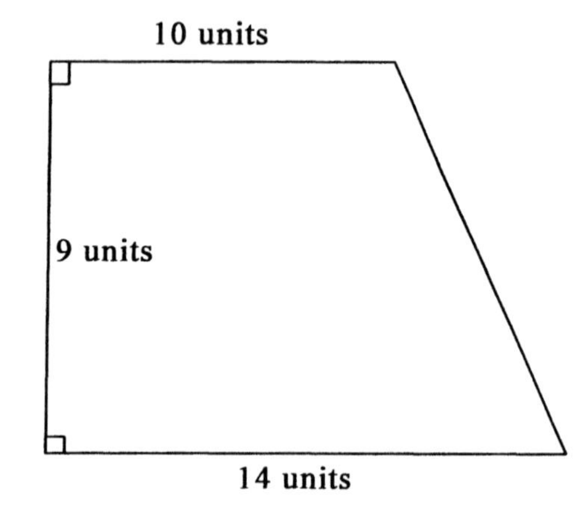

5. ABCD is a parallelogram. Find its area.

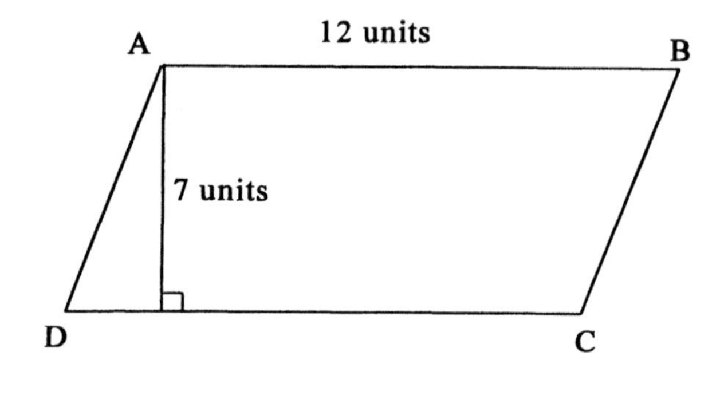

6. In the figure at right, $\overline{AB} \,||\, \overline{CD}$; CD = 5 units; Area $\triangle$ACD = 15 sq units. Find area $\triangle$BCD.

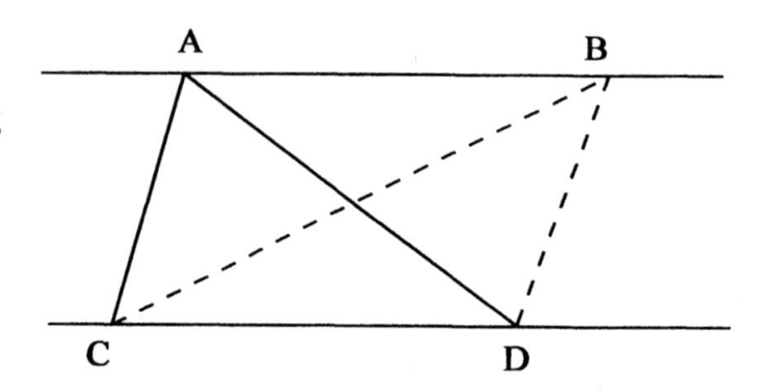

7. Find the approximate area of each figure by making the necessary constructions and measurements. Use metric units.

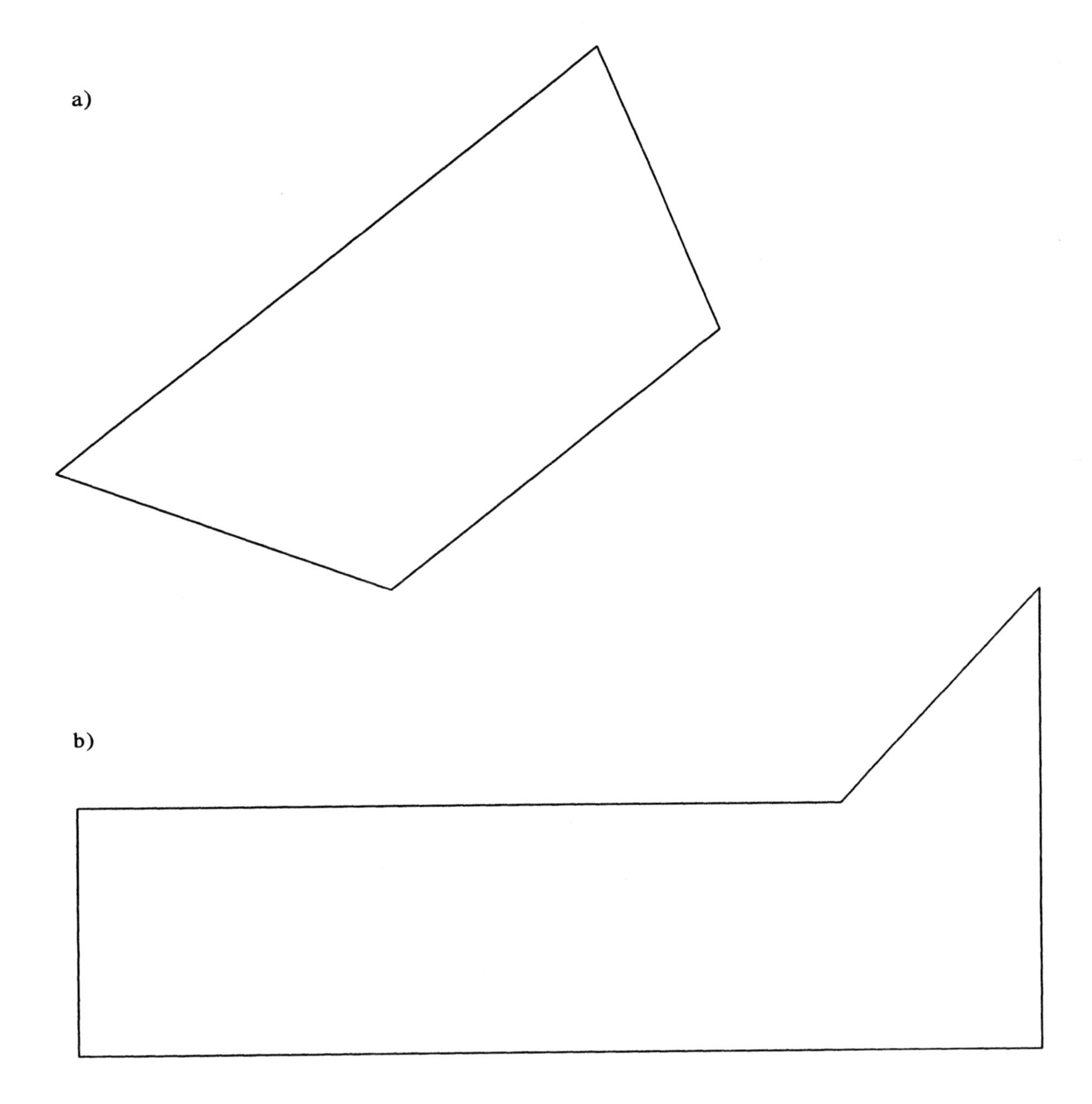

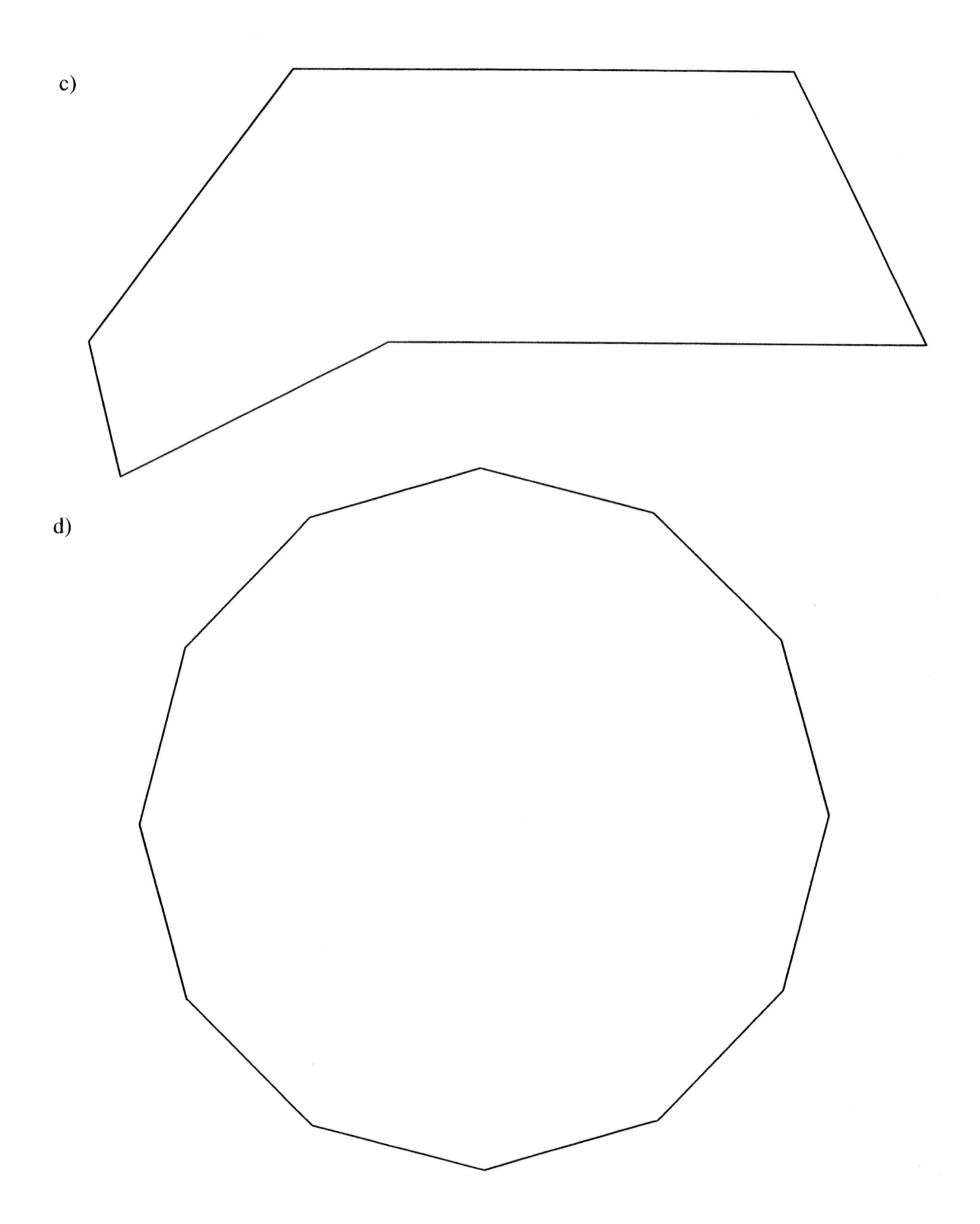
c)
d)

Section 41: Area of Other Figures

In the last set of exercises it should have become evident that once we know how to determine the areas of rectangles and triangles, we can use this information to find areas of parallelograms, trapezoids or any polygons.

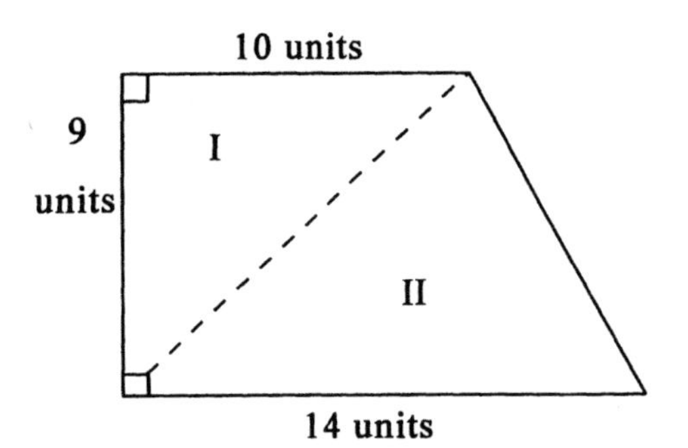

Example 1: Reconsider Exercise 4 on page 192.

To determine the area of this trapezoid, we separate the region into two triangles as shown above.

Area I $= \frac{1}{2}(9)(10)$ and

Area II $= \frac{1}{2}(9)(14)$.

So Area of the Trapezoid is $\frac{1}{2}(9)(10) + \frac{1}{2}(9)(14) = \frac{1}{2}(9)[10 + 14]$ (Distributive property).

This can be generalized into a formula: Area of Trapezoid $= \frac{1}{2}a(b_1 + b_2)$, where b_1 and b_2 are the parallel bases and a is the altitude to those bases.

Example 2: Reconsider Exercise 5 on page 192.

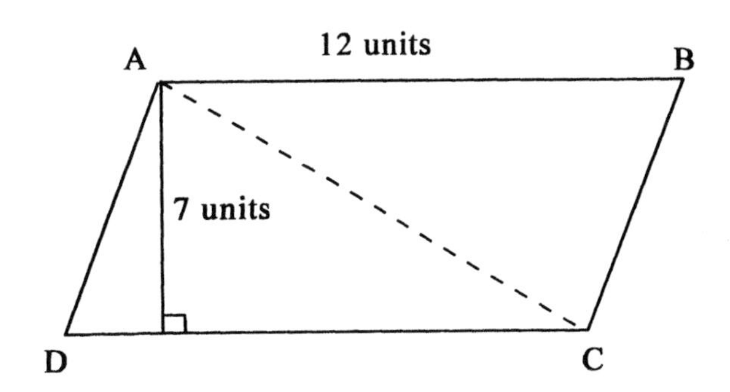

If we draw diagonal $\overline{AC}$ as shown then

$$\begin{aligned}\text{Area } ABCD &= \text{Area } \triangle ACD + \text{Area } \triangle ABC \\ &= \frac{1}{2}(7)(12) + \frac{1}{2}(7)(12) \\ &= 7(12).\end{aligned}$$

Example 3: Reconsider Exercise 3 on page 192.

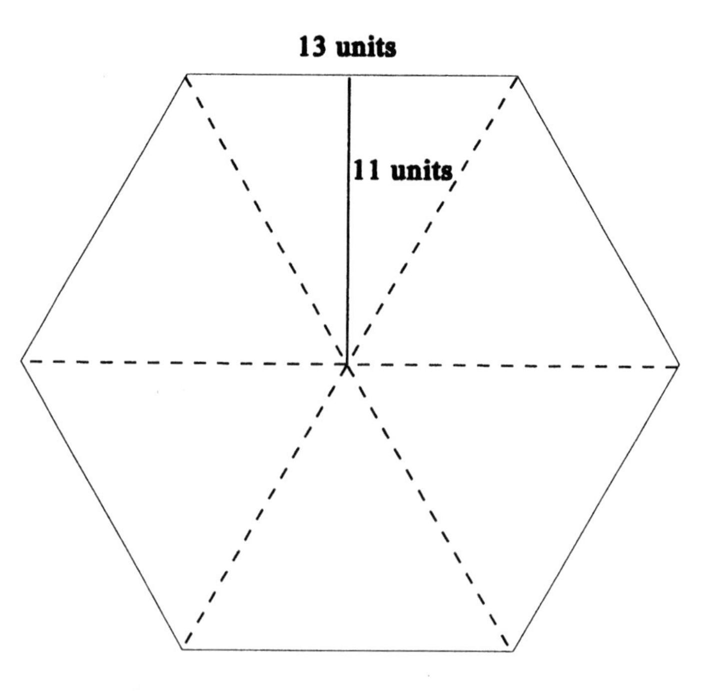

Each side of this regular hexagon is 13 units, and the distance from P (the center) to the sides is 11 units. Hence the area of each triangle into which the polygon has been separated is $\frac{1}{2}(13)(11)$. Since the triangles are congruent, and there are six of them, the area of the hexagon is $6(\frac{1}{2})(13)(11)$.

In general, the area of a regular polygon is $n(\frac{1}{2})(s)(t)$ where n is the number of sides, s is the length of a side, and t is the distance from the center to each side. Since ns represents the perimeter of the polygon, we can also say that the area is $\frac{1}{2}pt$, where p is perimeter and t is the distance from center to sides.

This last example suggests an explanation for the formula we use to determine the area of a circle. Actually deriving the formula involves the concept of limit, which is beyond the scope of this course and certainly beyond the level of your future students. But the simplified explanation should be within the grasp of sixth or seventh graders.

First notice that the area of inscribed polygons come closer and closer to the area of the circumscribed circles as the number of sides of the polygons increases:

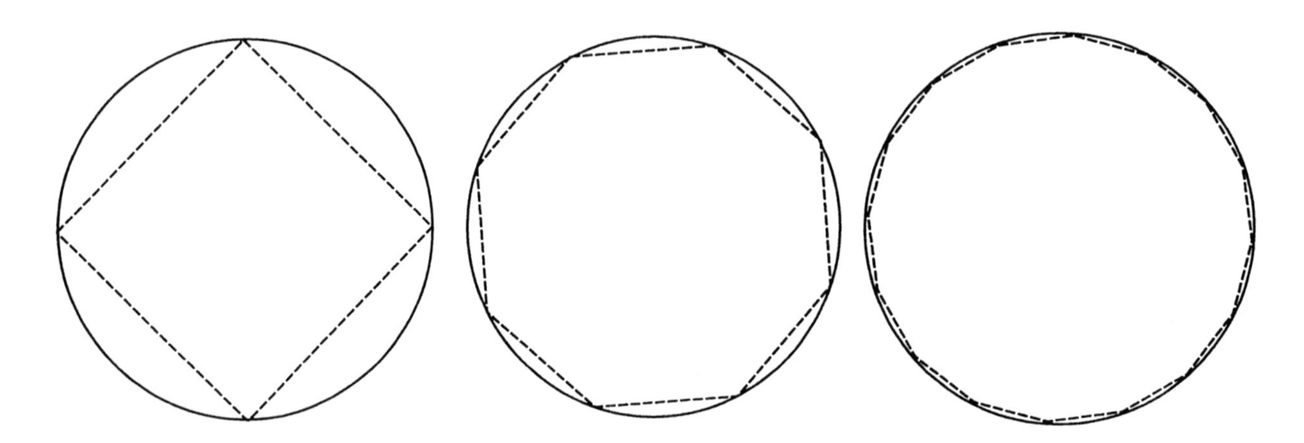

Suppose we could make the sides of the polygon just one point long; then the area of the polygon would equal the area of the circle! Now recall that this area would be $\frac{1}{2}pt$, where p is the perimeter of the figure and t is the distance from the center to sides. Since the perimeter of a circle (circumference) is $2\pi r$, and the distance from the center to "sides" is r (radius), we have:

$$\text{Area of circle} = \tfrac{1}{2}pt = \tfrac{1}{2}(2\pi r)r = \pi r^2.$$

A sector is a portion of a circular region which is bounded by two radii and an arc.

Examples:

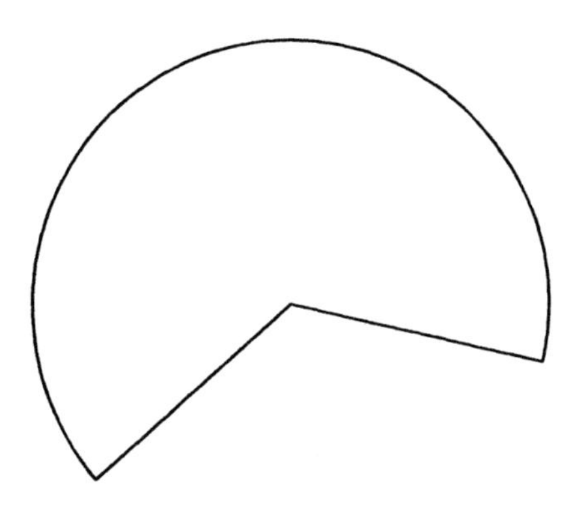

To determine the area of a sector, we must find out what fraction of the circular region it is. Since the angle measure of a circle (with vertex at the center) is 360 , the measure of the angle formed by the two radii of the sector, divided by 360, represents the fraction we need.

Example: Find the area of the sector shown at the right:

150°

6 units

This sector represents $\frac{150}{360}$ of a circle of radius 6 units.

Thus the Area of Sector ≈ $\frac{150}{360}(36\pi) \approx 15\pi \approx 47$ sq units.

Homework: Section 41

1. Use the information given to find the area of the washer (shaded area) below:

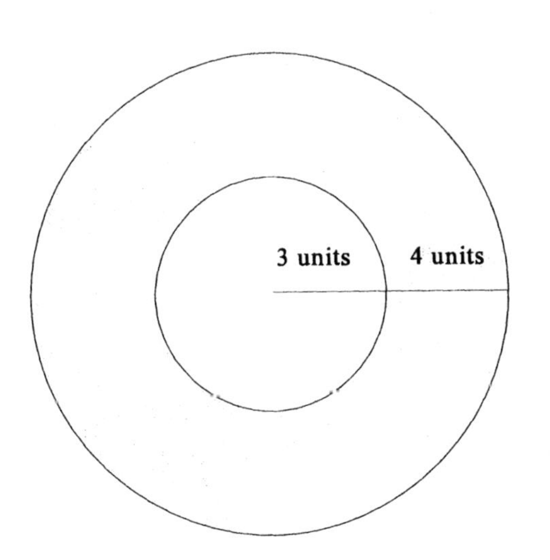

2. In the picture below, a circle is inscribed in a semi-circle. (The diameter of the circle is the radius of the semi-circle.) The radius of the circle is 3 units. Find the shaded area.

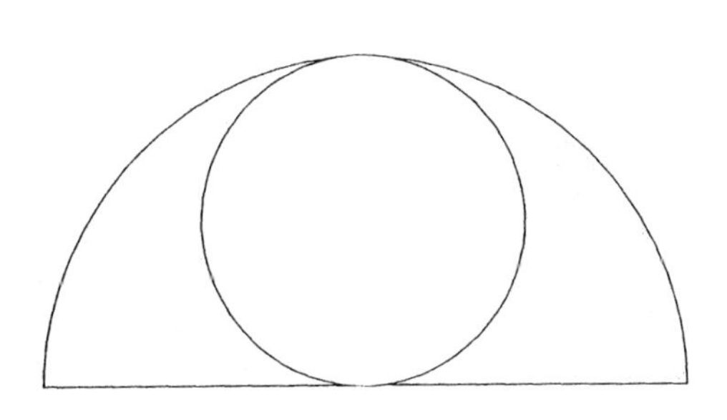

3. Find the area of the sectors by making the necessary measurements.

a)

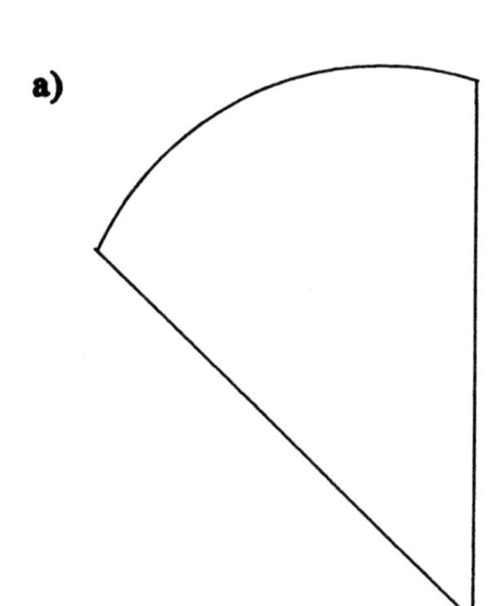

b)

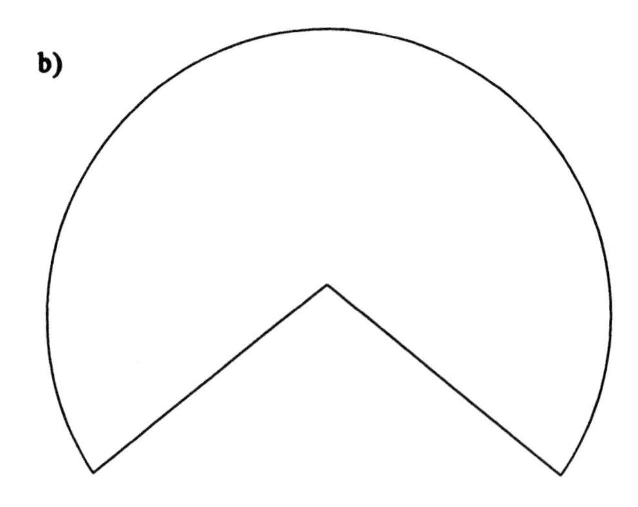

4. The outside circumference of a washer is 22 cm, and the inside circumference is 10 cm. What is the area of the washer?

5. You have a circular rose garden of diameter 20 ft.
   a) One pound of weed-killer covers 10 sq. yd. How many pounds of weed-killer will be needed to cover the garden?
   b) Bricks are 8 inches long. How many bricks will be needed to build a wall, six bricks high, around the garden?

6. a) Construct a right triangle with hypotenuse congruent to this segment:

   ____

   b) Circumscribe a circle about the triangle.

Notice that the hypotenuse of the triangle is the diameter of the circle! This suggests an easy way to construct a right angle: Construct a circle; construct a diameter; pick any point on the circle except an endpoint of the diameter and name it P; construct the segments from P to each endpoint of the diameter --- these segments are perpendicular!

Section 42: The Pythagorean Theorem

One of the best-known and most useful facts in all of mathematics is known as the Pythagorean Theorem. It was first proved thousands of years ago by Greek mathematicians and, through the centuries, it has served as the cornerstone of applied geometry. The theorem states:

> If a and b are the lengths of the legs of a right triangle and c is the length of the hypotenuse, then $a^2 + b^2 = c^2$.

Proof: Let $\triangle RST$ be any right triangle with sides as shown in the picture below:

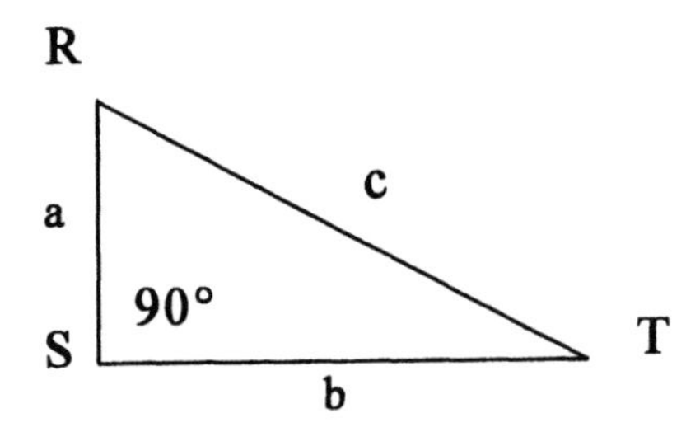

Extend rays $\overline{SR}$ and $\overline{ST}$ and copy the segments as shown:

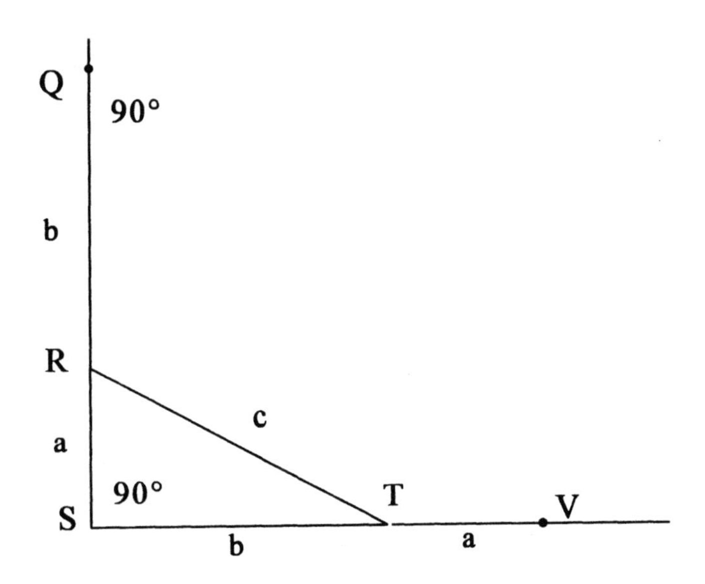

Now we construct the perpendicular to $\overline{RS}$ at Q and the perpendicular to $\overline{ST}$ at V.

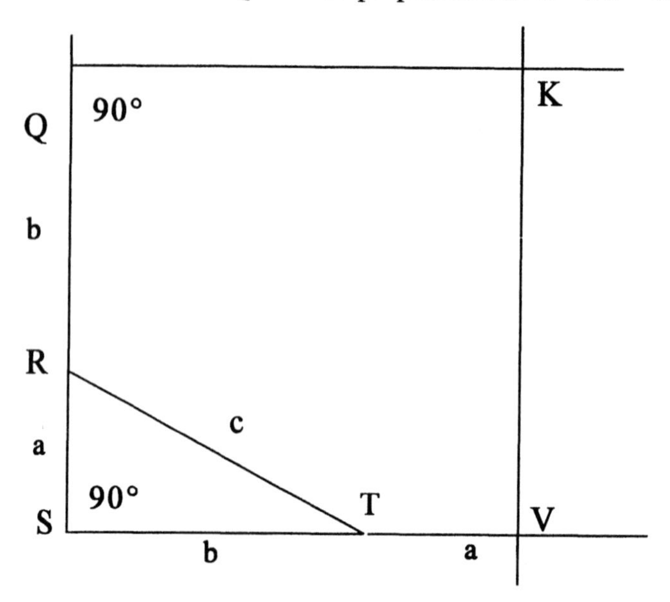

It is easy to show that QSVK is a square with sides of length a + b. So we can label points P and M such that QP = KM = a and PK = MV = b. We now construct segments $\overline{RP}$, $\overline{PM}$, and $\overline{MT}$.

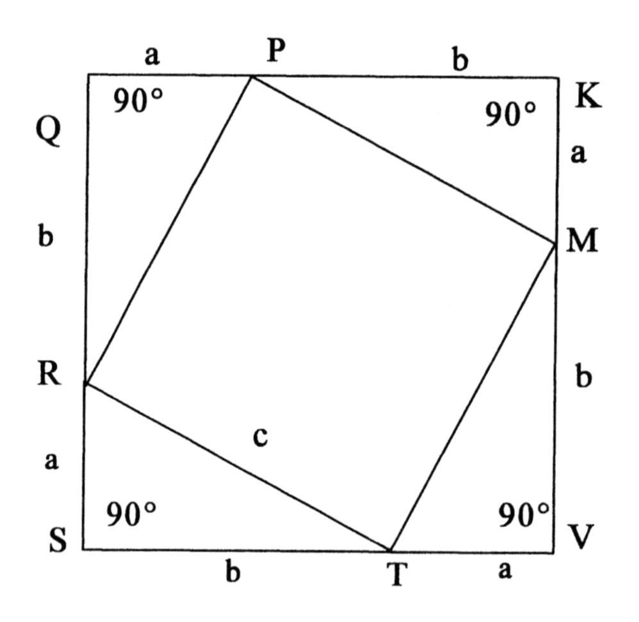

The four triangles at the corners of QKVS are congruent by SAS, so
RP = PM = MT = RT = c,
$m\angle QRP = m\angle KPM = m\angle VMT = m\angle STR = x$, and
$m\angle QPR = m\angle KMP = m\angle VTM = m\angle SRT = y$.

Now consider the figure below with markings to represent all this information:

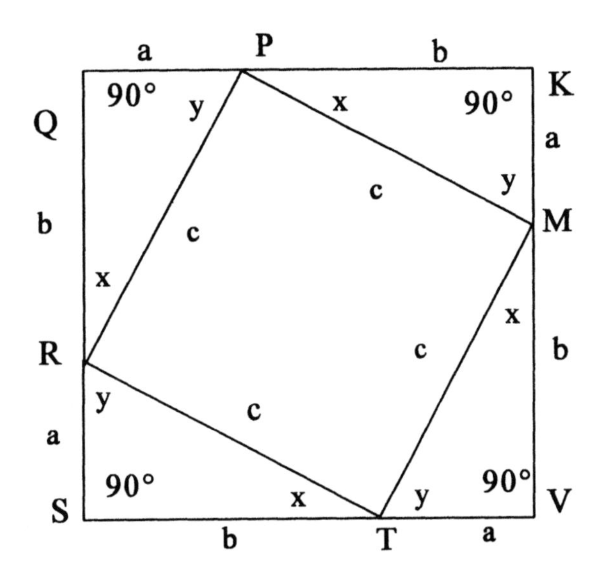

Since $m\angle S = 90°$, it follows that $x + y = 90°$ (why?). Therefore, $m\angle PRT = 90°$ (why?). By the same reasoning, $m\angle RPM = m\angle PMT = m\angle MTR = 90°$. Hence RPMT is a square (all sides equal and all right angles).

Since QSVK is a square with sides of length a + b, we know that

area QSVK = (a + b)(a + b).

But the area of QSVK can also be determined by adding the areas of the four triangles and the small square PMTR. The four triangles each have area ½ab and the area of PMTR is c^2, so we have:

$$\text{Area QKVS} = 4(\tfrac{1}{2}ab) + c^2$$
$$(a + b)(a + b) = 2ab + c^2$$
$$a^2 + 2ab + b^2 = 2ab + c^2$$
$$a^2 + b^2 = c^2$$

The beauty of this proof lies in its simplicity. It is amazing that a theorem as important as this can be proved using only the most basic concepts of geometric figures and area.

Example 1: A 13-foot ladder is leaning against a wall with the foot of the ladder 5 ft from the wall. At what height is the ladder resting against the wall?

First we draw a picture:

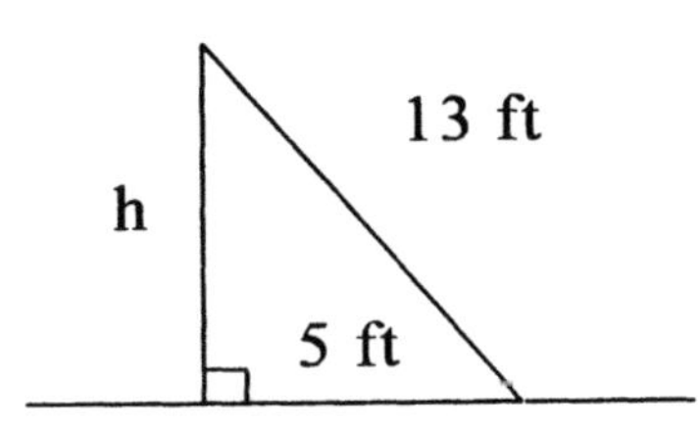

The height h can be found by using the Pythagorean relationship:

$$h^2 + 5^2 = 13^2$$
$$h^2 = 169 - 25$$
$$h^2 = 144$$
$$h = 12$$

Hence the ladder is resting 12 ft up the wall.

Example 2: A camper hikes 4 miles due south from his tent. Then he turns and hikes 3 miles due west. How far is he from the tent?

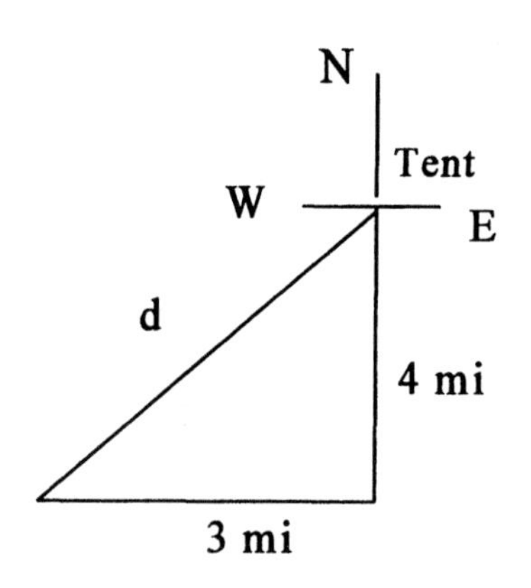

Again, we begin with a picture.

The distance d can be found by:

$$d^2 = 4^2 + 3^2$$
$$d^2 = 25$$
$$d = 5$$

Therefore the hiker is 5 miles from his tent.

Example 3: In the figure at right, find x.

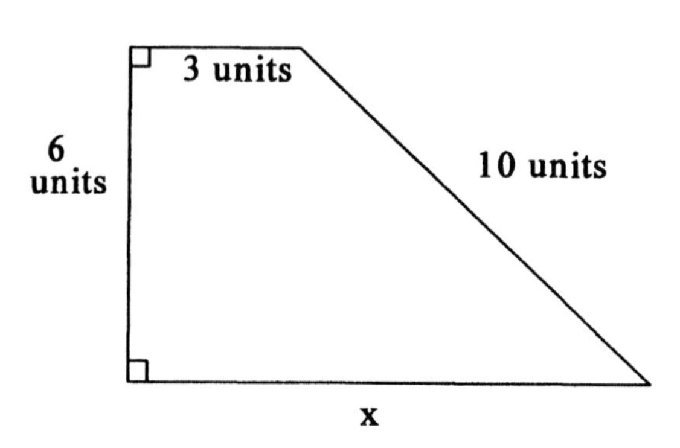

If we construct the segment shown below, then the Pythagorean theorem can be used to find y. Now we use $x = y + 3$.

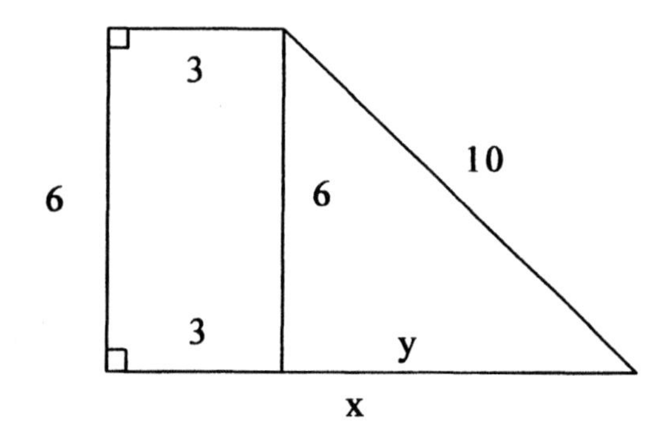

$$6^2 + y^2 = 10^2$$
$$y^2 = 100 - 36$$
$$y^2 = 64$$
$$y = 8$$

So $x = 8 + 3 = 11$ units.

Example 4: Find x.

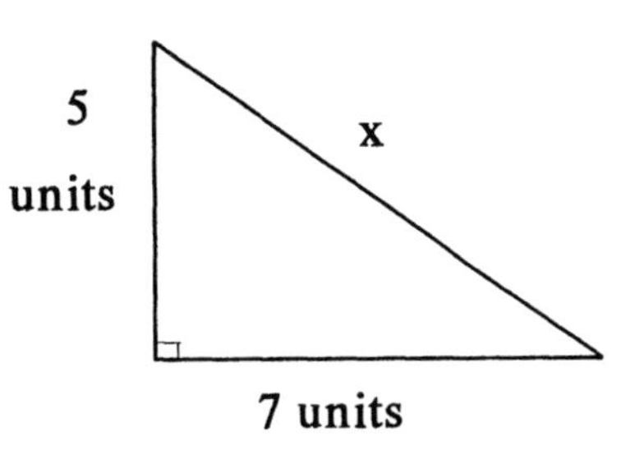

By the Pythagorean Theorem, $x^2 = 5^2 + 7^2$,
$x^2 = 74$.

In the three previous examples we had $h^2 = 144$, $d^2 = 25$, and $y^2 = 64$. It was obvious that h = 12, d = 5, and y = 8. But $x^2 = 74$ is an entirely different matter.

We know that x is the number whose square is 74, but it isn't at all obvious what that number is. We say that x is "the principle square root of 74", or $x = \sqrt{74}$. But these names tell us very little about the nature of the number. Is it a fraction? Is it a decimal? How "big" is it? Since

$$8^2 = 64$$
$$x^2 = 74$$
$$9^2 = 81$$

it is clear that x is about half-way between 8 and 9. When we check, $(8.5)^2 = 72.25$; so x is bigger than 8.5. Next, we find that $(8.6)^2 = 73.96$. This is <u>very</u> close to 74, but still not exact. Testing 8.61, we get $(8.61)^2 = 74.1321$. Now we know that $\sqrt{74}$ is between 8.6 and 8.61, but we still do not know its <u>exact</u> decimal representation. The fact is, $\sqrt{74}$ cannot be written as a terminating or repeating decimal. It is an infinite, non-repeating decimal--we can write it with as many decimal places as we choose, but whenever we stop, the result is only an <u>approximation</u> of $\sqrt{74}$. Such numbers are called "irrational numbers." They have a place on the number line and they can be approximated to as many significant digits as we choose, but they can never be written as fractions. In geometry, where all real-world measurements are also approximate, irrational numbers should not cause any problems. We understand that $\sqrt{5} \approx 2.2$ simply means that $(2.2)^2$ is closer to 5 than $(2.1)^2$ or $(2.3)^2$. (Note: $(2.1)^2 = 4.41$; $(2.2)^2 = 4.84$; $(2.3)^2 = 5.29$.)

Recall that π is also an irrational number. Actually, there are many more irrational than rational numbers, but in elementary mathematics the only ones we usually encounter are π and square roots.

In order for us to add or multiply irrational numbers, we usually round them off to a reasonable number of significant digits and proceed with decimal arithmetic. Sometimes, however, it is helpful to simplify square roots by using this relationship: if x and y are positive, then $\sqrt{xy} = \sqrt{x}\sqrt{y}$.

So $\sqrt{27} = \sqrt{9}\sqrt{3} = 3\sqrt{3}; \sqrt{225} = \sqrt{49}\sqrt{5} = 7\sqrt{5}; \sqrt{108} = \sqrt{36}\sqrt{3} = 6\sqrt{3}$. Note, also, that irrational numbers enjoy all the same nice properties as the rational numbers--commutativity, associativity, etc.

So $3\sqrt{7} + 5\sqrt{7} = (3 + 5)\sqrt{7} = 8\sqrt{7}$.

One particularly useful application of the Pythagorean Theorem is called the "30-60-90 Theorem." It gives a relationship among the sides of a triangle whose angles measure 30°, 60°, 90°.

Consider an equilateral triangle, $\triangle ABC$, with altitude AM.

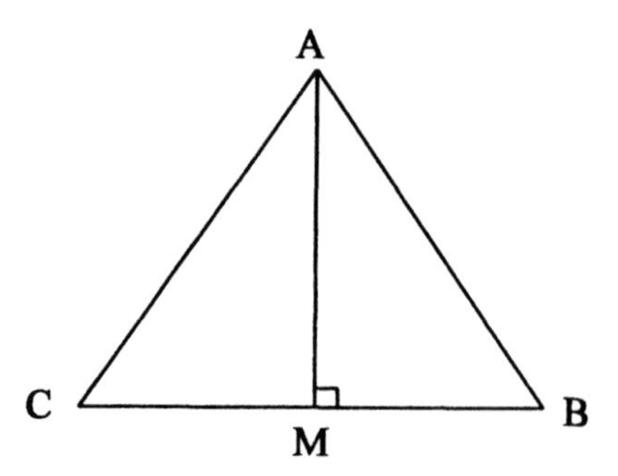

Since $m\angle B = 60°$ (each angle of an equilateral triangle measures 60°) and $m\angle AMB = 90°$ (definition of altitude), it follows that $m\angle MAB = 30°$. It is easy to show, by ASA or SAS, that $\triangle BAM \cong \triangle CAM$. Therefore CM = MB. This means that $MB = \frac{1}{2} CB = \frac{1}{2}AB$.

Now consider $\triangle MAB$:

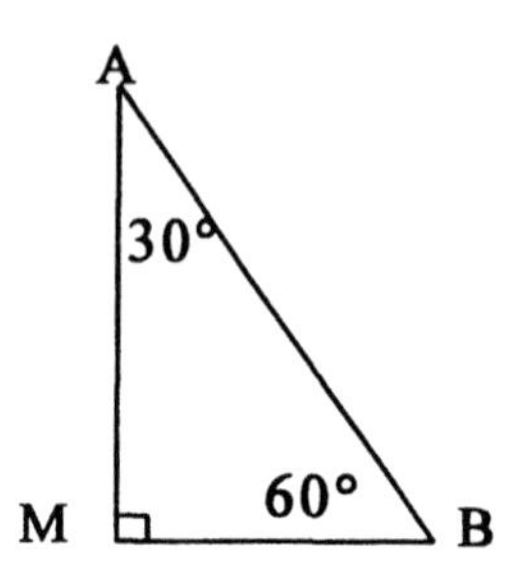

We have just noticed that $MB = \frac{1}{2}AB$, so we have shown:

In a 30°- 60°- 90° triangle, the side opposite the 30° angle is half the hypotenuse.

Now notice that if $MB = \frac{1}{2} AB$, then $AB = 2(MB)$. So the Pythagorean relationship for $\triangle MAB$ can be written as:

$$(MB)^2 + (AM)^2 = (2(MB))^2$$
$$(MB)^2 + (AM)^2 = 4(MB)^2$$
$$(AM)^2 = 3(MB)^2$$
$$AM = \sqrt{3}\ (MB).$$

This can be generalized to:

In a 30°- 60°- 90° triangle, the side opposite the 60° angle equals $\sqrt{3}$ times the side opposite the 30° angle.

Putting these two facts together we have:

In a 30°- 60°- 90° triangle, if the side opposite the 30° angle is x, then the hypotenuse is 2x and the side opposite the 60° angle is $\sqrt{3}x$.

Homework: Section 42

I. 1. A ship sails 7 miles due East then it turns and sails due North until it is 25 miles from its starting point. How far North did the ship sail?

2. A 32-foot telephone pole is held in place by a support wire which is attached to the top of the pole and to a stake in the ground 24 ft from the bottom of the pole. How long is the wire?

3. Find area.

4 units

9 units

$3\sqrt{5}$ units

7 units

4. Find x.

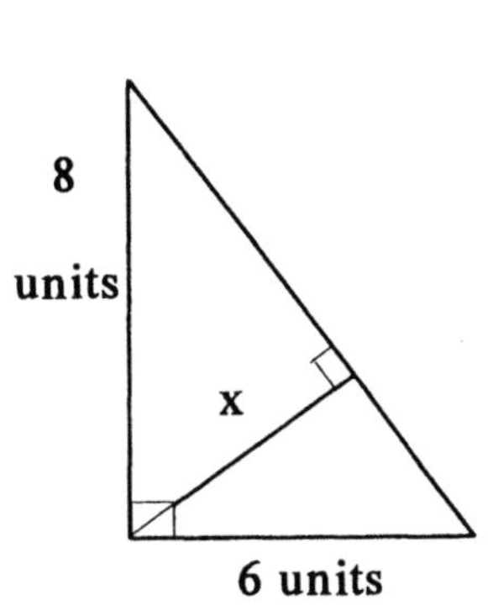

5. Find x.

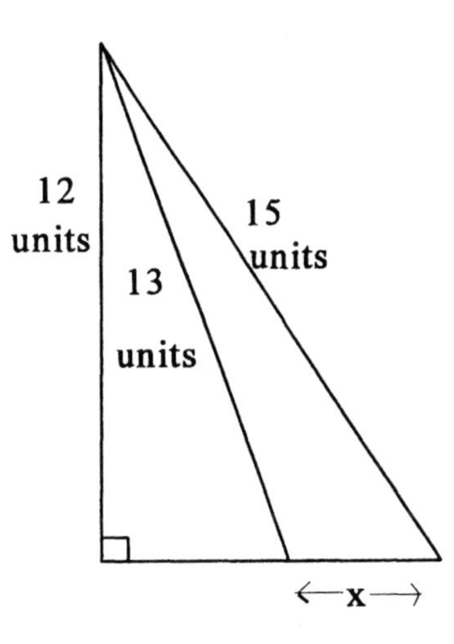

6. Find area.

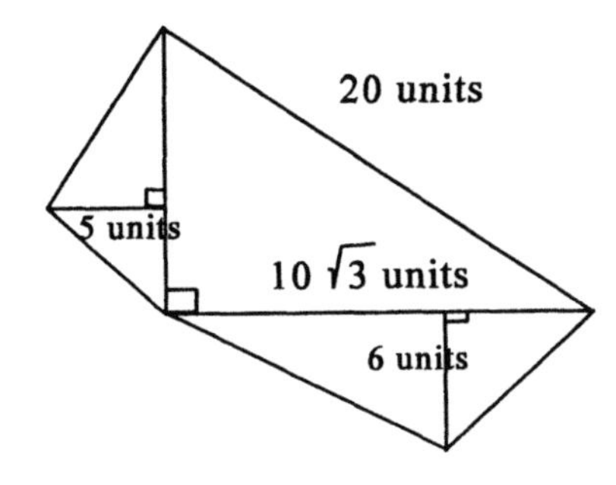

7. Find area.

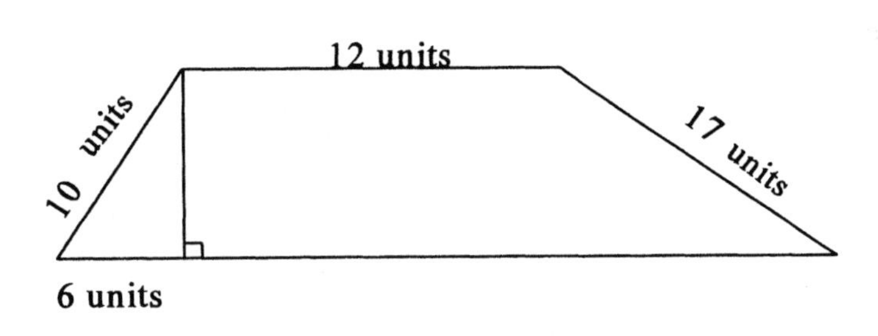

8. The altitude of an equilateral triangle is 6 cm. What is the area of the triangle?

9. A rhombus has sides of length 12 units and one angle of 150°. What is its area?

10. The hypotenuse of a right triangle is 50 ft and one leg is 14 ft. What is its area?

11. The diagonals of a square are 16 inches. What is the area of the square?

12. The vertex angle of an isosceles triangle measures 120°, and the equal sides are 10 inches. What is the area of the triangle?

II. For each figure

a) Make the <u>fewest</u> measurements needed to find the area of the figure. (Since every measurement has a possible error, using the fewest number of measurements will give us the best approximation of the area.)

b) Trace or Xerox each figure, cut it out, and bring it to class.

1.

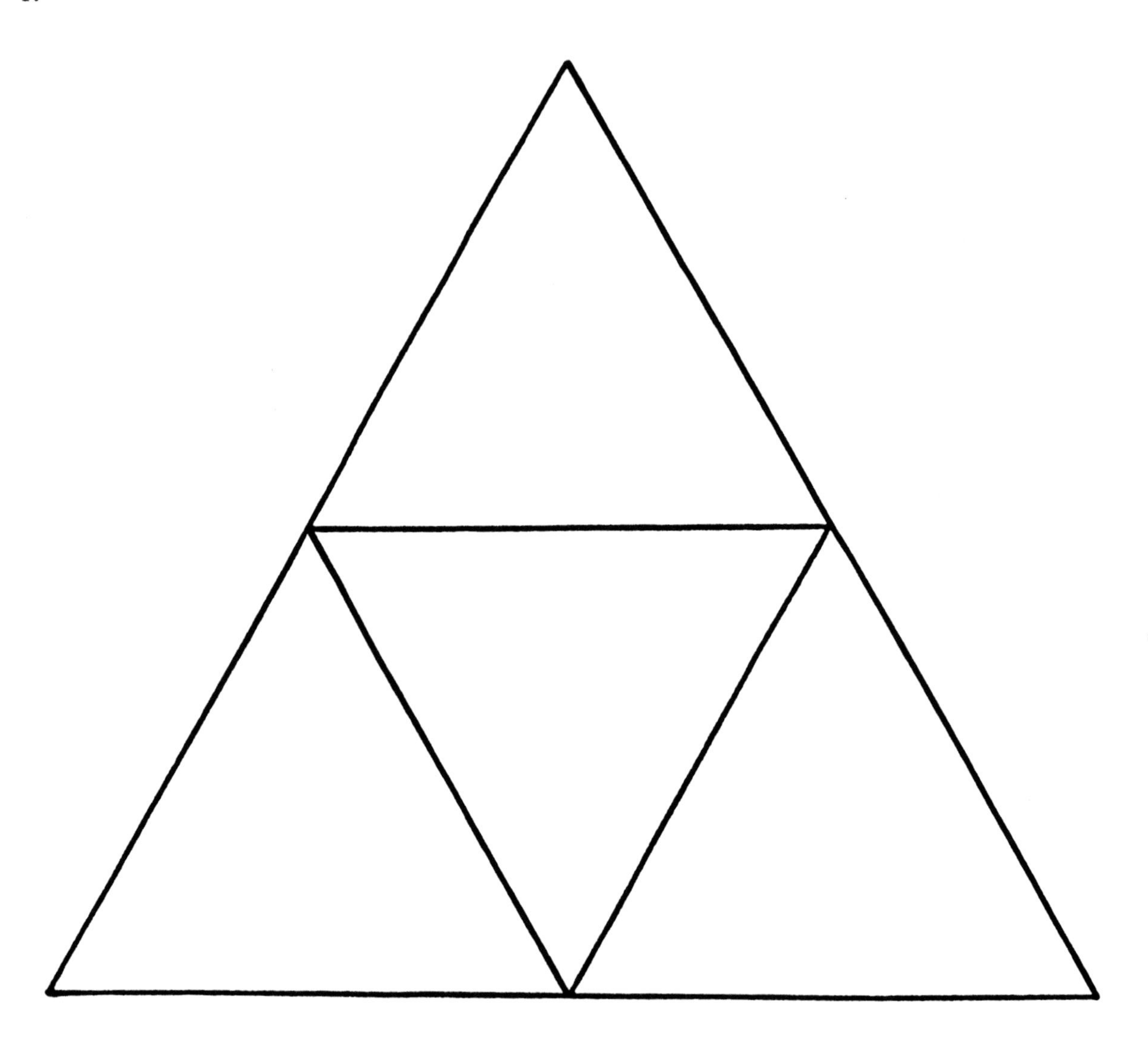

2.

3.

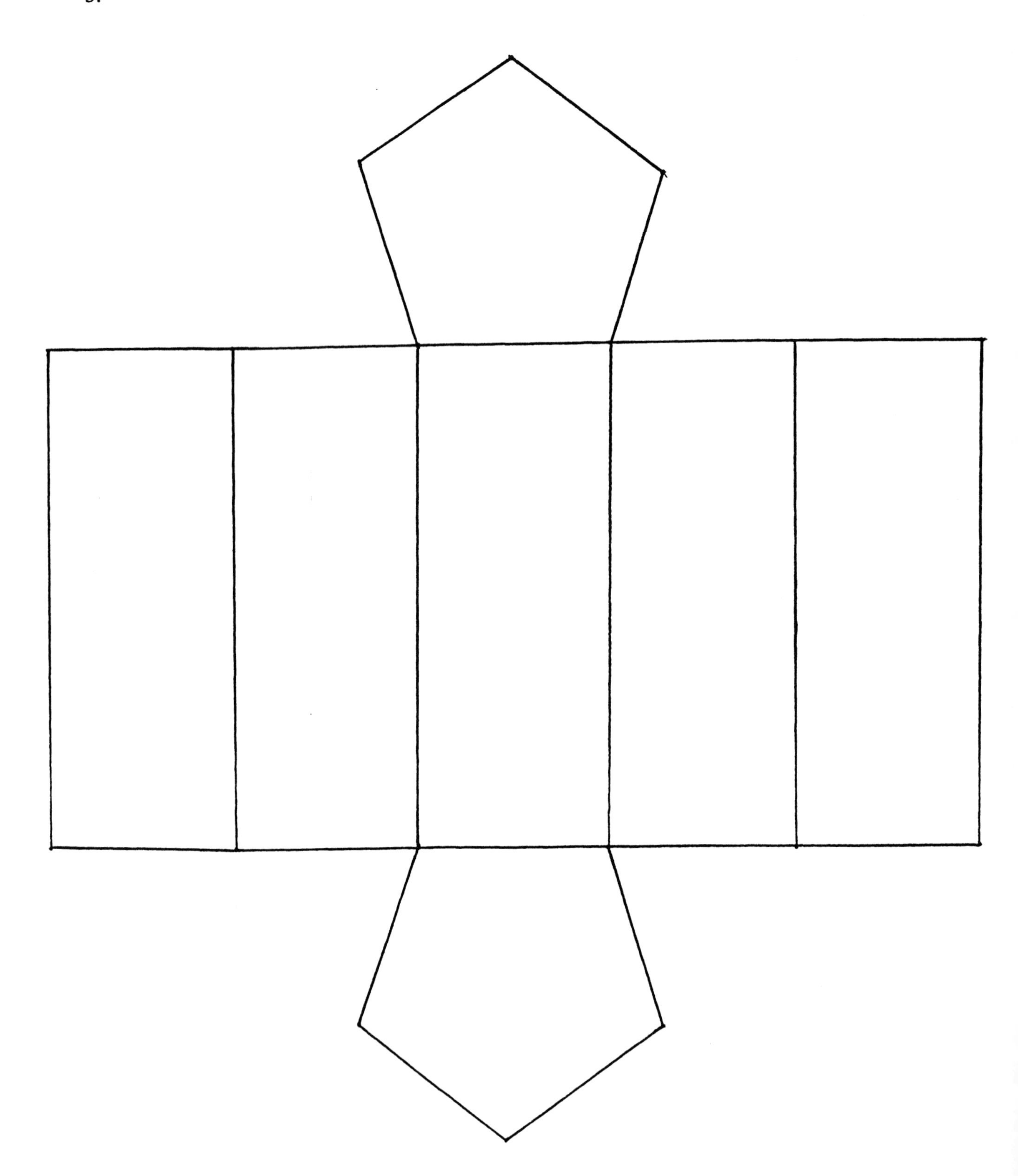

4.

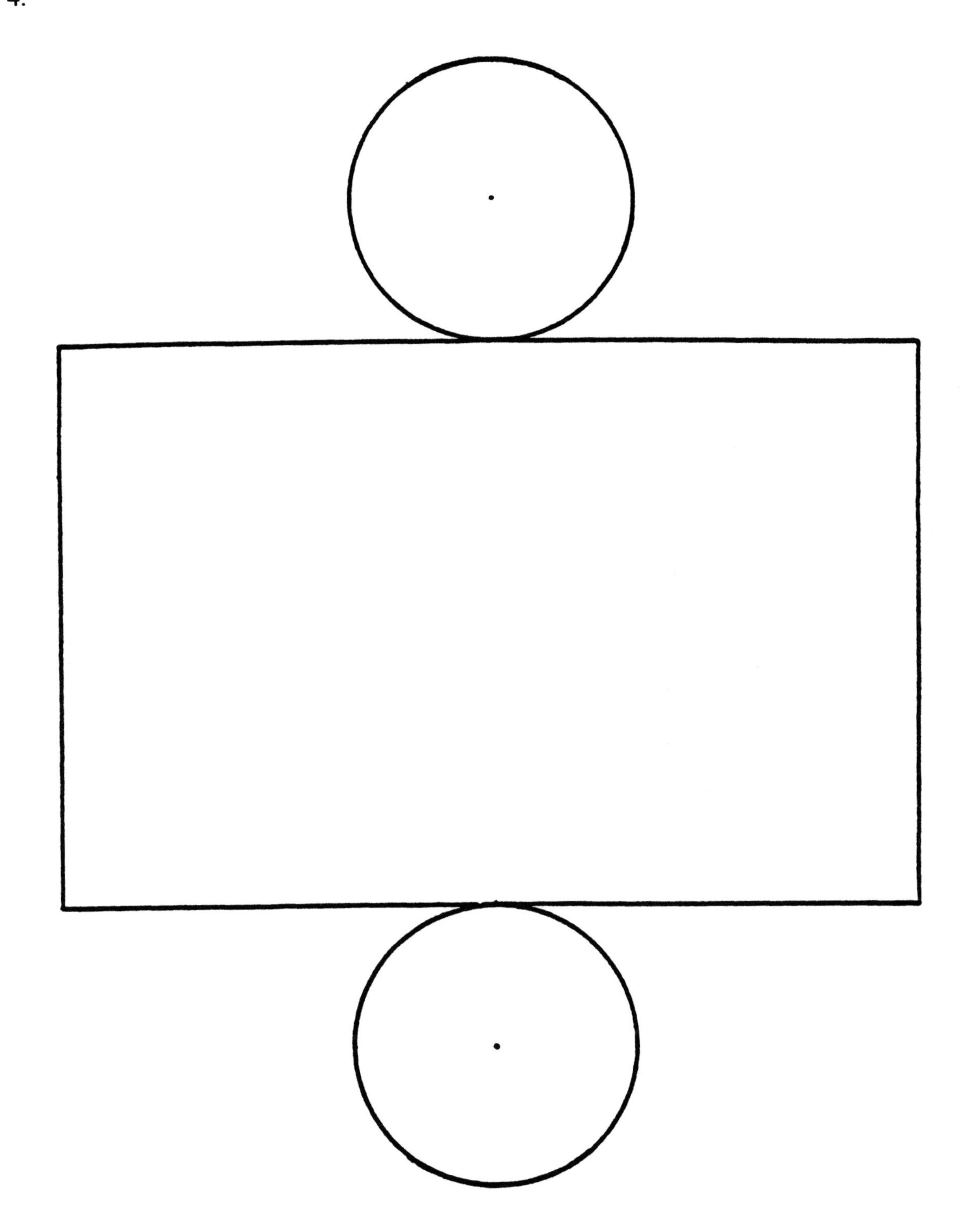

5.

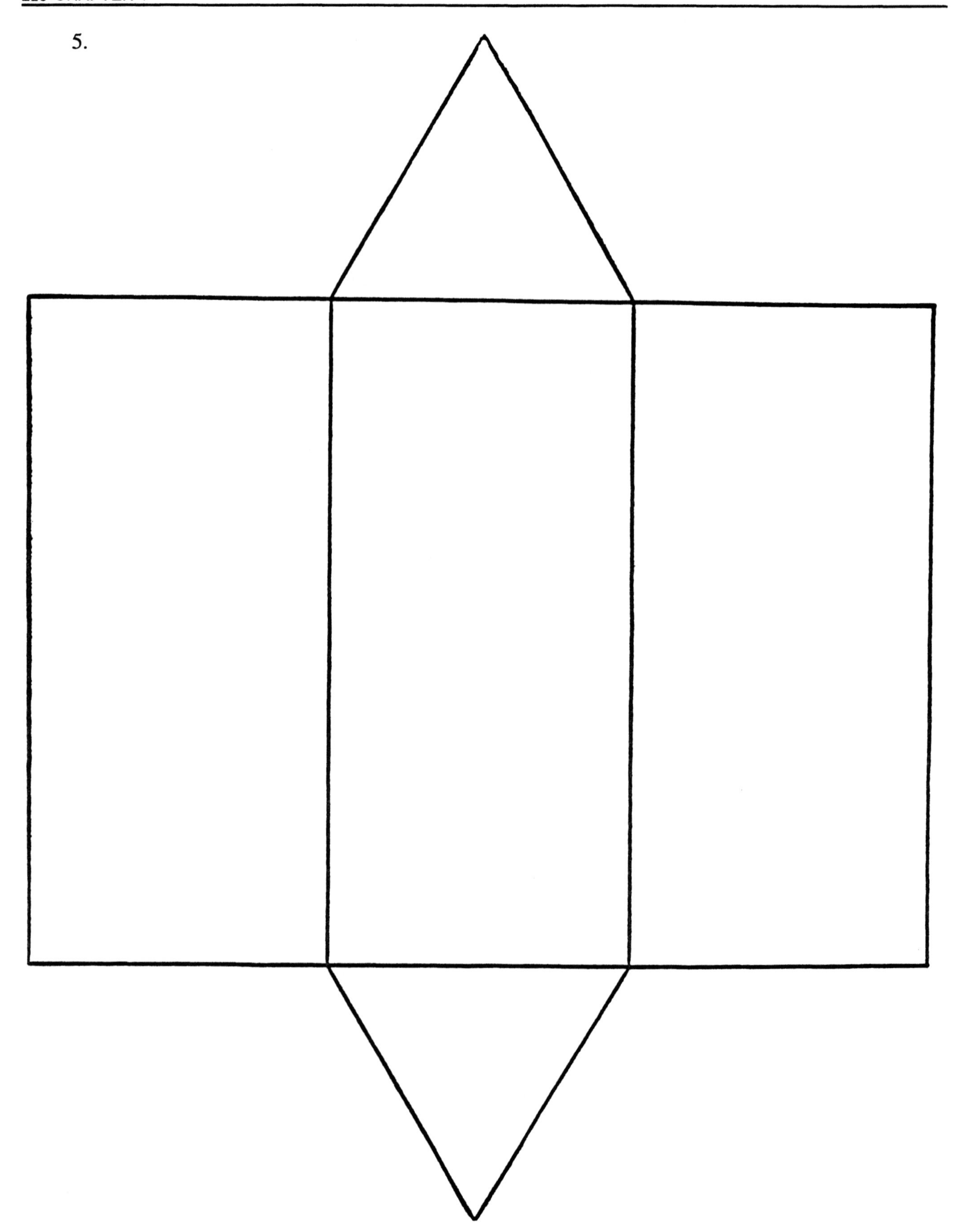

6.

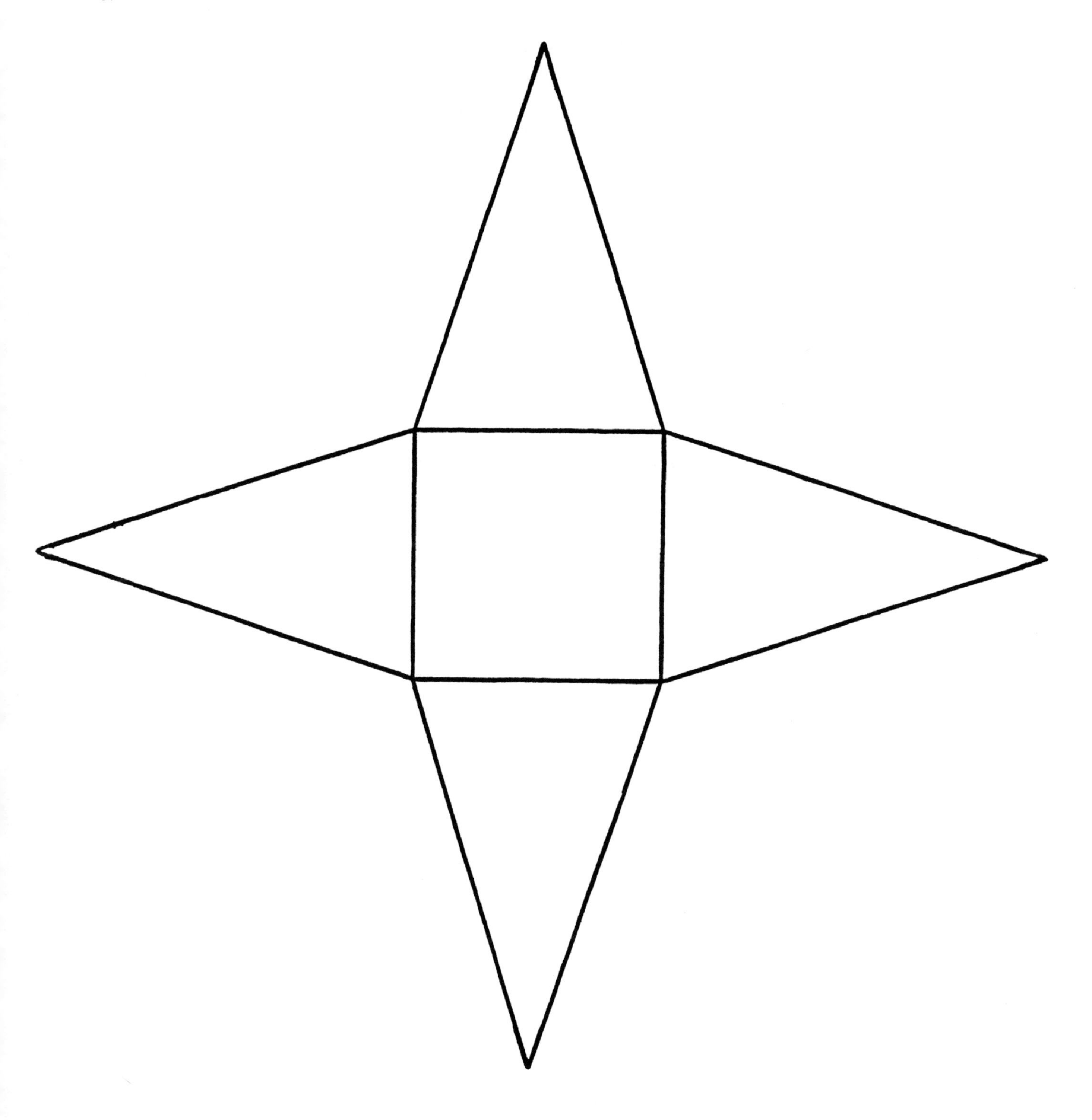

7.

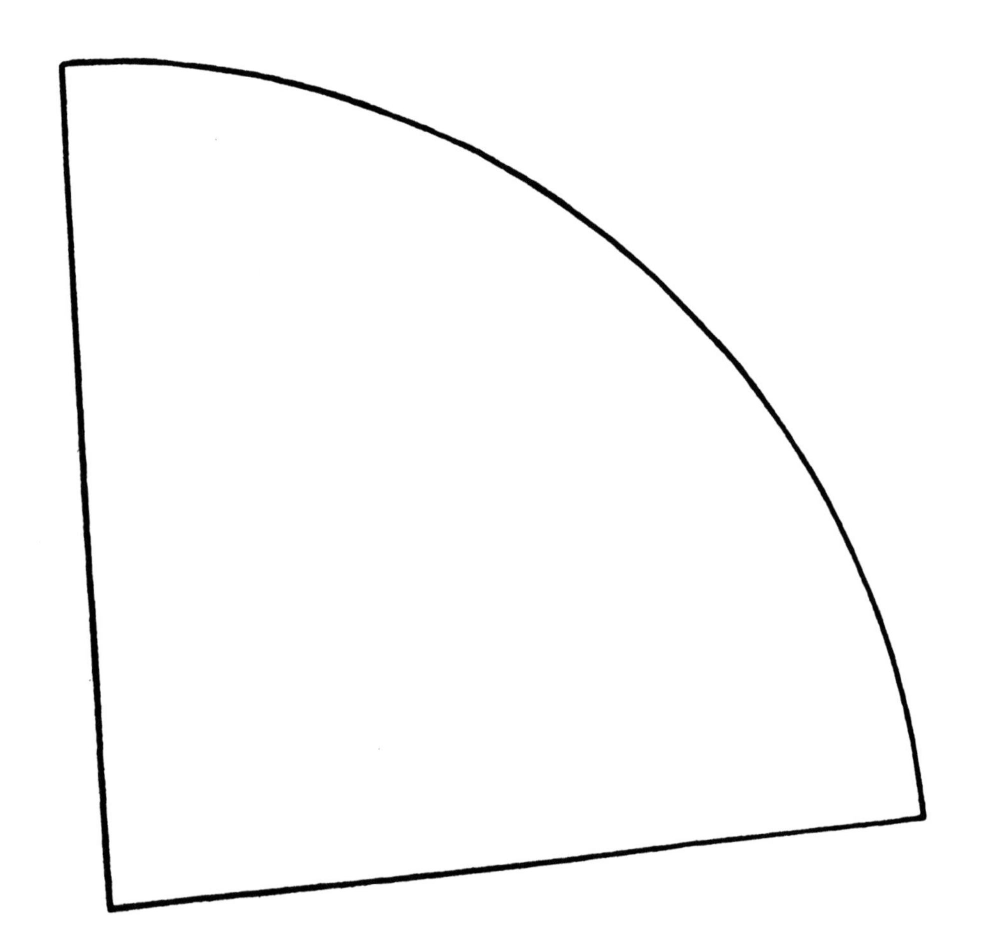

8.

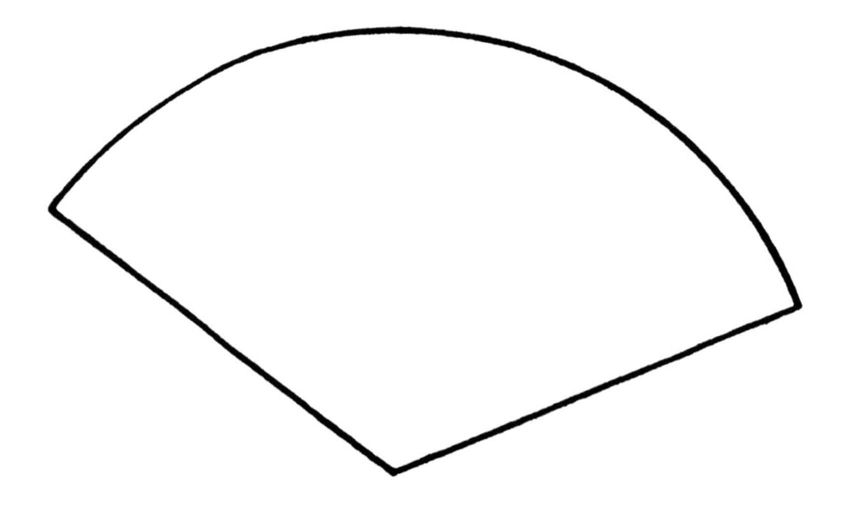

9.

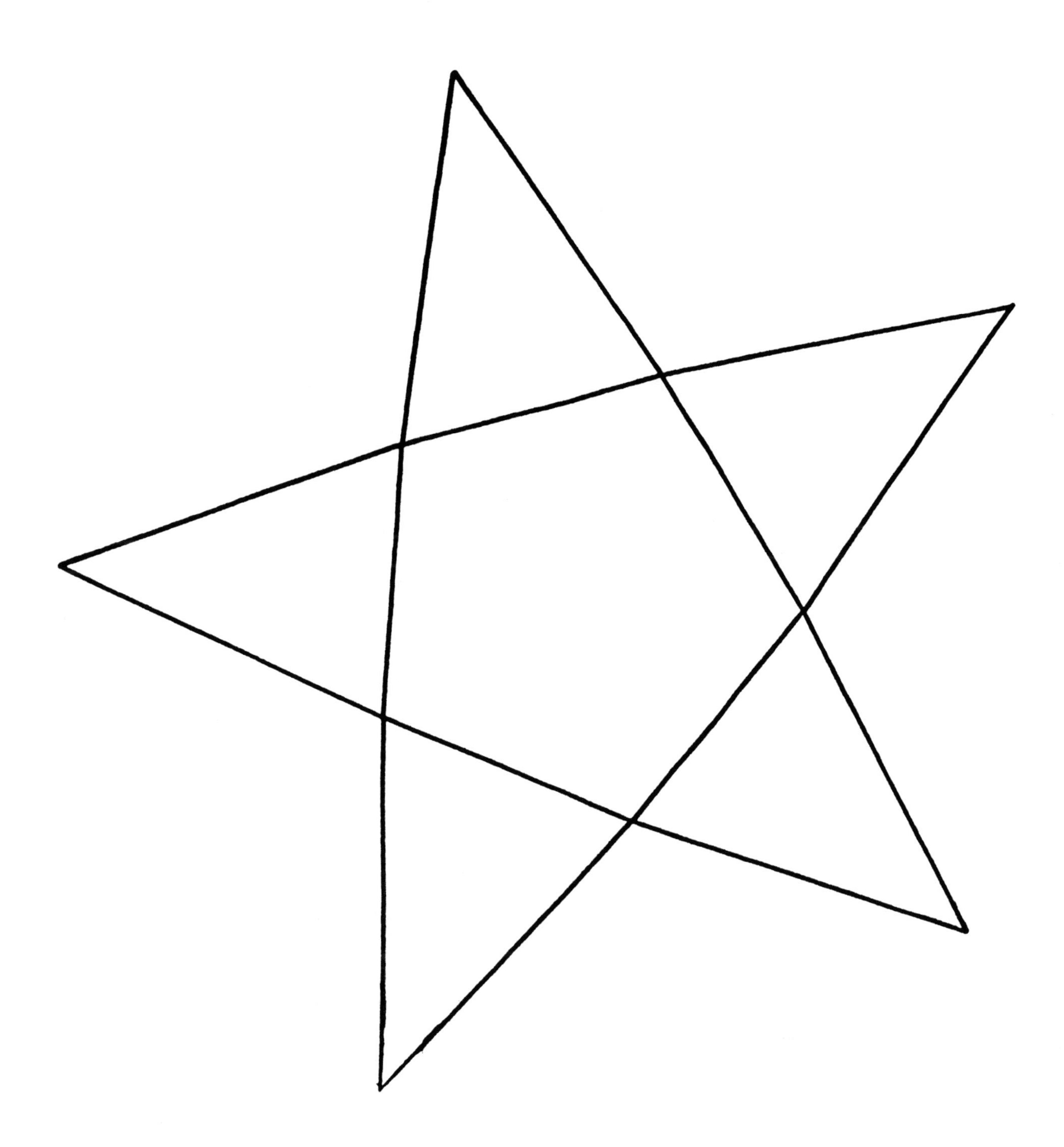

Section 43: Surface Area

If the cut-out figures from pages 206-213 were folded along their edges, they would form familiar three-dimensional figures. Hence they can be considered as "patterns" for the prisms, cylinders, pyramids, or cones. The surface area of a three-dimensional figure is the area of its pattern. Some figures, such as spheres, cannot be made from a simple pattern; hence their surface area must be determined in other ways.

A prism consists of two congruent n-gon bases and n rectangles. If the bases are rectangles, then the prism is simply a box.

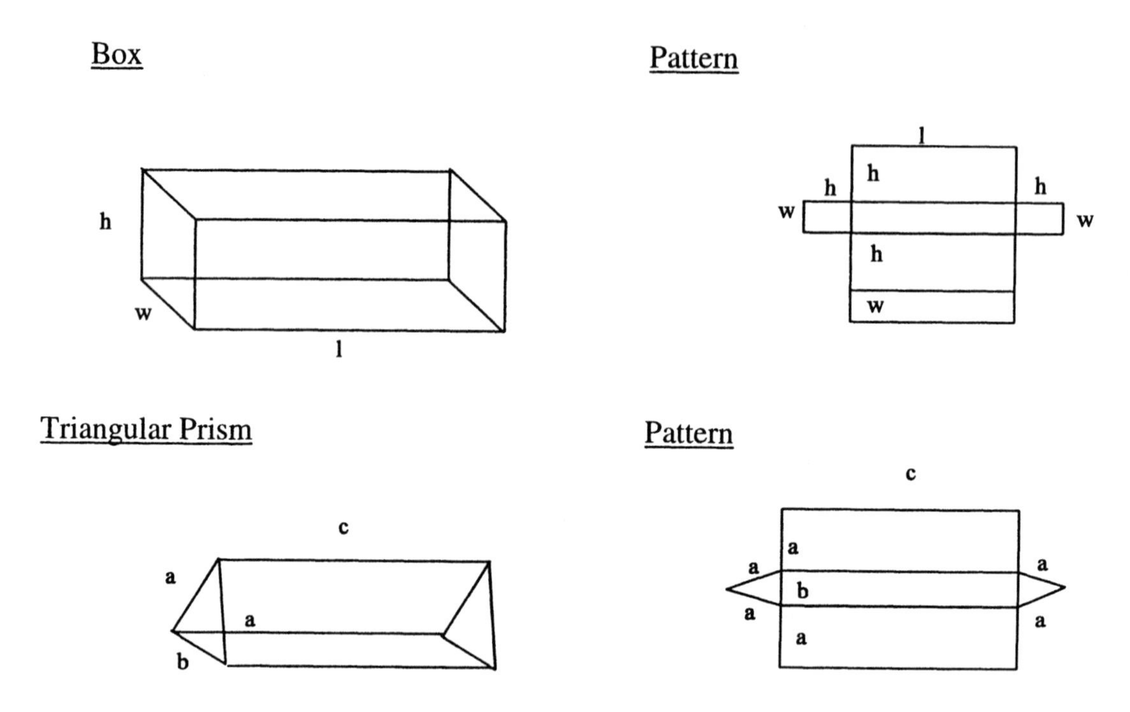

The figure on page 208 is the pattern of a pentagonal prism.

A cylinder can be thought of as a prism with circular bases and only one rectangular "side."

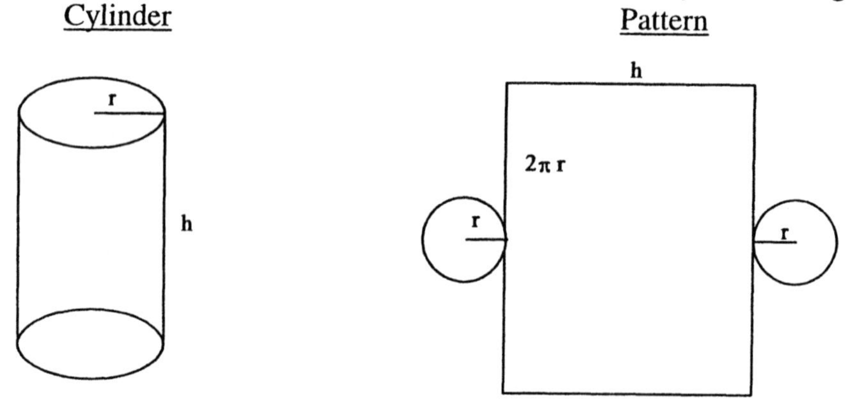

Example 1: Construct the pattern for a cylinder which is to be 6 cm long and have bases of radius 2 cm.

We must begin with a rectangle 6 cm long and $2\pi r = 4\pi$ cm wide. (Notice that when the pattern is formed into a cylinder, the width of the rectangle must fit the circumference of the bases.) Then we simply add the circular bases tangent to the rectangle along each width. (Tangent means intersecting in exactly one point.)

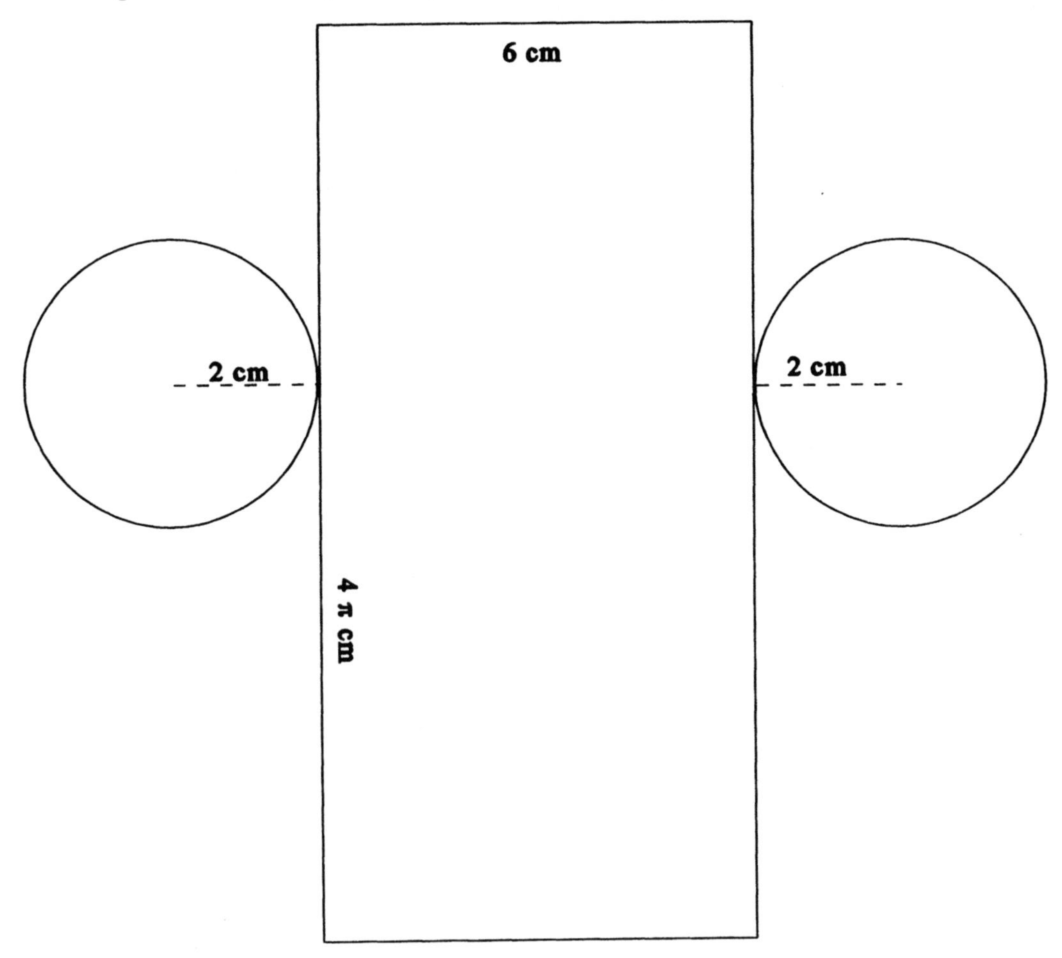

Pyramids have a single polygonal base and triangular sides (adjacent to each base) which meet at a common point called the apex. The distance from the apex to the base is the height of the pyramid.

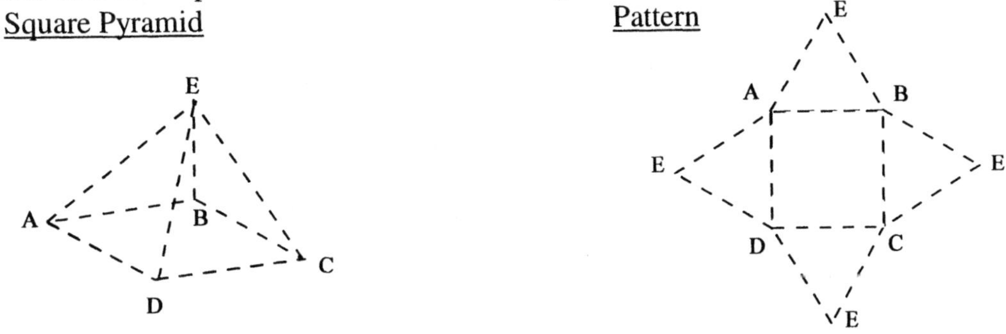

The figure on p. 206 is the pattern for a triangular pyramid, and the figure on p. 213 is the pattern for a pentagonal pyramid.

Example 2: Construct a square pyramid with a base of 6 cm on each side and height 4 cm.

Clearly, we begin with a 6 cm by 6 cm square. Next, we must know the altitude of triangular sides. (It is not 4 cm!)

To find the altitude, imagine a segment drawn from the apex of the pyramid to the center of the base. This is the height of the pyramid h. Now imagine the perpendicular segment from the center to one of the sides; call it b.

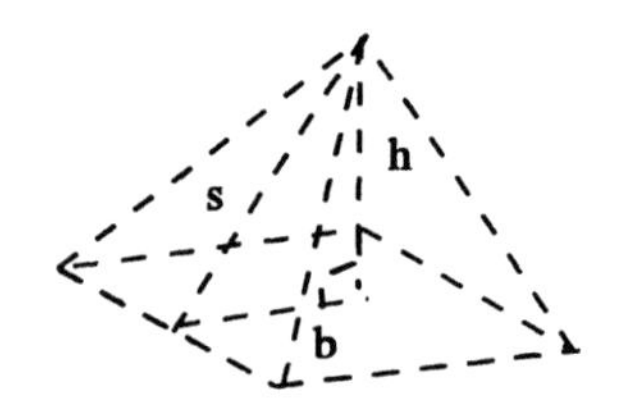

The segment s which forms a right triangle with h and b is the altitude of the triangular sides of the prism. It is called the 'slant height' of the pyramid, and can be found by the Pythagorean equation:

$$h^2 + b^2 = s^2$$
$$4^2 + 3^2 = s^2$$
$$25 = s^2$$
$$5 = s$$

Therefore the altitude of the triangle must be 5 cm. Hence the pattern is:

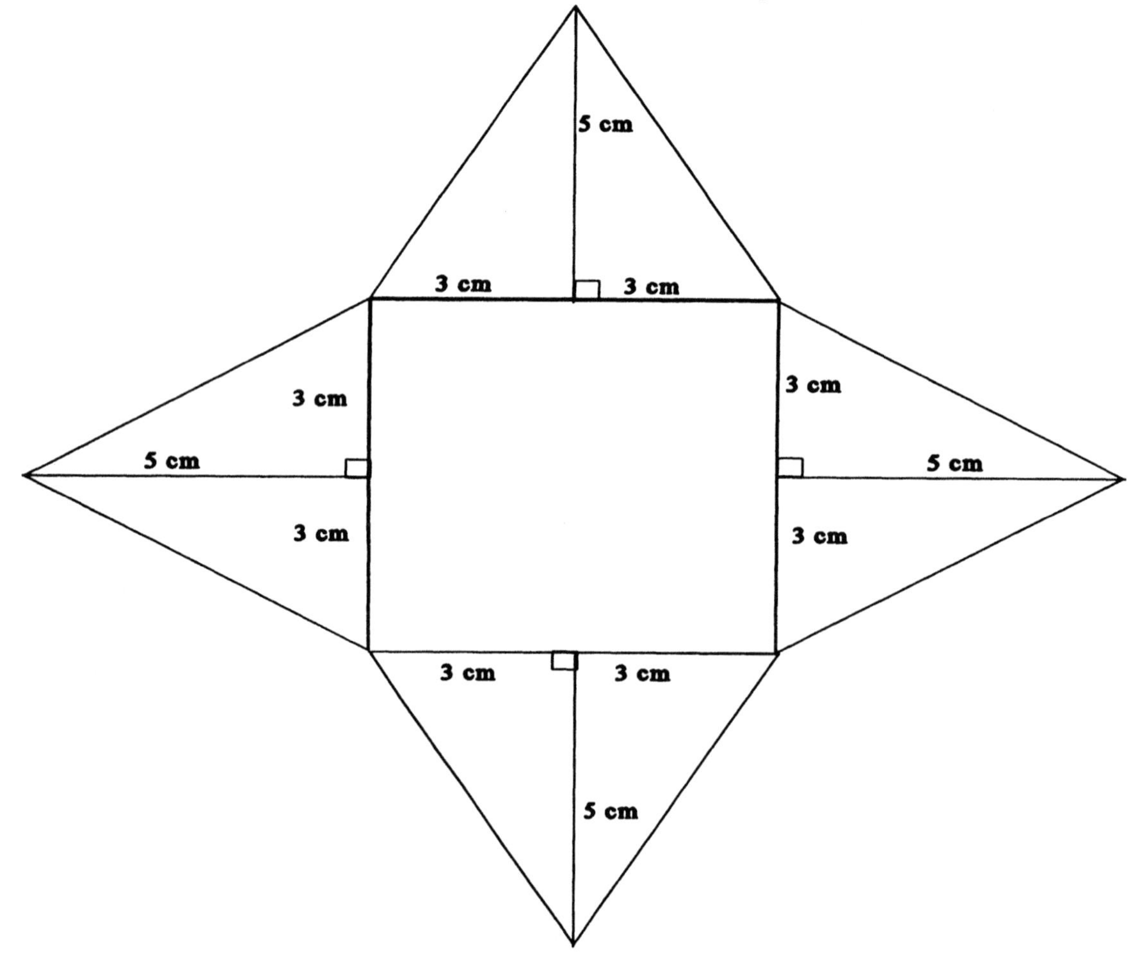

Homework: Section 43

1. Construct the pattern for a cylinder which has diameter 2 inches and is 5 inches long. What is the surface area of the cylinder?

2. Construct the pattern for a hexagonal prism which is 10.0 cm long and whose bases have sides of 3.0 cm each. What is the surface area of the prism?

3. Construct a pyramid which is 7 cm high and has a square base with sides 4 cm long. What is the surface area of the pyramid?

4. Points A and B are 5 units apart. How many lines in a given plane containing A and B are 2 units from A and 3 units from B?

5. An isosceles triangle has a base of 10 and two sides of 13. What other base can an isosceles triangle with equal legs of 13 have and still have the same area?

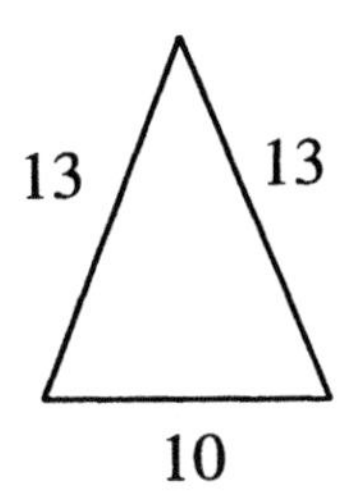

6. The hypotenuse of an isosceles right triangle $4\sqrt{2}$ units. How many square units are in the area of the triangle?

7. Express the perimeter of the triangle ABC as a decimal rounded to the nearest hundredth.

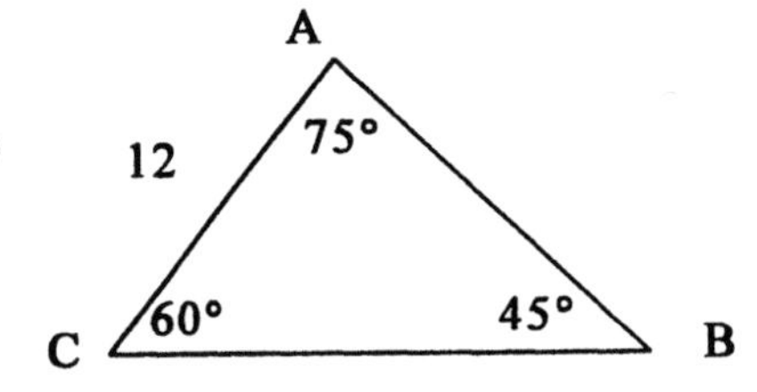

8. The dart board has a 7" radius with each region of the dart board 2" wide, except the bull's eye, which has a 1" radius. If a dart is thrown at random and is guaranteed to hit the board, what is the probability that it will hit the ring worth 50 points?

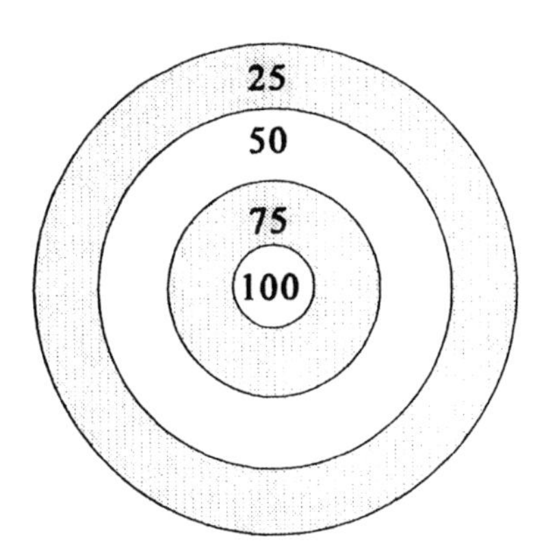

Section 44: Surface Area II

A cone is similar to a pyramid, but its base is a circle and its "side" is a sector.

Cone

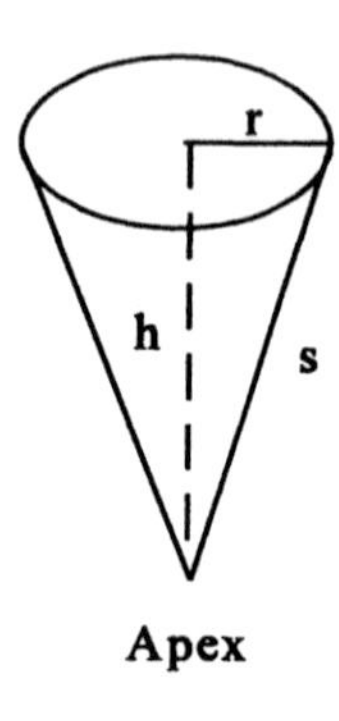

Pattern (for "side")

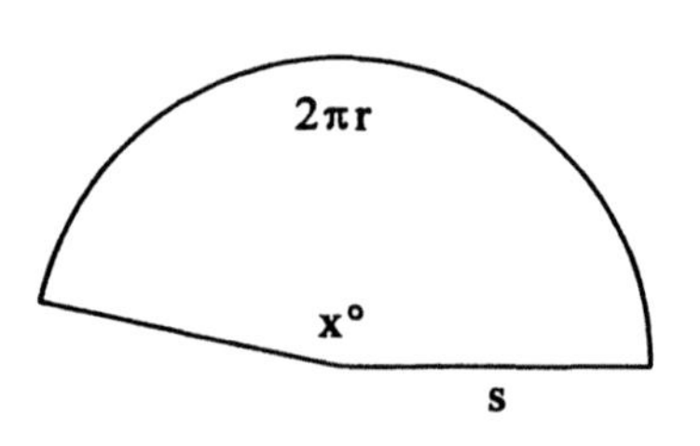

Notice that the radius of the sector is the slant height of the cone, and the length of the arc of the sector is the circumference of the base of the cone. Since the sector is x/360 of the circle from which it comes, its arc is x/360 of the circumference of the parent circle. So we have:

$$2\pi r = \frac{x}{360}(2\pi s)$$

$$r = \frac{x}{360}s$$

$$\frac{r}{s} = \frac{x}{360}$$

Example 1: If a cone were made from the pattern below, what would be the diameter of the base and what would be the height?

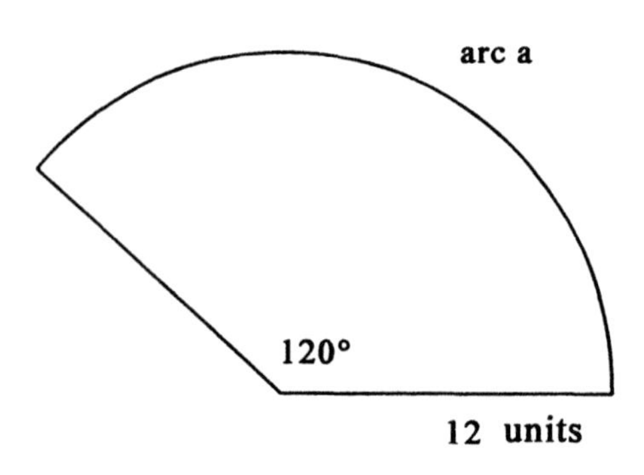

Remember that arc a represents 120/360 of the circumference of the parent circle of the sector; so arc $a = \frac{120}{360}\,2(12)\pi = \frac{1}{3}\,(24\pi) = 8\pi$ units. But arc a also represents the circumference of the base of the cone; so if d is the diameter of the base, then arc a = πd units. Thus $\pi d = 8\pi$, which implies d = 8. Therefore the diameter is 8 units.

To find the height of the cone, we use the Pythagorean theorem on this right triangle:

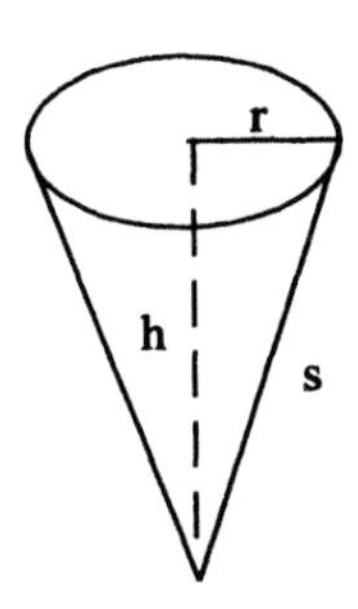

We know that s = 12 and r = 4, so

$$r^2 + h^2 = s^2$$
$$h^2 = 144 - 16$$
$$h^2 = 128$$
$$h = \sqrt{128} \approx 11$$

Thus the height of the cone is $\sqrt{128}$ units or about 11 units.

Example 2: Make a pattern for a cone which will be 8 cm high and have a base of diameter 12 cm.

First, we note that the cone will look like the picture at the right: So by the Pythagorean Theorem,

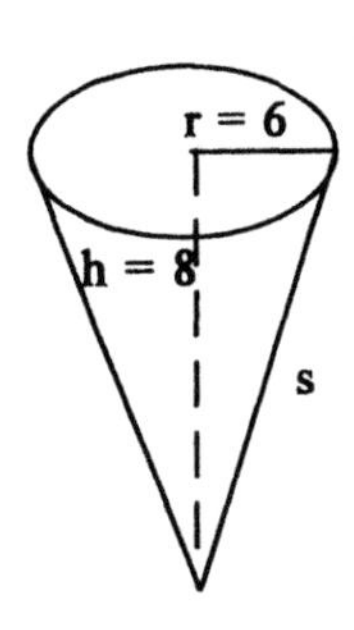

$$s^2 = 8^2 + 6^2$$
$$s^2 = 100$$
$$s = 10.$$

Now recall that the pattern for this cone must have the measurements shown at the right--where r/s = x/360. So

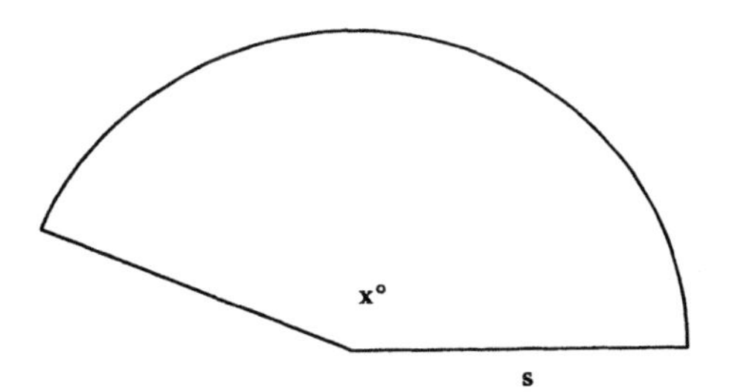

$$\frac{6}{10} = \frac{x}{360}$$

$$10x = 2160$$
$$x = 216°.$$

Therefore the pattern for the cone is:

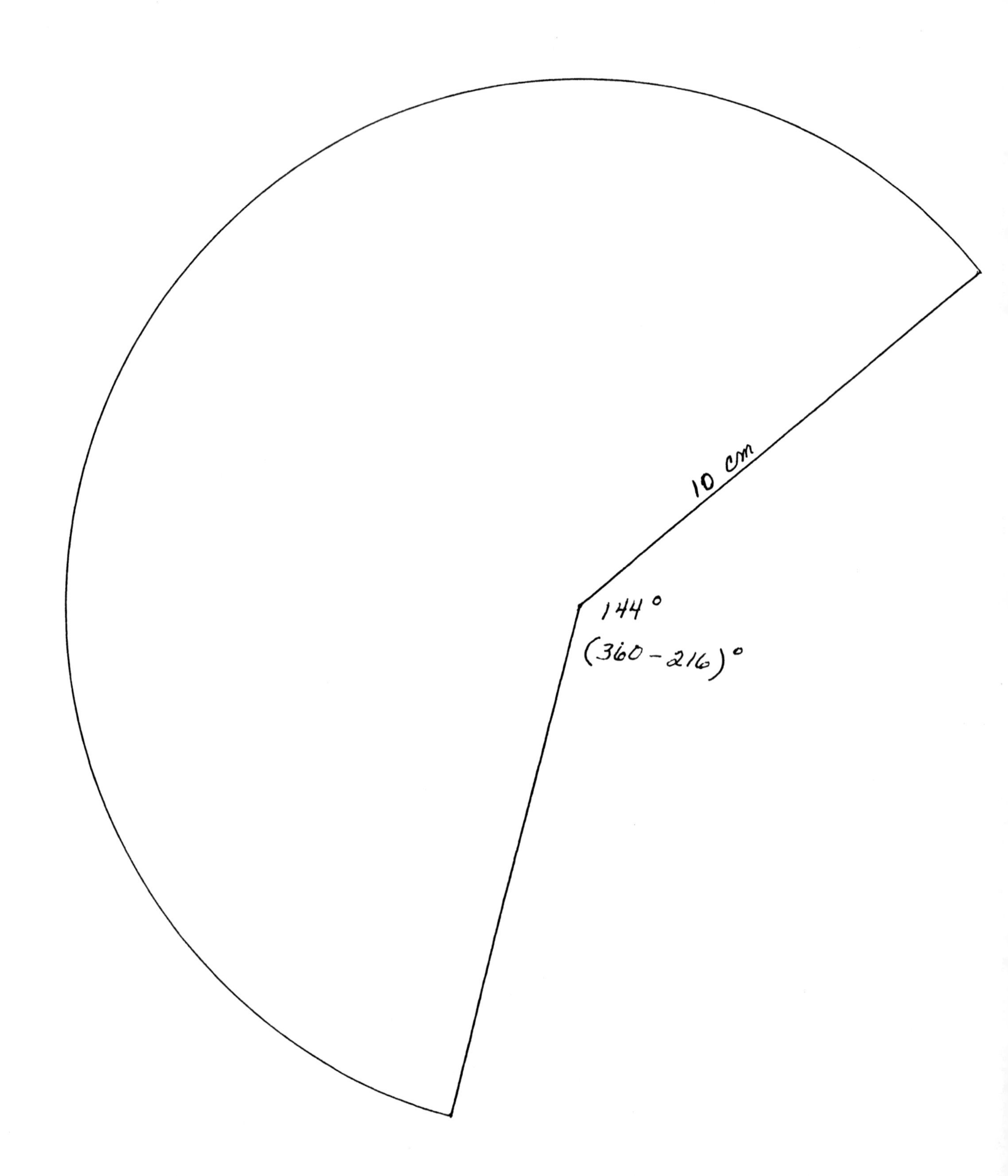

Example 3: A cone-shaped cup is 12 cm high and 10 cm in diameter. How much paper was used to the make the cup?

The question is asking for the surface area of the cone--the area of the pattern. So our first job is to determine the measurements of the pattern.

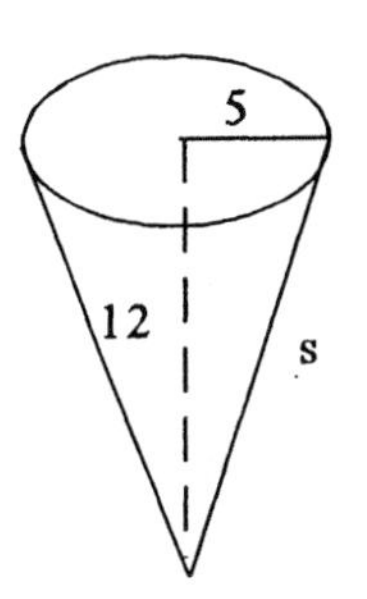

Using the figure at the right, we have:

$s^2 = 5^2 + 12^2$
$s^2 = 169$
$s = 13.$

Thus the radius of the pattern should be 13 cm, which implies that the circle from which the pattern is cut has area $(13)^2\pi$ sq cm. Now we must determine what fraction of this circle makes up the sector. The fraction is x/360 (where x is the angle formed by the two radii of the sector). And since x/360 = r/s = 5/13, the sector is 5/13 of its parent circle. Therefore, the area of the sector is:

$$* \qquad \frac{5}{13}(13)^2\pi = 5(13)\pi = 65\pi \text{ sq cm}$$

Therefore about 200 sq cm of the paper was required to make the cup.

Notice the equation above marked *. If we substitute r and s for their numerical values, the result is:

$$\text{Area of sector} = \frac{r}{s}(s)^2\pi$$
$$= rs\pi.$$

In reasoning through a particular problem we have derived a very simple formula for finding the area of any sector. This is the way that most mathematics has been developed. In trying to solve a real-world problem, someone notices a pattern or generalization which can be used to solve other, similar problems. Such formulas are convenient and time-saving--if we know them or know where to find them. But please remember, if you understand the concepts and are willing to think, the problems can be solved even though you don't know the formula.

The last figure whose surface area we will consider is the sphere. A sphere is the set of all points which are a given distance (radius) from a given point (center). There is no elementary way to explain why the surface area of a sphere is what it is. We can only give the formula, note that it depends on more sophisticated mathematics than we can discuss here, and ask students to accept it.

Surface Area of Sphere = $4\pi r^2$, where r is the radius.

Homework: Section 44

I.

1. Cone-shaped cups are 7 cm in diameter and 12 cm high. Make a pattern for one of these cups. How much paper is needed to make such a cup?

2. If a cone were made from the pattern at the right, how high would it be? What is its surface area?

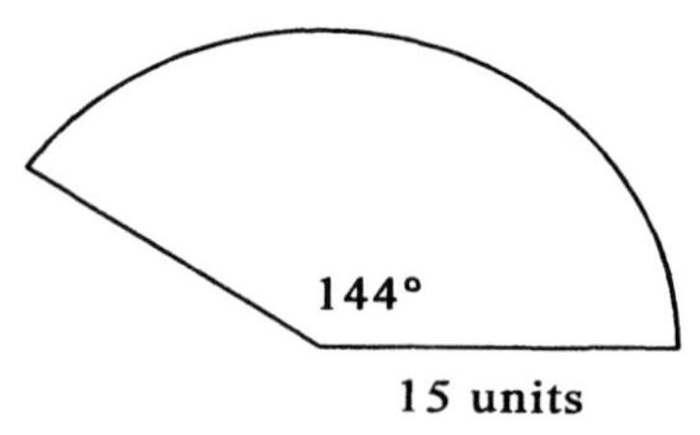

3. A basketball is 30 cm in diameter. What is its surface area?

4. A spherical space balloon has circumference 60π ft and is made from aluminum sheeting. If the sheeting weights 2 oz per square foot, how much does the balloon weigh?

5. A cardboard box is 4 ft long, 2 1/2 ft wide, and 1 1/2 ft high. Cardboard costs .5¢ per sq foot. What is the cost of enough cardboard to make 500 boxes?

6. A cylindrical storage tank is 15 ft high and 20 ft in diameter. The inside of the tank (bottom and sides) is to be coated with a liquid sealer. If a can of sealer covers 100 sq ft, how many cans will be needed?

7. The base of a prism is an equilateral triangle with sides each 2 inches. If the prism is 6 inches long, what is its surface area?

8. Cone-shaped snowball cups are 4 inches high and 6 inches in diameter. How many sq inches of paper is needed to make such a cup?

9. The top of the Superdome is a semi-sphere of radius 200 ft. If a gallon of paint covers 200 sq ft, how many gallons will be needed to paint the top of the Dome?

10. The Pentagon building in Washington, D.C. has 60 ft sides, is 90 ft high, and has 8 stories. The distance from the center of the building to the outside walls is 40 ft. What is the total floor space in the building? If half of the outside walls are windows, how much window space is there in the building?

11. You know that the wheels on your bike are 20 inches in diameter. How could you use the bike to find the distance from Griffin Hall to Maxim Doucet Hall?

12. What is the surface area of a cylinder which is 5 inches high and has a diameter of two inches?

13. A pyramid has a regular hexagonal base 3 cm on each side and slant height 5 cm. How much paper does the pattern use? What is the height of the pyramid?

II.

1. On poster paper (or light cardboard), construct patterns for these figuress:
   a) A prism (open at one base) with square base 6 cm on each side, and 8 cm high.
   b) A pyramid with square base 6 cm on each side, and 8 cm high.
   c) A cylinder (open at one end) with base of diameter 10 cm, and 8 cm high.
   d) A cone with base of diameter 10 cm, and 8 cm high.

2. Fold and tape the patterns at their edges to form the respective three-dimensional figures.

3. Fill the pyramid with table salt and empty it into the prism. Repeat until the prism is full. Record the number of pyramids-full required to fill the prism.

4. Fill the cone with salt and empty it into the cylinder. Repeat until the cylinder is full. Record number of cones-full needed to fill the cylinder.

Section 45: Volume

Recall that volume is a three-dimensional measure; it is given in cubic units or units of liquid capacity. Suppose the pattern shown below were folded and taped to form a rectangular prism (box):

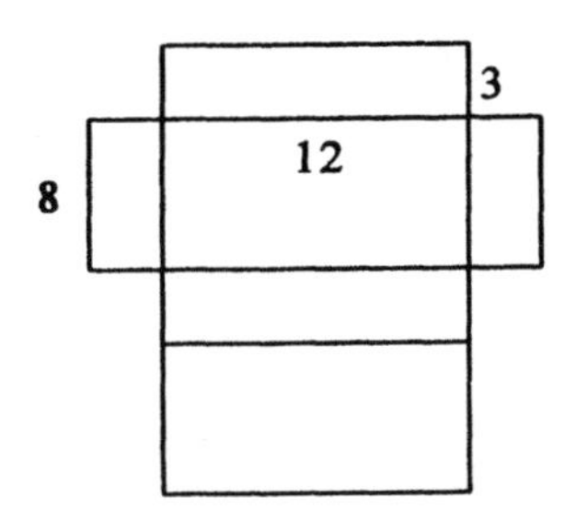

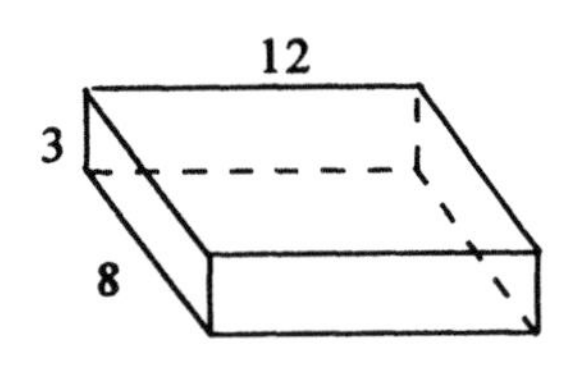

There are numerous properties of the box which can be measured: length is 12 units; width is 8 units; height is 3 units; bottom of the box covers 96 units; 312 units of paper were required to make the pattern. The box would hold 288 cubic units of sand. Length, width, and height are linear measures; area of bottom and total surface area are square measures; and the amount of sand the box can hold is a measure of capacity or volume (cubic).

It is easy to understand how we calculate volume for a rectangular prism. Imagine that you have a supply of blocks which measure 1 unit on each edge. (Hence each block is 1 cubic unit.) Now imagine that you begin filling the box described above by putting a layer of blocks on the bottom. How many blocks would there be in that layer? Clearly, there would be 96--8 rows with 12 blocks in each row. (And this of course represents the area of the bottom of the box--96 square units.) Now, in order to fill the box with blocks, how many layers are needed? Obviously three, since the height of the box is 3 units. Hence the box would hold 96 x 3 = 288 cubic units.

This can be generalized into a formula for finding the volume of any prism or cylinder since all of these figures can be considered as stacks of uniform layers:

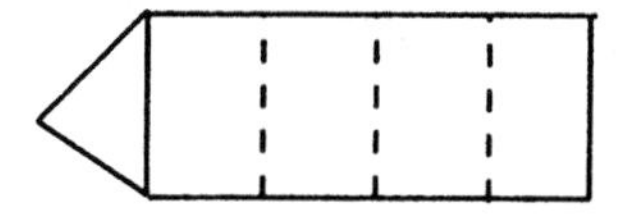

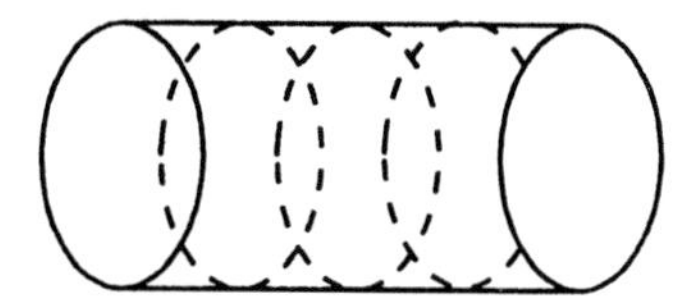

Volume = Area Base x Height

In the last set of exercises you collected some data which suggests a relationship between the volumes of pyramids and prisms of the same base and height and cones and cylinders of the same base and height. Explaining why these relationships exist, or actually deriving formulas for the volumes of cones and pyramids, requires concepts and techniques from calculus, so we will accept them on faith and the evidence from your experience.

$$\text{Volume of Cone or Pyramid} = \frac{1}{3}\ (\text{Area of Base}) \times \text{Height}$$

Deriving or explaining the formula for the volume of a sphere is also beyond the scope of this course, so we accept it, too, on faith:

$$\text{Volume of Sphere} = \frac{4}{3}\pi r^3, \quad \text{where r is the radius}$$

Homework: Section 45

1. Sand costs $30 per cubic yard. A sandbox is 5 ft long, 3 ft wide, and 15 inches high. How much will it cost to fill it with sand?

2. A feeding trough for cattle is shown at right. How many gallons of feed will it hold? (The bases of the prism are isosceles triangles.)

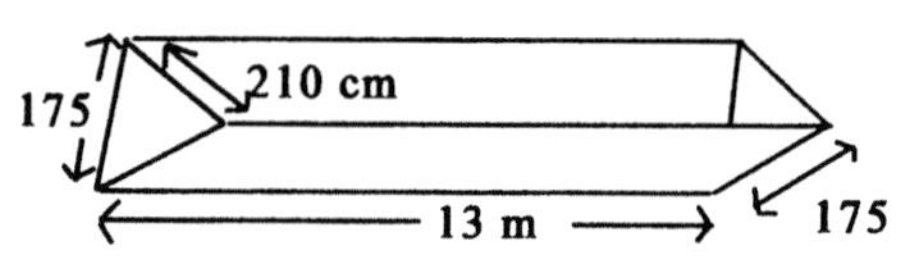

3. The Great Pyramid has a square base 750 ft on each side, and it is 500 ft high. How many cubic feet of stone were used to build it? (Assume the structure is solid stone.) If a company charges 5¢ per sq foot to clean the stone, how much will they charge to clean the entire surface of the Pyramid?

4. Three tennis balls fit exactly into a can. If each ball is 6 cm in diameter, how much empty space is in the can?

5. A cone-shaped cup is 12 cm high and 10 cm in diameter. How many of these cups can be filled from a two-liter bottle of Coke?

6. Copper weighs 4.2 grams per cubic cm. A solid copper sphere is 28 cm in diameter. How much does it weigh?

7. Beer cans are 14 cm high and 6 cm in diameter. When Joe College drinks a six-pack, how many quarts of beer does he consume?

8. A can holds 750 ml of tomato juice, and it is 15 cm high. About how many sq cm of tin were used to make the can?

9. A rectangular aquarium is 75 cm long, 60 cm wide, and 50 cm high. A solid glass ball 40 cm in diameter is lying at the bottom, and the tank is filled with water. How many gallons of water are in the tank?

10. The base of a circus tent is a regular octagon with 36 ft sides. The center pole which holds up the tent is 40 ft and is 30 ft from each side. How many sq ft of canvas were used to make the tent? What is the floor space in the tent?

11. These four boxes are all the same length and all of them have a girth of 36 inches. Which has the greatest volume? Which has the least volume?

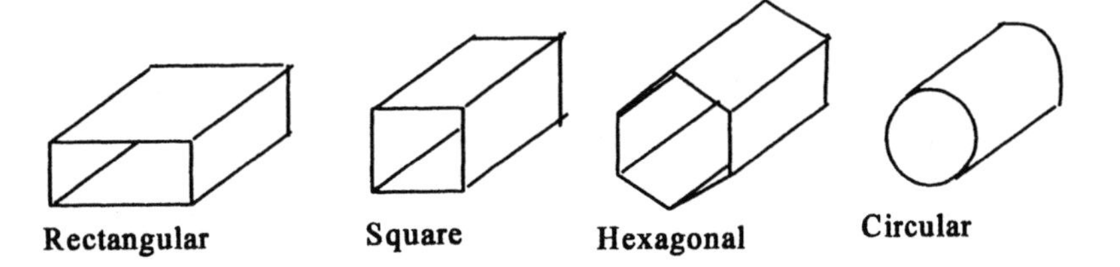

12. ABCD is a parallelogram.
Diameter of circle is 8 units.
Find shaded area.

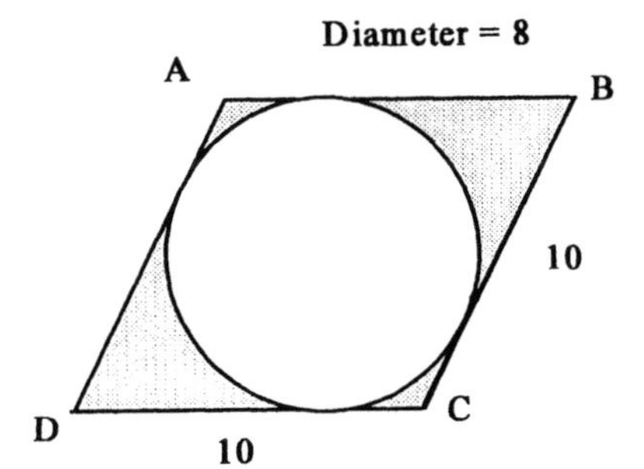

13. RSTQ is a square. Find shaded area.

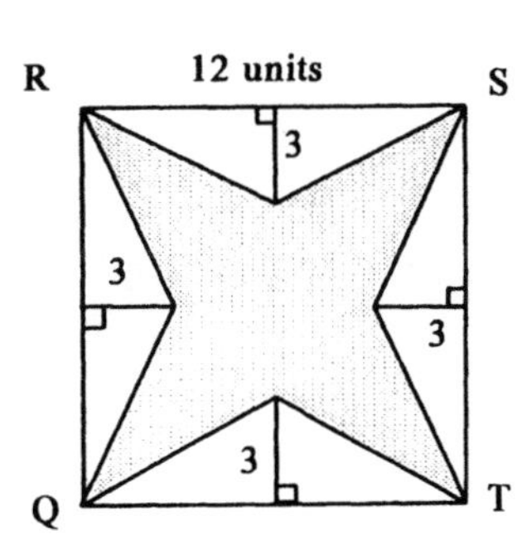

14. ABCD is a trapezoid.
Area ABCD is 50 sq ft.
Find the area of $\triangle$BCD.

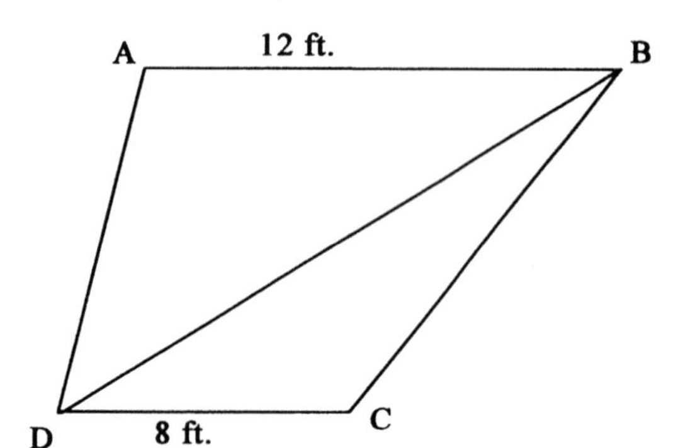

15. Metal washers like the one below are cut from sheets of iron which weigh 8 grams per sq cm. About how much does a washer weigh?

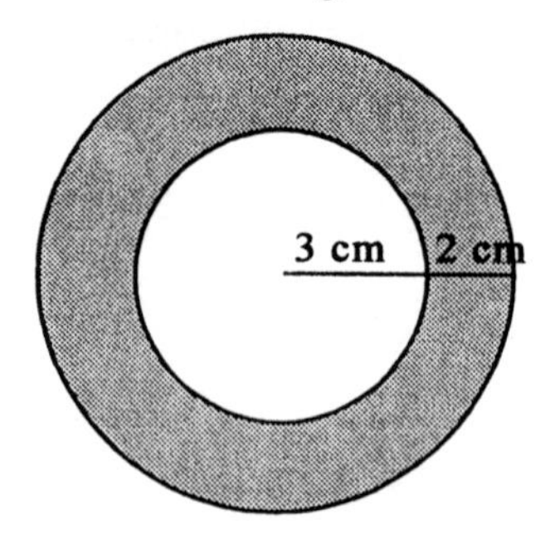

16. A right triangle has legs of 7 cm and 24 cm. What is the length of the altitude from the right angle to the hypotenuse?

17. The spokes of a bicycle wheel are 14 inches. About how many revolutions does the wheel make when the bike goes 100 yards?

18. How much paper would be needed to make a conical cup 8 cm in diameter and 9 cm high? How many fl oz of water would the cup hold?

19. A cylindrical water tower is 12 ft high, 10 ft in diameter, and open at the top. The entire tank, insidea dn out, is to be painted. If a gallon of paint covers 200 sq ft, how many gallons will be needed? How many gallons of water will the tank hold?

20. A piece of cardboard is 5 ft 8 in long and 3 ft 9 in wide. A box is made out of the cardboard by cutting out 6-inch squares from the corners and turning up the sides. Pecans weight 9 pounds per cubic foot. About how much will the box weigh when it is filled with pecans.

21. Make a pattern for a prism which has equilateral triangles with sides of 5 cm as bases, and is 9 cm long. What is the area of the pattern? What is the volume of the prism?

Chapter VI

Section 46: Collecting a Set of Data
Activity 1 - Part A

Figure 1

1)

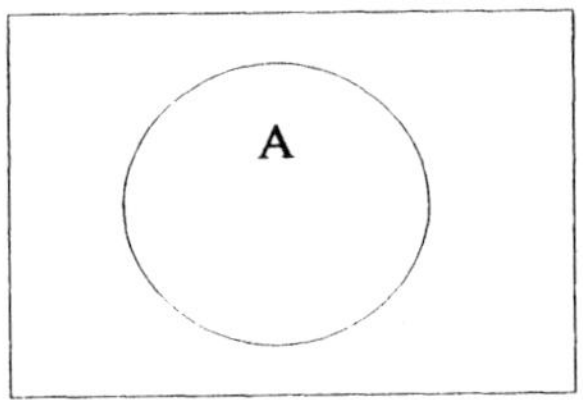

A

A: I ate breakfast today.

2)

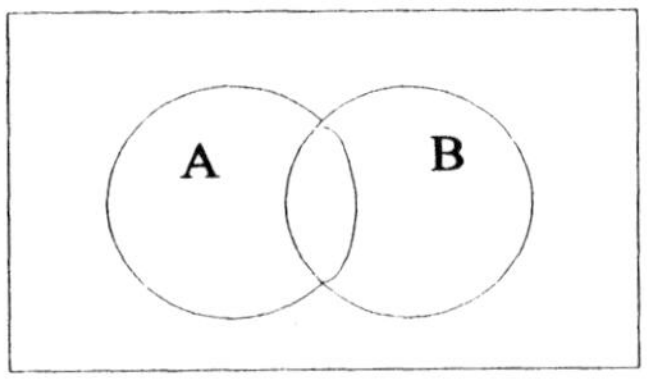

A B

A: I have a brother.
B: I have a sister.

3)

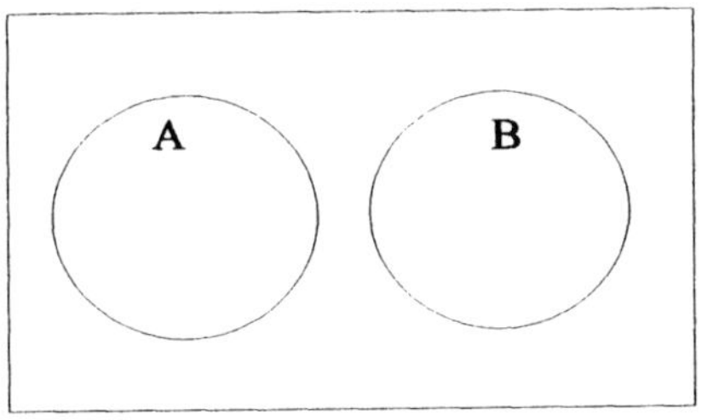

A B

A: My birthday is in May, June, July or August.

B: My birthday is in January, February, March, or April.

4)

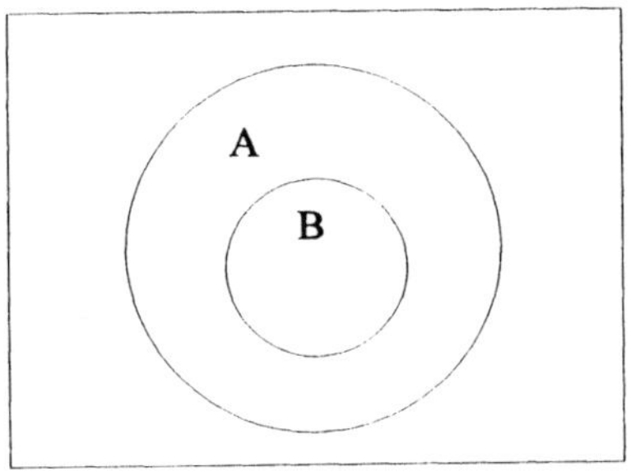

A B

A: I am majoring in Education.

B: I want to teach junior high.

5)

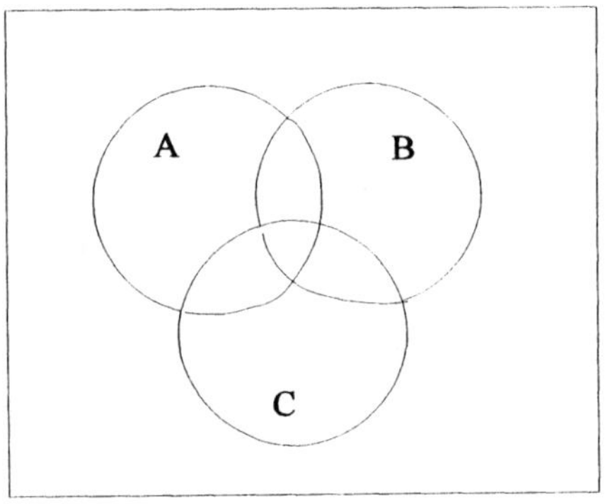

A B C

A: I am less than 22 yrs old.

B: I own my car.

C: I have a job.

6)

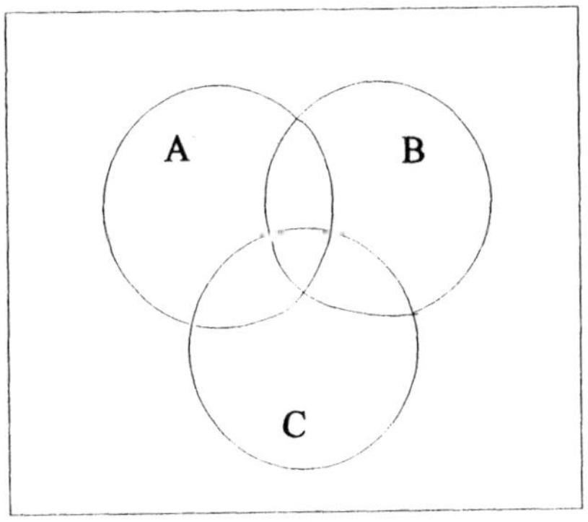

A B C

A: My telephone number ends in an even number.

B: My social security number ends in an even digit.

C: The number of letters in my first name is an even number.

Directions: The diagrams shown in Figure 1 are drawn on the blackboard. Think of each section of the diagram as a room which contains people as described by the headings. Notice that some rooms do not have specified headings but implied headings. Put yourself into the appropriate "room".

Rules: 1) You must sign each diagram.
2) You may only sign a diagram once.

Group Discussion: Discuss the thought process you used in placing yourself into a room. (Did it get easier or harder as you progressed form chart to chart? Why do you think this is happening? How did you know where to put yourself?

Activity 1 - Part B

Use the class diagrams created from Activity 1 Part A indicated to work the following but first change each diagram so that it contains the number of people in each region instead of the names of each person.

Diagram 3:

1) Why are the circles drawn without an intersection?
2) Describe the people not in either circle?

Diagram 4:

1) Could this diagram have been drawn another way?
2) How many people are majoring in Education and want to teach junior high?
3) How many people are majoring in Education and do not want to teach junior high?

Diagram 5:

1) How many people does the diagram represent?
2) How many people are less than 22 years old?
3) How many people own their car?
4) How many people have a job?
5) How many people are younger than 22, own their car and have a job?
6) How many people are at least 22 years old?
7) How many people are at least 22, don't own their car and don't have a job?
8) How many people don't have a job?
9) How many people are younger than 22 or own their car? (Why is this not just the sum of the answers to #2 and #3?)
10) How many people have a job and own their car but are 22 or older?
11) How many people are younger than 22 or own their car but do not have a job?

Diagram 6:

1) How many people have telephone numbers which end in an even digit, social security numbers which end in an even digit and have an even number of letters in their first name?

2) How many people have telephone numbers which end in an even digit and social security numbers which end in an even digit?

3) How many people have telephone numbers which end in an even digit and have an even number of letters in their first name?
4) How many people have social security numbers which end in an even digit and an even number of letters in their first name?
5) How many people have social security numbers which end in an even digit?
6) How many people have telephone numbers which end in an even digit?
7) How many people have telephone numbers which end in an odd digit?
8) How many people have an even number of letters in their first name?
9) How many people have an odd number of letters in their first name?
10) How many people does the diagram represent?

Discussion

Venn diagrams are a type of pictorial representation of the relationship between sets. Venn diagrams are named after John Venn (1834-1923), an Englishman who was the first to put them into general use. In these diagrams, a rectangle is commonly used to represent the "universe" or set of all possibilities for the problem in question. Within the universal set are usually circular shapes representing the other sets of interest. There is no requirement that circles must be used - any closed figure will work.

If the figures within the universe cross, then the two sets contain some of the same objects. For example, in Activity 1, the sets "I have a brother" and "I have a sister" were drawn so that the two circles would intersect forming a third region. Anyone who signed their name in this region should have both a brother <u>and</u> a sister. Remember, Rule 2 states, "You may only sign a diagram <u>once</u>."

Suppose one group's chart looks like this:

Figure 2

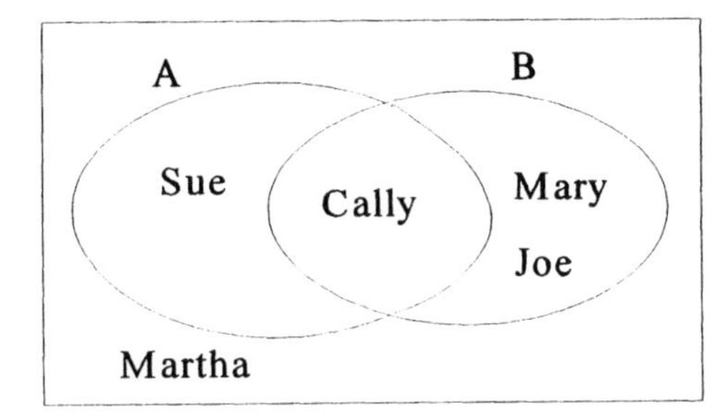

A: I have a brother.

B: I have a sister.

What do we know about this group? Individually, we know that Sue has a brother but not a sister, Mary and Joe each have a sister but no brother, Cally has both a brother <u>and</u> a sister and Martha has no brothers and no sisters. We do not know <u>how many</u> brothers or sisters anyone has except Martha. Mary could have three sisters and still be in set B. Sue could have 9 brothers and still be in set A. Cally could have 2 brothers and 1 sister and yet she is still a member of both sets A and B. However, Martha must be an only child!

As the number of sets of interest in the universe increase, the diagrams become more complicated in appearance and interpretation. The activity stopped at a diagram using three intersecting sets but the numbers of sets that could be used is actually unlimited.

Let's look back at Figure 2. Suppose we want to know how many people in the group have a brother <u>or</u> a sister. Four people fit this description. Sue has a brother, Mary and Joe have a sister and Cally has both. We include Cally because she would answer "yes" to the question, "Do you have a sister?" along with Mary and Joe. In an "or" statement, the minimum requirement is to satisfy one part of the statement. Notice that the new set of Sue, Cally, Mary, and Joe is made from all the members of set A together with all the members of set B.

Most of the time, we are more interested in knowing how many objects are in the sets rather than which objects are in the sets. For this reason, Venn diagrams often contain only the number of objects that are in each region. Thus, Figure 2 is usually drawn as

Figure 3

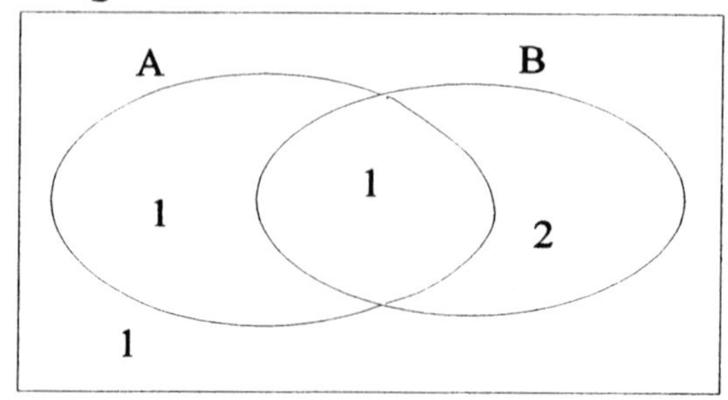

and the number of objects in set A or set B would be 1 + 1 + 2 or 4 as before. Notice, we do not add the 1 in the intersection twice - this 1 represents Cally and she is only one person.

Homework: Section 46

Use a Venn diagram of your own creation to help answer the following. You will need to reverse the processes you used in the previous activities.

1) Jane conducted a survey like the one in a diagram 2. She questioned 45 random people. These are the results she obtained:
   30 people had a brother
   19 people had a sister
   12 people had both a brother and a sister
   a) How many people surveyed by Jane did not have a brother nor a sister?
   b) How many people surveyed by Jane had a sister but not a brother?

2) Marty conducted a survey like the one in diagram 5. He surveyed 50 classmates and obtained the following results:
   17 are less than 22 yrs old
   27 own a car
   32 have a job
   11 are less than 22 and have a job
   7 are less than 22 and own a car
   16 own a car and have a job
   4 are less than 22, have a job, and own a car

   How many of Marty's classmates are 22 or older, don't own a car and don't have a job?

3) Mrs. Morgan surveyed her class of 20 students on whether they liked apples or bananas. Six students did not like either. Seven students liked bananas but not apples and 4 students liked apples but not bananas. How many students liked both apples and bananas?

4) Keith was doing inventory at his tackle shop and recorded the following information on some lures he found in a box on a shelf. Twelve lures had blue glitter on them, 3 lures had blue glitter and green dots, and 1 lure had blue glitter, green dots and red stripes. Nine lures had only green dots and 9 lures had only red stripes. Twenty-one lures had either green dots or red stripes but no blue glitter. Six lures had blue glitter and red stripes but no green dots. There were four lures that were solid gold. If all the lures in the box have been described, how many did Keith count?

5) Look at the two diagrams drawn below.
   a) Write a sentence which describes the relationships shown in each diagram.
   b) Do the diagrams represent different situations or just two ways of drawing the same situation? Explain.

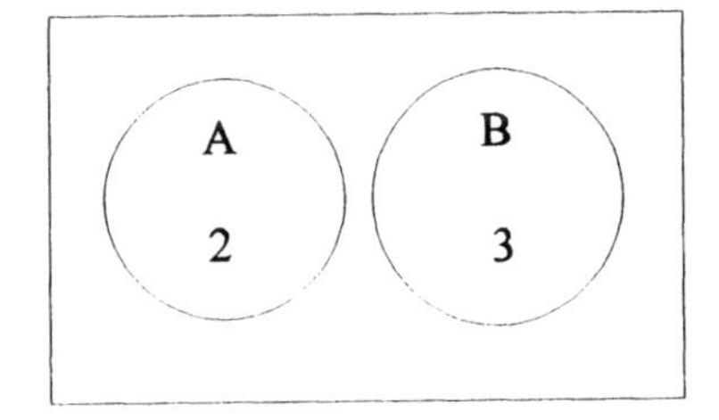

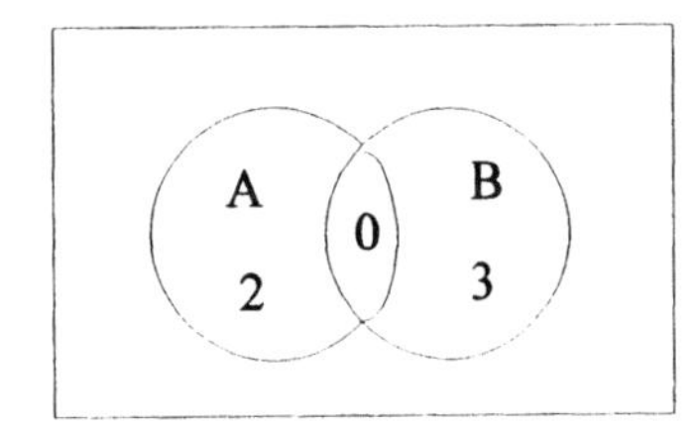

   c) When translating information into Venn diagrams involving 2 circles, the circles can be drawn as intersecting or nonintersecting. Are there any situations which require the use of a particular setup? Are there any situations in which one of the setups would be inappropriate? Give examples or explain.

6) Data collection: Ask 5 **adults** for the following information:
   1) What color are your eyes?
   2) How tall are you?
   3) How many brothers and sisters (total) do you have?

   Bring your results to class tomorrow. You will combine your results with the other members of your group for analysis.

Section 47: Methods of Organizing Data

Data collection and analysis is an important topic in the elementary classroom. Children in kindergarten can collect data quite easily from their own classmates. Examples: What color are your eyes? How many teeth are you missing? Do you like spinach? Notice that the answers to the questions are all different types of responses. It seems appropriate then for there to be different ways to represent the responses pictorially. In this section, we will discuss various pictorial representations of different types of data.

Data which involves several non-overlapping categories is often put into a circle graph or pie chart. For example, with the data collected on eye color, we have one major topic - eye color - separated into many nonintersecting categories - blue, brown, green, hazel, etc. To make a pie chart, one circle is separated into the special categories using percentages of the total population which belong to that category. For example, if 25% of the people surveyed have blue eyes, then the "slice" of the circular pie for blue eyes should represent 25% of the area of the circle. This slice is quite easy to draw since it represents 1/4 of the circle. What happens if we have 15% with blue eyes? Then it is a bit more difficult to separate the circle into a slice which represents 15% of the area. However, a quick review from our geometry course will remind us that the angle we need to draw for this sector is also 15% of a complete revolution - or 15% of 360°. Therefore, if we make a sector with an angle of 54° from the center of our pie chart, we will create a section which will contain 15% of the area of the circle. (See Figure 4)

Figure 4

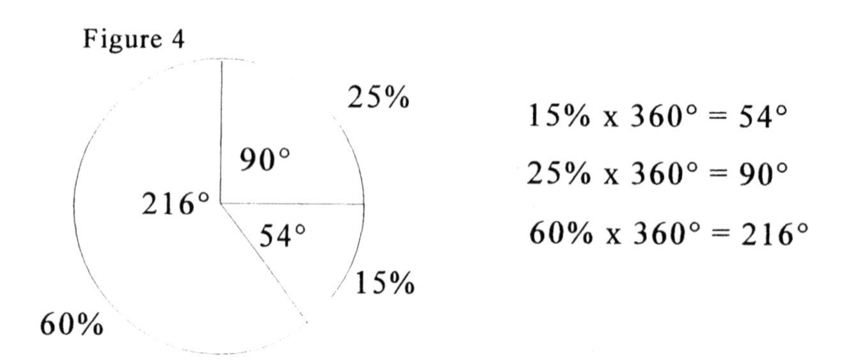

Another way data can be displayed is with a bar graph or histogram. Histograms can be easily constructed if the data is first organized into a frequency table. A frequency table is made by putting the data into categories or intervals and tallying the number of data pieces in each category or interval.

Example 1: A group of 15 students were asked to pick their favorite ice cream flavor - chocolate (C), vanilla (V), or strawberry (S). The responses were:

C C C. V S

S V S C S

V S S C S

Organized into a frequency table, the data would look like:

Flavor	Frequency (f)
Chocolate	5
Vanilla	3
Strawberry	7
	15

Example 2: The day of the month on which of 36 faculty members were born were recorded to determine if any patterns existed. These were the resulting days of the month:

3	9	12	27	28	7	10	18	28
12	27	6	8	12	30	14	14	17
20	25	8	20	30	26	16	18	30
30	11	18	4	6	19	2	12	22

As the number of responses increases, there is a tendency to put the data into intervals to speed up the organizational process. There is no particular rule that must be followed in separating the data into intervals. Generally, there should be between 5 and 20 intervals which cover the entire range of the data. Intervals must be constructed so that no piece of data will fall into more than one interval.

For the days of the month example, a possible frequency table could be:

Day of the Month	Tally	Frequency (f)
1-5	|||	3
6-10	~~||||~~ ||	7
11-15	~~||||~~ ||	7
16-20	~~||||~~ |||	8
21-25	||	2
26-30	~~||||~~ ||||	9
		36

Notice that the width of the intervals (the number of possible days in each) is the same. Also, the largest value (30) is not the largest possible day but the largest value in our data set. The column marked "Tally" is a useful organizational tool when, as in this case, the original data set is not given in numerical order.

From the frequency tables, a bar graph is usually made by putting the interval or categories on a horizontal axis and the frequencies on a vertical axis. Bar graphs can also be made with the intervals on the vertical. Each interval is represented by a rectangle whose height is determined by the frequency.

All rectangles should be the same width since the intervals are also of equal widths. However, this is not a necessity. Below are two bar graphs for the two previous examples.

Both examples represent discrete data or data that is not a measurement. It would not make sense to have the bars on the graphs touch. Data that is continuous, like a measurement, is often shown with the rectangles touching. In those cases, it is impossible for a given data piece to be exactly on the line - remember all measurements are approximate. If, for example, the measurements are given to one decimal place, then the intervals can be made to two decimal places to guarantee that no data is "on the line". A bar graph where the rectangles touch is called a histogram.

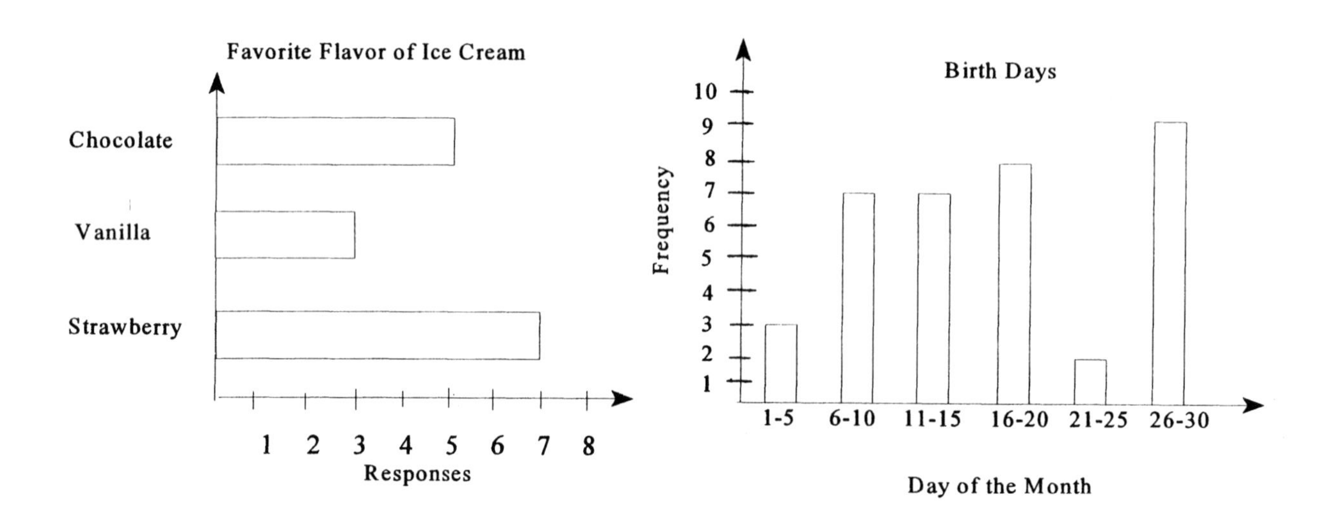

Example 3: The heights, in cm, of Mrs. Gruber's third grade class are as follows:

124.0	120.3	115.7	135.4	126.9
123.8	119.0	121.5	132.8	138.9
120.5	122.1	126.3	130.0	115.2
125.6	131.4	123.7	124.2	128.5
118.8	123.6	127.2	124.9	128.3

A possible frequency table and histogram could be:

Height (cm)	Tally	f
114.95-118.95	|||	3
118.95-122.95	||||	5
122.95-126.95	|||| ||||	9
126.95-130.95	||||	4
130.95-134.95	||	2
134.95-138.95	||	2
		25

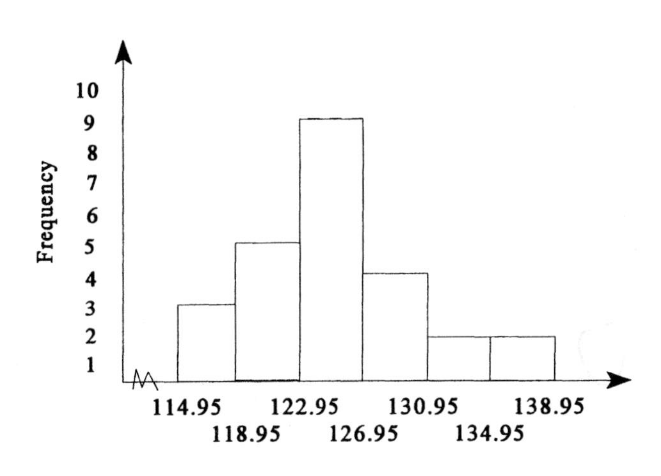

In this case, we make the intervals "touch" in the frequency table so that when we transfer the information to the histogram, the rectangles will also touch. The horizontal axis for continuous data is basically the same as the real number line. Therefore, the broken line at the beginning indicates that we are not starting at the zero for this set of data.

A frequency polygon or broken-line graph is another type of graph that can be made to help organize a set of data. Frequency polygons are often used in newspapers and magazines to report stock prices, trends in consumer prices and other data that evolve over time. Some frequency polygons can be formed from histograms. By joining consecutive midpoints of the tops of the rectangles with line segments and connecting the first and last rectangles to points on the horizontal axis where the next midpoints would appear, a polygon is formed using the horizontal axis as a side. Below is the frequency polygon for the heights of Mrs. Gruber's third grade class. The histogram is shown by the dotted lines but is not normally drawn in the frequency polygon.

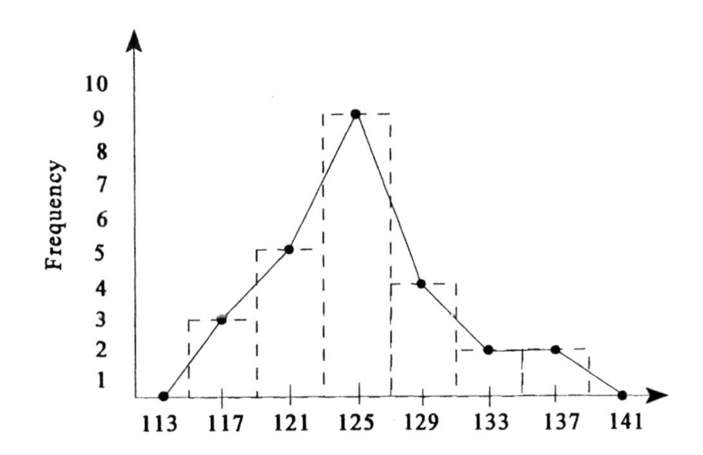

Homework: Section 47

1. Assignment to turn in:
   Cut out or photocopy five different frequency polygons from a current newspaper or magazine. Write a paragraph describing the similarities and differences you find in your graphs.

2. Use the data collected by your group in Section 46 Problem 5:
   a) Construct a pie chart for the data collected on eye color.
   b) Construct a frequency table and bar graph for the data collected on siblings.
   c) Construct a frequency table, histogram and frequency polygon for the data collected on heights.

3. Katie Jackson claims her average in Geography class is 93. Mrs. Jackson knows that Katie has only 4 grades and she remembers that the first three grades were 82, 75, and 85. Mrs. Jackson also knows that Katie's teacher never gives a grade over 100 and never gives bonus points. Katie was grounded for two weeks for lying to her mother. How did Mrs. Jackson know that Katie was lying?

4. 4, 6, 8, 10, ...
   a) If the pattern continues, what could the next two numbers be? Explain how you found them.
   b) Without finding all the ones in between, explain how to find the tenth number in the pattern. (A formula is acceptable.)

5. 2, 4, 8, 16, ...
   a) If the pattern continues, what could the next two numbers be? Explain how you found them.
   b) Without finding all the ones in between, explain how to find the tenth number in the pattern.

6. 1, 3, 5, 7, ...
   a) If the pattern continues, what could the next two numbers be? Explain how you found them.
   b) Without finding all the ones in between, explain how to find the tenth number in the pattern.

7. John and Mary Smith are trying to decide on a girl's name for their unborn child. They will choose a first name and a different middle name. They have a few names that they like.
   a) How many ways can they use both Marie and Kellie in forming a name?
   b) If they like Marie, Kellie, and Donna, how many different way can they name their daughter using two of these?
   c) How many ways can they use two of the three (Marie, Kellie, and Donna) if they don't care which is the first name and which is the middle name?

Section 48: The "Center" of the Data - Part I

Results of opinion polls and surveys are not always given in a pictorial manner. Often, it is more convenient to report the "average" response. Government surveys generally report on the "average" family. Communities looking for potential residents report the "average" cost of a new home in their area. Companies seeking new employees will advertise the "average" salary paid for a particular position. Teachers report grades for students using an "average." Many questions begin to arise. Are these averages all the same sort of number? Do we compute these averages in the same way? Is there a different meaning attached to different averages? If the averages are different, how do we know which one to use?

For most students, an average is simply the grade they make in a class. Unfortunately, this is but a small part of what an average is all about. In fact, the Merriam-Webster Dictionary defines average in several ways including "a ratio of successful tries to total tries, "being about midway between extremes", and "being not out of the ordinary."

To report an average for a given set of data, we try to assign to that data a single value that will best describe the overall flavor of the data set. In this sense, the average will not be an extreme value - very high or low, but somewhere in the middle. The average tries to level out the data. Previous discussions of average have shown a concrete way to represent this leveling process to younger children. Example 1 should refresh your memory.

Example 1: Five students each own a different number of CD's. What is the average number of CD's owned by these students?

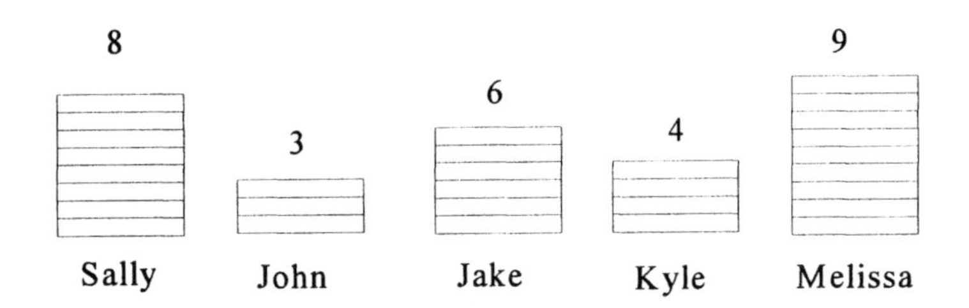

To level out the stacks, we first approximate what the middle (average) value should be. Suppose we use 5 as our average. Then, we should take a few CD's from the stacks that have more than 5 and put them in the smaller stacks. There are many ways in which this can be done. For this example, let's take 2 CD's from Sally's stack and put them in John's stack. That will leave Sally with 6 CD's and John with 5 CD's. Then, we can put 1 CD from Jake's stack into Kyle's stack leaving them both with 5 CD's.

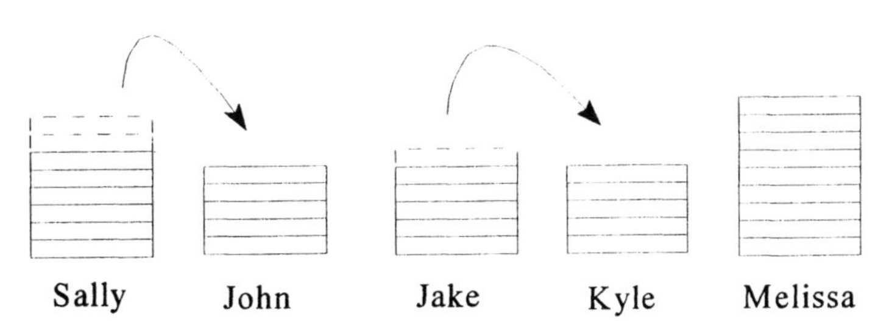

So now our stacks look like this:

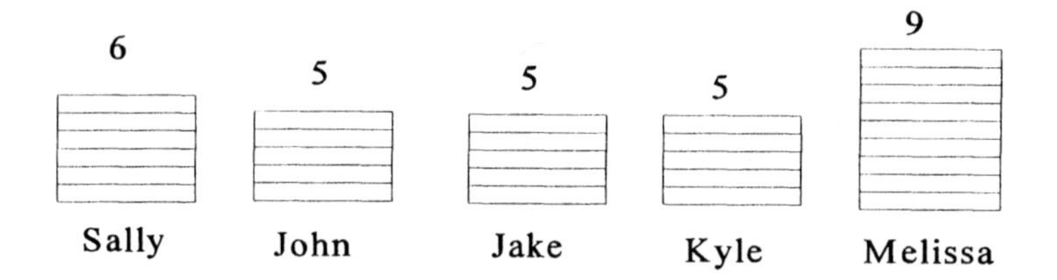

At this point we should be able to see that our estimate of 5 is too low. Melissa and Sally still have over 5 CD's while all the others have exactly 5. So, we try to increase our estimate. Since these numbers are small, we might only increase our estimate by 1 or 2 CD's. Could we level out the stacks so that there are 6 CD's in each stack? Yes, look at Melissa's stack. She has 3 CD's over the estimate of 6. There are 3 stacks that are under the estimate by 1 CD. So, share Melissa's extras with John, Jake, and Kyle.

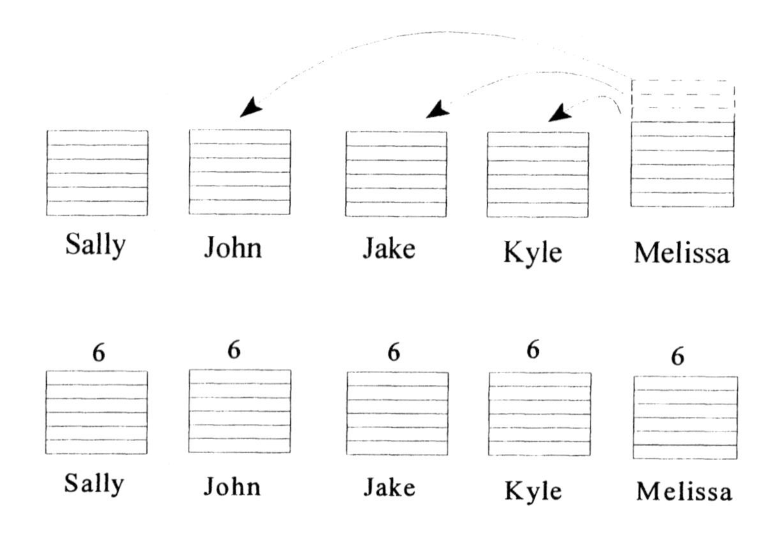

The final result is that all the stacks have 6 CD's in them. Therefore, the average number of CD's owned by these students is 6 CD's.

There are actually three different types of averages that can be used to describe the center of a set of data. The first is the arithmetic average or the mean. The mean is generally what most people think of when someone says "average." To find the mean number of CD's, we add the number of CD's that each student owns and divide the total by the number of students. In general, add the data values together and divide by the number of values.

Thus, the mean number of CD's owned is: $\frac{8+3+6+4+9}{5} = \frac{30}{5} = 6$ So in our concrete example, we actually found the mean. We use the symbol $\bar{x}$, read as "x-bar," to represent the mean.

Example 2: Caroline spent 4 hrs cleaning her room on Monday, 2 hrs on Tuesday, 1 hr on Wednesday, 5 hrs on Thursday, and 30 min on Friday. How many hours a day did she average cleaning her room?

One important note about averages; the data must all be the same type of measurement or number. (Why?) Therefore, before we can compute Caroline's average, we must change her 30 min to ½ hr. So, the mean or average number of hours a day Caroline spent cleaning her room is:

$$\bar{x} = \frac{4+2+1+5+\frac{1}{2}}{5} = \frac{12\frac{1}{2}}{5} = 2\tfrac{1}{2} \text{ hrs}$$

The second type of average is called the <u>median</u>. The median is the value that is physically in the middle of a set of numbers when the numbers are put in order from smallest to largest (or largest to smallest). The median separates the data into two equal sets - exactly half are less than or equal to the median and exactly half are greater than or equal to the median. For example, the data on Caroline's hourly cleaning would be ordered as

½ 1 2 4 5

and since 2 is directly in the middle, 2 hrs would be her median time. There is no common symbol for the median.

Example 3: The batting average for nine players on a baseball team are listed below. Find the median batting average for the team.

0.333 0.250 0.125 0.111 0.228

0.125 0.286 0.300 0.250

First, the numbers need to be reordered.

0.111 0.125 0.125 0.228 0.250 0.250 0.286 0.300 0.333

Then, the value that is in the middle of the set, 0.250, can be obtained. Notice that values that occur more than once must be repeated in the list. This is true for any averaging process.

Example 4: Milo completed four rounds of golf with the following scores: 85, 80, 89, 78. Suppose Milo wants to give his average using the median. First, he orders the scores.

78 80 85 89

Then, he notices that there is no score which falls in the middle. What should he do?

Milo's problem will occur every time there is an even number of data values. To avoid saying there is no median, in these cases, the median will be the arithmetic average or mean of the two numbers which are in the middle. In Milo's case, the mean of 80 and 85 is 82.5. Therefore, we say that the median score is 82.5.

As the size of the data set increases, it becomes difficult to find the middle value just by looking. For that reason, we will use a shortcut. If there are **n** pieces of data, then the median will be in the $\frac{n+1}{2}$th position when the data are ordered. Notice that for an even value of **n**, this shortcut will yield a fraction. The fractional position is where the arithmetic average of the two middle positions falls. For example, with four data pieces, the median's position is $\frac{4+1}{2}$ or 2 ½; indicating you need to find the arithmetic average of the values in position 2 and position 3.

The third type of average that we will discuss is the mode. The mode is simply the value in the data set that occurs the most often.

Example 5: Karen has five test scores: 57, 64, 64, 82, 90. The mode of this set of scores is 64 because it appears more than any other value.

Obviously, as in example 5, this type of averaging could lead to several problems. What happens if no value is repeated, like Milo's golf scores? Or what if more than one value is repeated, like in the batting averages? Convention has determined that if no value is repeated then there is no mode. If more than one value is repeated the same number of times, then there is more than one mode. So in the case of the batting averages, there would be two modes -- 0.125 and 0.250.

Now that all three types of averages have been discussed, the question is "Which one do I use?" The answer is not a simple one. In many cases, all three measures will be basically the same value so it would not matter. In other cases, there will be a great difference in the values. Then there are cases where the data collected is not numerical in matter but simply categorical - how do you find the mean of a favorite flavor of ice cream? But you could report the flavor most often chosen; i.e., the mode. In other words, whichever value you choose to use must have meaning! Typically, the mean is used in reporting average grades, ages, scores, heights, and other measurements. The median is more commonly used in average incomes, home prices, and car prices. (Why do you think this is so?) The mode is used mainly for categorical data where high or low frequency of response is most important.

Homework: Section 48

1. The LA State Field Archery Association held its annual indoor shoot recently. A perfect round has a score of 300. The scores of all male competitors for one round are given.

300	300	300	300	300	299	299	298	296	293
300	297	292	299	297	299	299	298	299	290
284	284	281	279	276	273	299	298	300	300
297	278	295	297	288	288	298	292	299	295
281	281	285	293	262					

   a) What was the modal value for this set of data?
   b) Find the mean and median of these scores. Which value seems most appropriate to use? Why?

2. Use the data collected by your group in Section 46 Problem 6 to answer the following. (Use the "most appropriate" measure in your response.)
   a) What is the "average" height of the people surveyed?
   b) What is the "average" number of siblings?
   c) What is the "average" eye color?
   d) Write a few sentences explaining which measure you used and why.

3. If the mean, median and mode are all equal for the set $\{90, 130, 100, 80, K\}$, then what is the value of K?

4. Must any of the average measures - the median, the mean, or the mode - be an actual data piece? If so, which one(s)?

5. ABC Printing, a small print shop, has seven employees. Their individual yearly salaries are $16000, $50000, $16000, $25000, $18000, $50000, and $16000.
   a) Find the average salary for the company using each of the three measures.
   b) Which best describes the "center" of this data?
   c) Which would you be most interested in if you were an employee of this company?

6. The sum of nine consecutive even numbers is 180. What is the result when the mean of these numbers is subtracted from the median?

7. a) Find the mean of these test scores: 62 74 74 63 89 74 62 74
   b) Notice that $\dfrac{62(2) + 63(1) + 74(4) + 89(1)}{8}$ will give the same mean.

   Will this process work every time? Explain why or why not.

8. 0, 3, 6, 9, ...
   a) If the pattern continues, what could the next number be? Explain how you found it.
   b) Without finding all the ones in between, explain how to find the ninth number in the pattern.

Section 49: The "Center" of the Data - Part II

As our discussion on data analysis proceeds it will become increasingly necessary to use more and more mathematical notation. We use these notations or symbols merely as a form of shorthand. The first symbol we need is the Greek letter sigma Σ. Mathematically, Σ means summation. So Σx means to total all the x values. When calculating a mean, each data item is considered to be a value of x. Let n be the number of items of data and then

$$\bar{x} = \frac{\Sigma x}{n}.$$

Technically we should write Σx as $\sum_{i=1}^{n} x_i$ where each value of x has it's own subscript. But then the notation becomes more complicated than the process and we are trying to make things less complicated.

Every example or homework problem shown thus far on averages has been with data that is not in a frequency table. This type of data is called ungrouped data. If the data is organized in a frequency table, then it is called grouped data. Suppose we want to find the average amount of time spent waiting in line at the supermarket but the only information we have is given in the form of a frequency table:

Time (min)	f
0.0-3.0	4
3.0-6.0	12
6.0-9.0	30
9.0-12.0	5
12.0-15.0	3
15.0-18.0	1
18.0-21.0	2

How can we find a mean? We don't know the original x's. The same problem arises in trying to determine a median or a mode. By condensing the data, we have lost some information - the individual values. To remedy this situation, we will use a representative value of x for each interval. The value we use is the midpoint of the interval or the mean of the upper and lower endpoints of the interval to represent the x's. Our new table looks like this:

Time (min)	f	x
0.0-3.0	4	1.5
3.0-6.0	12	4.5
6.0-9.0	30	7.5
9.0-12.0	5	10.5
12.0-15.0	3	13.5
15.0-18.0	1	16.5
18.0-21.0	2	19.5

Before we calculate the mean, look closely at the table. The frequency column indicates how many data pieces are in a particular interval. For example, there are 30 pieces of data in the interval 6.0 - 9.0 min. In order to find the mean, we cannot just add one value of 7.5. Likewise, there is only 1 value in the interval 15.0 - 18.0 that needs to be considered. If the mean is to be an accurate representation of the average, then we need to adjust our calculations appropriately. Instead of one value of 1.5, we need 4; instead of one 4.5, we need 12 and so on. Therefore, since multiplication is a shortened version of repeated addition, we can multiply each x by its frequency to get a total for each interval. Then, we can add these totals to get the final "sum of the x's."

Therefore, the formula for the mean of a set of grouped data becomes $\bar{x} = \frac{\Sigma fx}{n}$. Notice that n is the same as the sum of the f's.

Time (min)	f	x	fx
0.0-3.0	4	1.5	6
3.0-6.0	12	4.5	54
6.0-9.0	30	7.5	225
9.0-12.0	5	10.5	52.5
12.0-15.0	3	13.5	40.5
15.0-18.0	1	16.5	16.5
18.0-21.0	2	19.5	39
Total	57		433.5

$$\bar{x} = \frac{433.5}{57} \approx 7.6$$

Note: If at all possible, do not round until the very last step.

Because of the problem of not having all the information, the median and mode are not normally found when dealing with grouped data.

Homework: Section 49

1. Find the mean of the data from each of these previously discussed problems:
   a) The height of Mrs. Gruber's third grade class grouped data (Example 3, Section 47).
   b) The number of siblings ungrouped data (Problem 2, Homework: Section 47).
   c) The archery scores (Problem 1, Homework: Section 48). This data needs to be grouped prior to calculation. Start the first interval at 260.5 and make the intervals 10 units wide.

2. If the mean of x and y is 0 and y > 0, what is the relationship between x and y?

3. The mean of three numbers is 11. If one of the numbers is 9, what is the sum of the other two?

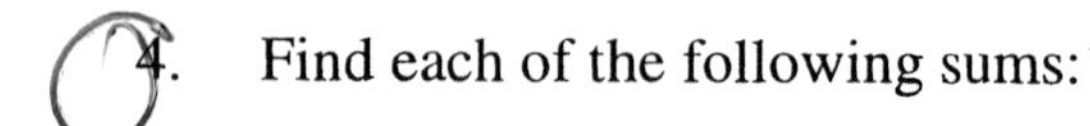

4. Find each of the following sums:

   a) $\sum_{n=1}^{10} 2n$ b) $\sum_{n=0}^{9} (2n+1)$ c) $\sum_{j=2}^{6} \frac{120}{j}$

5. John is looking for a job and hears that the mean monthly salary at a company is $1620. Upon further investigation, he discovers that the actual monthly salaries for employees are $960, $960, $960, $960,$960, $980, $2700, $2700, and $3400. Does the mean salary give John an idea of what his salary would be? Is there another number which would be better for John to know?

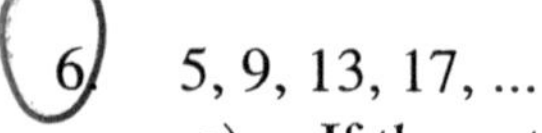

6. 5, 9, 13, 17, ...
   a) If the pattern continues, what could the next number be? Explain how you found it.
   b) Without finding all the ones in between, explain how to find the twentieth number in the pattern.

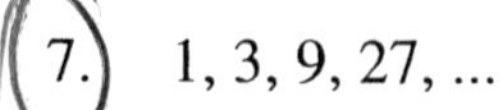

7. 1, 3, 9, 27, ...
   a) If the pattern continues, what could the next two numbers be? Explain how you found them.
   b) Without finding all the ones in between, explain how to find the ninth number in the pattern.

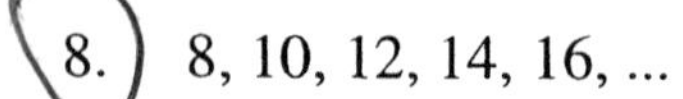

8. 8, 10, 12, 14, 16, ...
   a) If the pattern continues, what could the next two numbers be? Explain how you found them.
   b) Without finding all the ones in between, explain how to find the twelfth number in the pattern.

9. Lou owns a diner in which most of the tables are square and seat one person on a side. If he puts two tables together, he can seat six people. If three tables are placed end-to-end, eight people can be seated and so on.
   a) How many people can be seated at a table created from six square tables?
   b) How many people can be seated at a table created from t square tables?
   c) Can you come up with a rule to help Lou figure out how many tables he needs to seat n people?

Section 50: The "Spread" of the Data

Mr. Carlotti is in the process of choosing a student to represent his English class at the literary rally contest. He normally chooses the student with the highest average in the class. This year, there are three students with the same high average. Here are their grades on five tests.

Mark	87	99	99	100	95	$\bar{x} = \frac{480}{5} = 96$
Keisha	95	98	96	96	95	$\bar{x} = \frac{480}{5} = 96$
Ryan	100	100	92	92	96	$\bar{x} = \frac{480}{5} = 96$

Who should Mr. Carlotti choose? Why?

Mr. Carlotti's problem is a very common one. The statistic normally used does not provide enough information. In cases such as this, it is sometimes more useful to have information based on the consistency of performance or variability of the data. Mr. Carlotti would be wise to send the student whose grades fluctuate the least, i.e. Keisha.

Mathematically, there are a few measures which can be calculated to indicate amounts of variability or spread. The simplest measure of spread is called the range. The range is the difference between the highest and lowest values in the data set.

Example 1: Find the range of the scores for each student.

Mark: Range = 100 - 87 = 13 points
Keisha: Range = 98 - 95 = 3 points
Ryan: Range = 100 - 92 = 8 points

A lower range should indicate less spread. However, the range only deals with the extreme values and does not always give a true picture of the data.

A better measure to use when discussing variability of data is the standard deviation. The standard deviation is the average of each data point's distance or deviation from the mean of the data set. The formula for computing the standard deviation looks very complicated but really is just another version of an average. First, we need a way to compute deviation from the mean. Let x represent a data point and $\bar{x}$ represent the mean of the data set. Then $x - \bar{x}$ would be the amount of deviation. If this subtraction were calculated for each x in the set, then the average difference could be computed quite easily. However, look at what happens when this process is applied to Mark's grades.

Mark:

x	$x - \bar{x}$
87	87 - 96 = -9
99	99 - 96 = 3
99	99 - 96 = 3
100	100 - 96 = 4
95	95 - 96 = -1

The average of the deviations would be

$$\frac{-9+3+3+4+(-1)}{5} = 0$$

This will occur for every data set simply because of the basic definition of the mean. (Think about the leveling process). So, to remedy this situation, it is necessary to eliminate the positives and negatives. We don't really care in which direction the deviation occurs, we simply need to know the amount. One way of getting rid of the signs is to square each value. Then, the average of the squared deviations can be obtained.

Mark:

$(x - \bar{x})^2$
$(-9)^2 = 81$
$(3)^2 = 9$
$(3)^2 = 9$
$(4)^2 = 16$
$(-1)^2 = 1$

$$\frac{81+9+9+16+1}{5} = \frac{116}{5} = 23.2$$

The problem now is with the units on this result. The value obtained is 23.2 squared points. Squared points are not a useful unit in the context of this problem. However, in some problems, the value obtained at this step is useful. Therefore, it has been given a name -- the variance. For most problems, we need to convert our variance back into unsquared units. How do we do this? What is the relationship between y and y^2 ? $\sqrt{y^2} = |y|$. So, if we look at the square root of the variance, we should have a value which is in the desired units. (Do we need to worry about absolute value for our problem?) The new value is called the standard deviation and is denoted by s.

Mark: $s = \sqrt{23.2} \approx 4.8$ units

Note: Since the standard deviation is denoted by s, the variance is s^2.

Example 2: Compute the standard deviations for Keisha and Ryan.

Keisha:

x	$x - \bar{x}$	$(x - \bar{x})^2$
95	95-96 = -1	1
98	98-96 = 2	4
96	96-96 = 0	0
96	96-96 = 0	0
95	95-96 = -1	1

$$s^2 = \frac{1+4+0+0+1}{5} = \frac{6}{5} = 1.2$$

$$s = \sqrt{1.2} \approx 1.1$$

Ryan:

x	$x - \bar{x}$	$(x - \bar{x})^2$
100	100-96 = 4	16
100	100-96 = 4	16
92	92-96 = -4	16
92	92-96 = -4	16
96	96-96 = 0	0

$$s^2 = \frac{16+16+16+16+0}{5} = \frac{64}{5} = 12.8$$

$$s = \sqrt{12.8} \approx 3.6$$

Now, if we compare standard deviations, we see that Keisha's scores have the lowest average deviation with Ryan second and Mark third. The calculation of the standard deviation does require a bit of time. We can shorten it slightly by writing the algorithm as a formula.

$$\sqrt{\frac{\sum (x - \bar{x})^2}{n}}$$

The formula shown here is normally used when the data collected represents the entire population and not a sample of the population. Keisha's grades are her population of grades, but if we looked at the entire class as the population, then Keisha's grades would be a sample. When a sample is used, we want to be more conservative in our results. A smaller denominator will produce a slightly higher value of s. The formula for standard deviation of a sample is

$$s = \sqrt{\frac{\sum (x - \bar{x})^2}{n - 1}}$$

To maintain the difference between populations and samples, statisticians use μ (mu) and σ (sigma) for population statistics and $\bar{x}$ and s for samples.

	Sample	**Population**
Mean	$\bar{x} = \dfrac{\sum x}{n}$	$\mu = \dfrac{\sum x}{n}$
Variance	$s^2 = \dfrac{\sum (x - \bar{x})^2}{n - 1}$	$\sigma^2 = \dfrac{\sum (x - \bar{x})^2}{n}$
Standard Deviation	$s = \sqrt{\dfrac{\sum (x - \bar{x})^2}{n - 1}}$	$\sigma = \sqrt{\dfrac{\sum (x - \bar{x})^2}{n}}$

Note: Notation varies on calculators but many brands use σ_n and σ_{n-1} to distinguish between the two standard deviations where σ_n denotes a population and σ_{n-1} denotes a sample.

The formulas developed before are used for ungrouped data. As with the formula for the mean, when the data is given in the form of a frequency table, the formulas must be adjusted. Look at the calculation for the variance of Ryan's grades. Since he had two 100's, we see the deviation 100 - 96 = -4 twice. So, we could have written $2(100 - 96)^2$. Likewise for 92, we get $2(92 - 96)^2$. The contribution for each data value x is $f(x - \bar{x})^2$ where f is the frequency of that value. So, for grouped data, we need to include the frequency for each x since it represents the entire interval. Therefore,

$$s = \sqrt{\frac{\sum f \cdot (x - \bar{x})^2}{n - 1}}$$

Example 3: The weights of twenty newborn babies at County General Hospital are given in the frequency table. Find the mean and standard deviation of the weights of the newborn babies at County General.

Weight (oz.)	f
80.0-96.0	1
96.0-112.0	1
112.0-128.0	5
128.0-144.0	8
144.0-160.0	3
160.0-176.0	2
	20

Solution: The two formulas we will use are $\bar{x} = \dfrac{\sum f \cdot x}{n}$ and $s = \sqrt{\dfrac{\sum f \cdot (x - \bar{x})^2}{n - 1}}$

We will adjust the frequency table so that all necessary headings are included and then fill in the table. (All steps are shown here but some can be eliminated.)

Weight (oz)	f	x	fx	$x - \bar{x}$	$(x - \bar{x})^2$	$f(x - \bar{x})^2$
80.0-96.0	1	88	88	88-133.6 = -45.6	$(-45.6)^2 = 2079.36$	(1)(2079.36) = 2079.36
96.0-112.0	1	104	104	104-133.6 = -29.6	$(-29.6)^2 = 876.16$	(1)(876.16) = 876.16
112.0-128.0	5	120	600	120-133.6 = -13.6	$(-13.6)^2 = 184.96$	5(184.96) = 924.8
128.0-144.0	8	136	1088	136-133.6 = 2.4	$(2.4)^2 = 5.76$	8(5.76) = 46.08
144.0-160.0	3	152	456	152-133.6 = 18.4	$(18.4)^2 = 338.56$	3(338.56) = 1015.68
160.0-176.0	2	168	336	168-133.6 = 34.4	$(34.4)^2 = 1183.36$	2(1183.36) = 2366.72
	n = 20	$\sum$fx = 2672				$\sum f(x - \bar{x})^2 = 7308.8$

$$\bar{x} = \frac{\sum f \cdot x}{n} = \frac{2672}{20} = 133.6 \text{ oz.}$$

$$s = \sqrt{\frac{\sum f \cdot (x - \bar{x})^2}{n - 1}} = \sqrt{\frac{7308.8}{19}} = \sqrt{384.67368\ldots} \approx 19.6 \text{ oz.}$$

Homework: Section 50

1. Using the grouped archery scores (Homework: Section 49, problem 1c), find the standard deviation of the scores.

2. If all the data values in a set are equal, what is the mean and standard deviation of the set?

3. If every value in a data set is decreased by 10, what happens to the mean and standard deviation of the data set?

4. Hilda surveyed 8 seniors to determine how many pairs of shoes each owned. These are the results: 5, 7, 12, 10, 12, 9, 5, and 4. Find the range and standard deviation for the number of pairs of shoes.

5. Rosco needs to buy a new machine which grinds fishing points. He has three potential machines available to try -- Machine A, Machine B and Machine C. Each costs the same and has the same service contract. Rosco decides to test the machines by running batches of 1000 points through each. The results of the testing are given in the table. Which machine should Rosco buy? (Show some mathematics to back up your answer.)

Number of Defects per Batch of 1000						
	Runs					
Machine	1	2	3	4	5	6
A	5	6	3	1	2	5
B	4	4	4	4	2	4
C	1	2	2	5	5	7

6. The test scores for Professor LeBlanc's history class are shown in the frequency table. Find the mean and standard deviation of the scores:

Test Score	90 - 99	80 - 89	70 - 79	60 - 69	50 - 59	40 - 49
Frequency	3	16	13	7	11	8

7. Results from a data collection activity show a mean of 32.6 and a standard deviation of 3.48 for a set of measurements.
   a) What measurement is one standard deviation above the mean?
   b) What measurement is one standard deviation below the mean?
   c) What range of measurements are within two standard deviations of the mean?

8. Becky had both a history test and a geometry quiz on Friday. She made an 83 on the history test and a 16 on the geometry quiz. The history class had a mean score of 75 with a standard deviation of 5. The geometry class had a mean score of 12 with a standard deviation of 3. Relative to the class scores, on which test did Becky do better? Explain.

9. 11, 7, 3, -1, ...
   a) If the pattern continues, what could the next two numbers be? Explain how you found them.
   b) Without finding all the ones in between, explain how to find the fifteenth number in the pattern.

10. 6, 12, 24, 48, ...
    a) If the pattern continues, what could the next two numbers be? Explain how you found them.
    b) Without finding all the ones in between, explain how to find the ninth number in the pattern.

Section 51: Standardized Scores

In the previous sections, we looked at two types of statistical measures -- measures of central tendency (mean, median, and mode) and measures of dispersion (range, variance and standard deviation). In some instances, such as standardized test scores, the measures reported do not fit into either of these two categories. Many standardized test scores are given as a measurement called a percentile. A percentile is one form of measure which deals with the position of a score if all the data are put in some order. Other measures of this type are quartiles and stanines. For each of these measures, the entire data set is divided into sets containing an equal number of elements. Scores are then given a rank based on the number of scores which are less than or equal to the score in question.

For example, if Martin scores in the 95th percentile on a given test, then about 95 percent of the other scores on the test are less than or equal to Martin's score. It also means that about 5% of the other scores are greater than or equal to Martin's score. Caution: Martin's 95th percentile score does not mean that Martin score 95% on the test.

Quartile ranks are a condensed version of the percentile ranks. The 25th percentile is the same as the first or lower quartile, the 50th percentile is the middle quartile and the 75th percentile is the upper or third quartile. If Kelly scores at the middle quartile, then she scored better than about 50% of the other students.

Example 1: Sixteen applicants for a military job were tested. Their scores were 53, 45, 45, 20, 39, 25, 42, 45, 45, 62, 57, 42, 42, 38, 52, 65. Bill's score was 45. What is Bill's percentile rank?

First, the scores must be put in numerical order:
20, 25, 38, 39, 42, 42, 42, 45, 45, 45, 45, 53, 57, 62, 62, 65

Since there are some scores that are the same as Bill's, by convention, half of those scores will be counted in the less than or equal to percentile and half will be in the greater than or equal to percentile. So, there are 7 scores lower than 45, 4 scores the same and 5 scores greater. We take half of the 4 equal scores, 2, and add to the 7 lower for a total of 9. Then, 9/16 = 56.25%. So, Bill's score is at about the 56th percentile.

What is the difference between the median and the 50th percentile? The difference is rather subtle. The median is the score which is exactly in the middle of the data set. The 50th percentile is the score which is greater than or equal to 50% of the data. Try to think of some cases where these values could be different.

Homework: Section 51

1. A skills test was administered to 20 pre-school students. These were their scores:
   9 3 4 5 5 12 7 12 12 7
   8 6 3 2 9 10 4 10 12 4
   a) What percentile does a score of 7 represent?
   b) What percentile does a score of 4 represent?
   c) What percentile does a score of 12 represent?
   d)What scores represent the first and third quartiles?

2. Teachers do not usually need to calculate percentiles but instead need to be able to interpret percentiles. Page 253 shows two real Stanford Achievement score sheets. Use them to answer these questions:

   a) Student A scored 41 out of 44 on the Computation section which relates to a score in the 95th percentile nationally. Student B scored 41 out 44 on the Language section which relates to only the 86th percentile nationally. Why is there a difference?
   b) Looking at the grade equivalence column and the national grade percentile bands for Student B, what would you tell the parents of this student about the results of this test? What additional information about the student would you like to have in order to answer that question better?
   c) Why, mathematically, is there a difference between the National percentiles and the Catholic percentiles? Look at Student A's Language Mechanics score as an example.

3. Professor Boudreaux assigns letter grades on her tests depending on the mean and standard deviation of the scores made by each class. Any score which is 2½ standard deviations above the mean is an A; scores which are between 1½ and 2½ standard deviations above the mean are B's; scores which are between 1½ and 2½ standard deviations below the mean are D's; scores which are more than 2½ standard deviations below the mean are F's; and all others are C's.
   a) If the mean is 68 with a standard deviation of 6 and a student makes an 81, what letter grade will the student receive?
   b) To what letter grade would a 52 correspond?

4. Jimmy's jean pocket contains 3 marbles, 2 pieces of gum, 4 bottle caps and a worm. Jimmy's mother decides to wash his jeans and checks his pockets by reaching in and grabbing one item. What is the probability that she grabs the worm?

5. The menu at Harry's Hamburger Haven features a Child's Choice section for children under 12 years of age. The Child's Choice consists of 1 entree and 1 drink. For entrees, a child may select a hamburger, a cheeseburger, or a grilled chicken sandwich. For drinks, the choice is cola, water, milk, or lemonade. Harry created his own order pads for his employees. He listed all the possible ways that a child could place an order consisting of both an entree and a drink. How many different ways of ordering did Harry list?

6. Marco has a box containing 60 marbles, each only one color: black, blue or green. Marco chooses one marble from the box. The probability that it is black is 1/4. The probability that is it green is 2/5. How many blue marbles are in the box?

7. How many three-letter passwords, such as ABC, can be formed using the letters below if the last letter must be C, one of the other two letters must be A and no letter can be used more than once in a password?
Possible letters: A B C D E F G

Achievement Test Scores for problem 2:

STUDENT A

TESTS	NO. OF ITEMS	RAW SCORE	SCALED SCORE	NATL PR-S	CATHL PR-S	GRADE EQUIV	AAC RANGE
Total Reading	94	81	711	90-8	81-7	PHS	HIGH
Vocabulary	40	37	730	93-8	85-7	PHS	HIGH
Reading Comp.	54	44	702	86-7	78-7	PHS	MIDDLE
Total Math	118	108	748	97-9	96-9	PHS	HIGH
Concepts of No.	34	33	794	99-9	99-9	PHS	HIGH
Computation	44	41	754	95-8	93-8	PHS	HIGH
Applications	40	34	722	91-8	89-8	PHS	HIGH
Total Language	60	49	697	87-7	76-6	PHS	MIDDLE
Lang Mechanics	30	25	705	90-8	75-6	PHS	HIGH
Lang Expression	30	24	690	76-6	69-6	PHS	MIDDLE
Spelling	50	34	677	62-6	41-5	9.3	MIDDLE
Study Skills	32	31	765	99-9	98-9	PHS	HIGH
Science	50	37	672	77-7	76-6	10.5	MIDDLE
Social Science	50	45	715	97-9	97-9	PHS	HIGH
Listening	45	36	681	84-7	79-7	12.0	MIDDLE
Using Information	68	62	730	97-9	97-9	PHS	HIGH
Thinking Skills	118	94	690	88-7	85-7	PHS	HIGH
Basic Battery	399	339	706	93-8	87-7	PHS	HIGH
Complete Battery	499	421	705	95-8	89-8	PHS	HIGH

NATIONAL GRADE PERCENTILE BANDS
1 10 30 50 70 90 99

STUDENT B

TESTS	NO. OF ITEMS	RAW SCORE	SCALED SCORE	NATL PR-S	CATHL PR-S	GRADE EQUIV
Total Reading	128	91	587	58-5	44-5	3.0
Word Stdy Skills	48	38	594	63-6	50-5	3.4
Vocabulary	40	26	576	48-5	34-4	2.8
Reading Comp.	40	27	595	57-5	48-5	3.0
Total Math	105	76	573	61-6	61-6	3.3
Concepts of No.	34	20	544	31-4	28-4	2.3
Computation	36	29	595	78-7	78-7	3.9
Applications	35	27	586	70-6	70-6	3.5
Language	44	41	639	86-7	82-7	5.4
Spelling	30	24	603	71-6	64-6	3.6
Environment	40	25	564	35-4	30-4	2.1
Listening	45	28	577	33-4	22-3	2.3
Basic Battery	352	260	584	62-6	55-5	3.1
Complete Battery	392	285	582	61-6	53-5	3.1

NATIONAL GRADE PERCENTILE BANDS
1 10 30 50 70 90 99

Section 52: Sampling

Activity 2

Each group has a sealed bag which contains some beads. **DO NOT LOOK IN THE BAG!** Each person in the group will take a random sample of 8 beads from the bag - returning their beads before the next person. **DON'T LOOK IN THE BAG!** The group will keep a record of the beads chosen in the chart below:

	Number of beads				
Color	Trial 1	Trial 2	Trial 3	Trial 4	Trial 5

As a group, predict the percentage of beads of each color that are actually in the bag. **DON'T LOOK IN THE BAG!**

Repeat the sampling process but choose 16 beads at a time instead of 8. **DON'T LOOK IN THE BAG!** Record your results below:

	Number of beads				
Color	Trial 1	Trial 2	Trial 3	Trial 4	Trial 5

As a group, do you want to make a new prediction on the percentages? If so, do it now.

Now, have one person from the group choose 40 beads at random and record the results. **DON'T LOOK IN THE BAG!** Record the results below:

Color	Number of beads

Change your predictions if necessary. When you are finished, put your final group predictions on the board for comparison with the other groups. Now you can look in the bag.

One of the main reasons for conducting surveys or taking samples is to try to get a picture of what is happening for a given process or event. We want to be able to report on the "big picture" by analyzing a smaller segment of the population in question. However, we must be very careful that every effort is made to take what is called a representative sample whose makeup is not influenced by the design of the sampling process. For example, if a pollster wants a true representation of what will happen at the voting polls, he should not ask questions of only Democrats in a particular region. Of course, the perfect solution would be to survey every person involved. However, the perfect solution is also the most impractical and often impossible. For that reason, most surveys are taken over as large a portion of the population as financially and practically possible. As the sample size increases, the closer the results get to the true population.

Project

At the beginning of the semester, you collected data from 5 adults. Was your sample random? Did you think at the time it was representative of the true population? Do you now think it was a good sample? Let's try those surveys again. This time ask 20 adults the same questions (eye color, height, siblings) but keep in mind the random sampling process. Once the data has been collected, answer the following questions:

1. What is the "average" eye color? Is this different than the original class data?

2. What is the "average" number of siblings? The class used the median for this originally - is this still good? Did you use something else? Why?

3. What is the mean height of those surveyed? Do you think it would be different if the men are separated from the women?

4. Construct a grouped frequency table for the heights. Calculate the grouped mean and standard deviation.

All surveys and results are to be turned in by the date given by your instructor. Be neat and organized!

Chapter VI Review

1. Cassie's class is making special treats for their siblings. The teacher collected the following information about the class: Twenty students have at least one brother, eighteen students have at least one sister, five students have no siblings and six students have both brothers and sisters or at least one of each. What is the minimum number of treats that could be made by this class if each sibling were to receive exactly one?

2. Becky opened a bag of multicolored candies and separated the candies by color. She discovered the following: 23 were brown, 18 were yellow, 12 were orange, 10 were green, 6 were red and 2 were blue. Create a pie chart that describes the color distribution of the candies.

3. Seventy-five students answered a survey on the number of hours per week they spent doing homework. The results are shown below. Find the mean and standard deviation of the number of hours spent studying.

No. of hrs. per week	No. of responses
0 - 3.55	20
3.55 - 7.10	26
7.10 - 10.65	7
10.65 - 14.20	9
14.20 - 17.75	8
17.75 - 21.30	5

4. Lee was looking through the recipe section of a monthly magazine and made note of the amount of fat in each of the recipes listed. The amount of fat, per serving, in grams for each recipe is shown:

18	3	9	10	7	10	3	5	16
7	4	4	10	6	2	0	5	5

   a) Find the mean number of grams of fat for the recipes listed.
   b) Find the modal number of grams of fat.
   c) Find the median number of grams of fat.

5. When James walked into the Burger Barn, the average height of the people in the room rose from 63 inches to 65 inches. If there were six people in the room before James walked in, how tall is James?

6. Keith took 3 exams this quarter. He had a mean score of 74, a median score of 73 and a range of 9. What were Keith's three scores?

7. At a grocery store, which statistic would be most helpful to the manager when reordering boxes of washing powder: mean, median or modal size of the box? Explain.

8. Carol complained to her teacher that her grade should be an A for the semester. She argued that since her average on 12 tests was an 89 and it was only one point from the lowest score for an A of 90, that she should have an A. Was Carol only one point from an A? What would you tell Carol?

9. -9, -6, -3, 0, ...
   a) If the pattern continues, what could the next two numbers be? Explain how you found them.
   b) Without finding all the ones in between, explain how to find the thirtieth number in the pattern.

CHAPTER VII

Section 53: Probability Review

Throughout our lifetime, we are faced with many questions to which we invariably answer "Probably!". When we say probably, we mean that the chance of occurrence of the questionable event is more likely than not to happen. Mathematically, we represent this occurrence as the fraction of time such an event will happen. Elementary school children need to be taught the relationship between the fraction of the whole and the probability of the event. With this in mind, the following problem set is a review of the basic concepts of probability.

1. A bag contains 5 green, 7 blue, 4 yellow and one black marble. One marble is chosen from the bag at random. What is the probability that it is:
   a) a blue marble
   b) a black marble
   c) a green or a yellow marble
   d) not a blue marble
   e) a purple marble

2. A pair of standard fair dice are rolled. What is the probability that the **sum** rolled is:
   a) even
   b) greater than eight
   c) nine or less than three
   d) seven and the first die shows a five

3. One card is drawn from a standard deck of 52 cards. What is the probability that it is:
   a) a black card
   b) a heart
   c) a red king
   d) a seven
   e) a face card

4. There are three children in a particular family. What is the probability that:
   a) all three are boys
   b) two are boys and one is a girl
   c) the last child is a girl
   d) the first two are boys

5. The two spinners shown below are each spun once. What is the probability that:
   a) the first lands on two and the second lands on two.
   b) the first lands on an even and the second on an odd.
   c) the sum of the spins is 5.
   d) the sum of the spins is even.

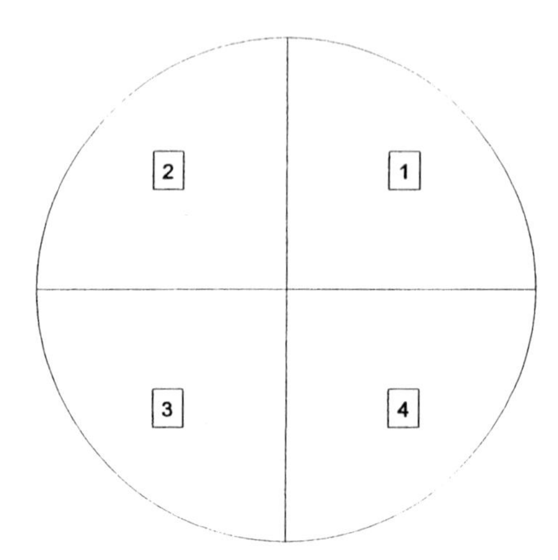

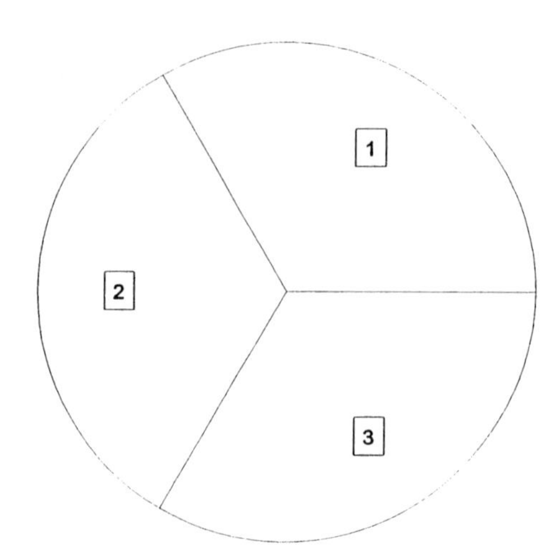

Recap of Section 53

A summary of some of the basic concepts used in answering the problems of section 53 follows. Let the symbol P(E) represent the probability of any particular event E.

a) P(E) is always between 0 and 1, inclusive.

b) The probability that E does not occur, written P(E′), can be found by subtracting P(E) from 1 or P(E′) = 1 - P(E).

c) The probability that one event E **or** another event F occur is the sum of their separate probabilities if the two events cannot both happen at the same time, or P(E or F) = P(E) + P(F).

d) When the two events, E and F can happen at the same time, we must be careful not to count anything twice. To guarantee this, we change the P(E or F) to P(E or F) = P(E) + P(F) - P(E and F) where P(E and F) is the probability that the two events both occur at the same time.

e) If E and F are independent events (one does not affect the other), then P(E and F) is the same as the product of their separate probabilities or P(E) · P(F). Cases where E and F are not independent will be discussed in the next section.

Section 54: Theoretical vs. Experimental

Activity 1

Each group has a pair of dice which are to be rolled **300** times. Each die in the pair has the same number of faces but the number of faces differs among the groups. As you roll your pair, record the sum of what is rolled on each die. It may be useful to first make a chart of all the possible sums before you begin rolling. Then, you need to only put a tally mark when a particular sum occurs. When you are finished collecting and recording your data, answer the following questions. Be prepared to share your results with the rest of the class.

1. What sum(s) occurred the most? What sum(s) occurred the least?

2. What sum(s) should have occurred the most? the least? What mathematical reason would support your claim?

3. Is there a relationship between the largest possible sum of your pair of dice and the number of faces on your dice? If so, what is it?

4. Compute the probabilities of each possible sum from the data you obtained in the experiment. (These are the experimental or empirical probabilities.)

5. Compute the probabilities of each possible sum using what you know about addition and what should have happened. (These are the theoretical probabilities.)

6. When the theoretical and empirical probabilities for a pair of dice are the same (or very close), then the dice could be considered "fair". Do you consider your dice to be "fair"? Explain.

Activity 2

Each group has a different size cylinder which is marked T and B for top and bottom.

1. Toss the cylinder to see how it lands. Did it land on its top, bottom or side?

2. Before you toss it again, let each member of the group predict how it will land. Did everyone make the same prediction?

3. Toss the cylinder 100 times and keep a record of how it lands. (Let one member be the tosser, one the recorder and one the counter).
   Put your results in the chart below:

   TALLY OF TOSSES:

TOP	BOTTOM	SIDE

4. Using your results: What is the probability of the cylinder landing on its top? What is the probability of the cylinder landing on its bottom?

5. Measure the height of the cylinder in centimeters. Measure the diameter of the cylinder in centimeters. What is the ratio of the height to the diameter as a decimal? Call this result R.

6. Collect the following information from all the other groups:
   R, the probability of landing on top - P(top), and the probability of landing on bottom - P(bottom)

R	P(top)	P(bottom)

7. Do you see any relationship between R and the two probabilities?

Section 55: Conditional Probability

Andrew and Ryan like to play a card game where each person chooses two cards from the deck at random. Yesterday, Ryan chose his cards first. He picked two aces. Andrew said "That's very hard to do!". Ryan said "I have the same chance of picking two deuces." Is Ryan correct?

Ryan's chance of choosing two aces depends on both the first card being an ace and the second card being an ace. The two picks can be thought of as two separate events. Let "the first card is an ace" be event A and "the second card is an ace" be event B. What we need to determine is P(A **and** B). However, events A and B may or may not be independent.

If the first card is returned to the deck before the second card is picked, then the events are independent. The probability that the first card is an ace is $\frac{4}{52}$ because there are 4 aces in the 52 cards. The probability that the second card is an ace is also $\frac{4}{52}$. The first card is back in the deck and it's as though the first pick never happened. In order to determine the probability of both events happening, we multiply the separate event probabilities. So, if the first card is returned to the deck, then the probability that both cards chosen are aces is $\frac{4}{52}$ times $\frac{4}{52}$ which equals $\frac{1}{169}$ when reduced.

If the first card is not returned to the deck before the second card is picked, then the second pick is affected by the first pick. The probability that the first card is an ace is still $\frac{4}{52}$. The probability that the second card is an ace is **not** $\frac{4}{52}$. If the first card is an ace, then there are only 3 aces left. Also, there are only 51 cards remaining in the deck. Therefore, the probability for the second card is now $\frac{3}{51}$. This new probability of picking an ace, $\frac{3}{51}$, is a type of probability called a conditional probability. A conditional probability is created when the sample space of the experiment is changed due to the previous occurrence of some known event. Our events A and B are still not independent, but since we have corrected B's probability to reflect A's occurrence, we can use the multiplication process like before. So, the probability of choosing two aces when the first card is **not** returned to the deck is $\frac{4}{52}$ times $\frac{3}{51}$ or $\frac{1}{221}$.

Going back to Ryan and Andrew, what is the probability that the two cards are deuces? Since there are the same number of deuces as aces in the deck, the probability is the same as for the aces. So, Ryan is correct!

Generalizing, the probability of the occurrence of an event F given that another event E has already occurred is symbolized as $P(F \mid E)$ and read "probability of F given E." The second time we computed the probability of the two aces, we actually multiplied P(A) times $P(B \mid A)$. In fact, we can always find P(E **and** F) by multiplying P(E) times $P(F \mid E)$. If the events are independent, then $P(F \mid E)$ will be the same as P(F). The extra knowledge that E has occurred does not change the probability of F.

Another way of determining the probability of E **and** F is through the use of a tree diagram. A tree diagram is created by drawing lines to represent the different possibilities for each event in an experiment. For multiple events, the lines begin to look like the branches of a tree. For example, suppose there are 10 slips of paper in a hat. Two slips will be chosen, without replacement of the first slip, at random from the hat. Seven of the slips are white and three are red. What is the probability that both slips chosen are red?

To create a tree diagram for this experiment, put a dot to represent the start. From the dot, two lines are drawn to represent the result of the first pick - either a red or white slip.

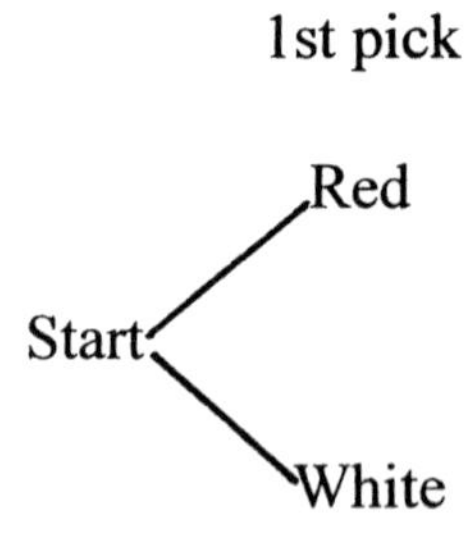

Next, the second pick is represented by extensions from **each** branch of the first pick.

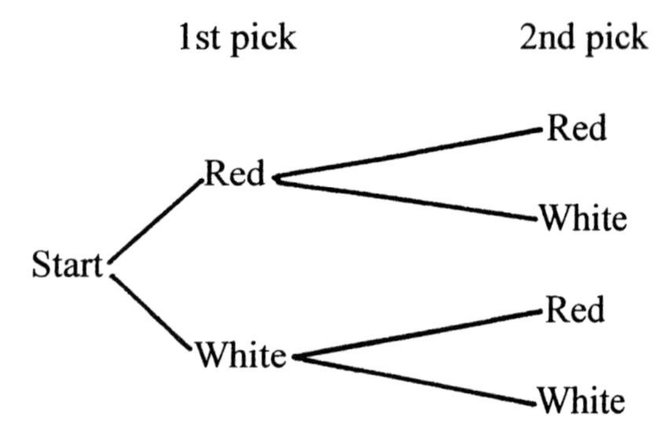

If you follow along each branch, you will be able to produce all the possible outcomes for the experiment:

Red and then Red	or	R and R
Red and then White	or	R and W
White and then Red	or	W and R
White and then White	or	W and W

The last step is to fill in the probabilities of each branch of the diagram. A necessary observation is that the second branches of this problem represent conditional probabilities since the first slip was not replaced.

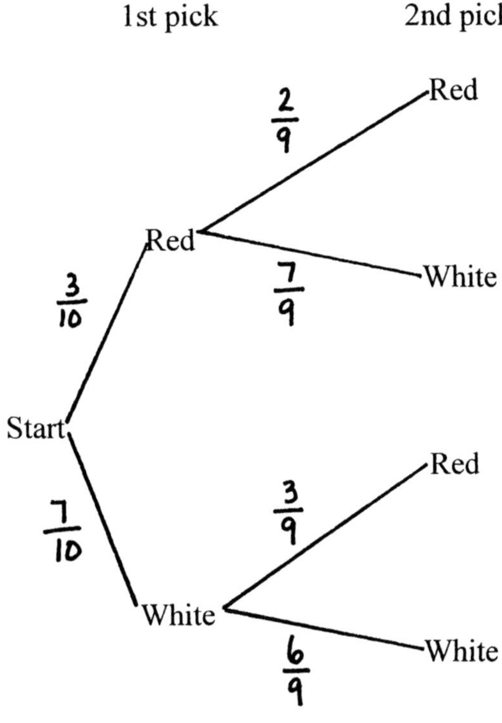

P(R and R) = P(R)P(R | R)

$$\frac{3}{10} \cdot \frac{2}{9} = \frac{6}{90}$$

P(R and W) = P(R)P(W | R)

$$\frac{3}{10} \cdot \frac{7}{9} = \frac{21}{90}$$

P(W and R) = P(W)P(R | W)

$$\frac{7}{10} \cdot \frac{3}{9} = \frac{21}{90}$$

P(W and W) = P(W)P(W | W)

$$\frac{7}{10} \cdot \frac{6}{9} = \frac{42}{90}$$

Which of these probabilities correspond to the original question - both red? Only the first case, P(R and R), will answer that question. So, the probability that both slips are red is 6/90 or 1/15.

What if we were interested in the probability of choosing a red in either position? Then, there are three outcomes which produce this result - P(R and R), P(R and W) and P(W and R). To obtain the final probability, we can look at the problem as an "**or**" situation - one or more of these outcomes must occur. Hence, we need to add the three probabilities together.

P(R and R) + P(R and W) + P(W and R)

$$= \frac{6}{90} + \frac{21}{90} + \frac{21}{90}$$

$$= \frac{48}{90}$$

$$= \frac{8}{15}$$

Homework: Section 55

1. Two cards are drawn from a standard deck without replacement. What is the probability of drawing:
   a) two clubs
   b) an ace and then a king
   c) a red ace and then a black ace
   d) the queen of hearts and then the jack of spades
   e) the queen of hearts and the jack of spades

2. A box contains 5 green, 3 blue and 2 yellow marbles. Two marbles are chosen at random, without replacement. What is the probability that:
   a) the second one chosen is blue given the first is green
   b) the second one chosen is blue
   c) there is at least one blue chosen
   d) both are yellow
   e) a green or a blue is chosen

3. Danielle chose a card from a deck after Cassie had already chosen a card. Cassie did not replace her card. What is the probability that Danielle chose:
   a) an ace if Cassie chose an ace
   b) an ace if Cassie chose a red ace
   c) an ace
   d) an ace if Cassie chose a red card

4. Harry's Hamburger Haven serves 5 types of burgers, 3 sizes of fries, 6 types of drinks and 4 types of pies. How many different orders could Mike place if he orders one burger, one fry, one drink and one slice of pie?

5. Danielle chose a number at random between 1 and 1000, inclusive. What is the probability that Becky guesses the number?

6. At Jenna's Horse Stables, 45% of the horses are mares and the rest are stallions. Twenty-five percent of the mares are brown. Fifty percent of the stallions are brown. A horse walks out of the stable at random. What is the probability that it is:
   a) a brown stallion
   b) not brown

7. A class contains several redheaded children. The ratio of redheaded children to non-redheaded children is 4 to 15. What is the probability of a child in the class being redheaded? Do you know how many children are in the class?

8. A pair of standard fair dice are rolled. What is the probability that the sum is:
   a) seven given an odd sum was rolled
   b) seven given the second die is 4
   c) seven given a prime sum is rolled

9. How many area codes exist if there are no restrictions on digits used in forming the code?

10. Keri has five dolls. She wants to take two of the dolls with her on a trip to the mountains. In how many different ways can she choose two dolls to take?

Section 56: Counting Techniques

We are now ready to extend our knowledge of probability by delving a little deeper into the subject. Our basic definition of the probability of some event happening relies on using simple counting techniques. We define the probability of a given event as the count of the number of ways the event could happen divided by the number of possible outcomes for the associated experiment.

Example 1: A bag contains 6 green, 2 yellow, 4 blue and 1 red marble. One marble is drawn from the bag. What is the probability that it is yellow?

How many different ways can the marble be yellow? 2 How many different marbles could be chosen? 13

Thus, the probability is $\frac{2}{13}$.

Sometimes, the counting is not so easy and more sophisticated counting techniques should be employed.

Example 2: Most people have three initials associated with their name. Suppose you and your teacher each write down a set of initials. What is the probability that the teacher writes the same set as you?

In how many ways can the two sets match? Only 1. How many different sets of initials exist? Here's where the counting becomes more difficult. A first reaction might be to list possible sets:

AAA AAB AAC AAD AAE . . .

It should be quickly apparent that this is not an efficient process. Let's look at this in another way. When you begin to write the first initial, how many choices of letters do you have?
Since there are 26 letters in our alphabet and every letter could be used, you have 26 choices. What about the second initial? Again, there should be 26 choices. You are allowed to use the same letter to start your middle name. So, each of the 26 first letters could be paired with one of the 26 second letters as in the chart below:

AA	BA	CA	DA	EA	. . .	ZA
AB	BB	CB	DB	EB	. . .	ZB
AC	BC	CC	DC	EC	. . .	ZC
. .	.	.	.	. . .	.	
. .	.	.	.	. . .	.	
. .	.	.	.	. . .	.	
AZ	BZ	CZ	DZ	EZ	. . .	ZZ

Therefore, we could make 26 times 26 or 676 sets of two letter initials.

To create our three letter set, each of the above sets could be matched with another one of the 26 letters in the alphabet:

(AA)A	(AB)A	(AC)A	. . .	(AZ)A
(AA)B	(AB)B	(AC)B	. . .	(AZ)B
(AA)C	(AB)C	(AC)C	. . .	(AZ)C
.	.	.	. . .	.
.	.	.	. . .	.
.	.	.	. . .	.
(AA)Z	(AB)Z	(AC)Z	. . .	(AZ)Z

This chart lists 676 possibilities using only the first column from the previous chart. We would need to repeat the process for each column from the previous chart for a total of 676 times 26 columns or 17576 sets of initials. Thus, the probability of a match between you and your teacher is $\frac{1}{17576}$.

Do you see a faster way to count the total possibilities? Try to come up with a faster method as you work the next activity.

Activity 1

Every country has an official flag. These flags sometimes have very simple designs and sometimes very complex designs. Many of the designs involve only stripes of varying colors. The flag of the United Arab Emirates has stripes as shown below:

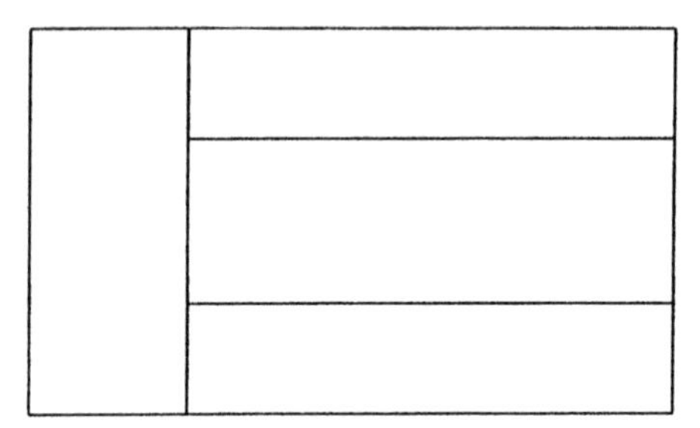

The colors used on this flag are black, green, red and white. Each stripe is a different color. Without looking at anyone else's design, color in the U.A.E. flag as you think it might look. (No color is used twice.)

The next flag design is for the nation of Zimbabwe. It looks something like this:

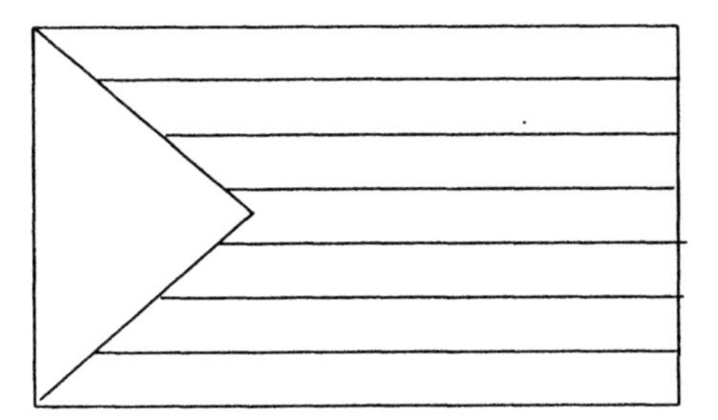

The colors used on this flag are black, green, red, white and yellow. The color of the triangle is not used again in the stripes. The stripes are colored symmetrically from the middle one. That is, the top and bottom stripes are the same color, the second and sixth are the same color and the third and fifth are the same color. The middle stripe is a different color from all the others so that all five colors are used on the flag. Color this flag as you think it looks.

The last flag design is for the imaginary country of Solvemeland. It looks like this:

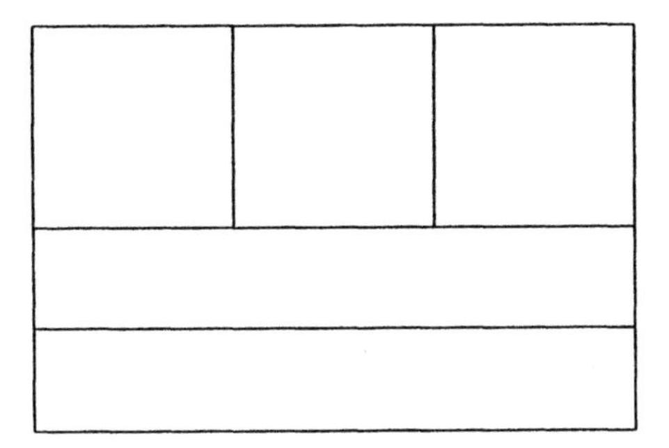

The colors used on this flag are the same as Zimbabwe: black, green, red, white and yellow. Each stripe is a different color and each color is used exactly once. Color this new flag as you think it should look. When all members of your group are finished, you will begin to compare your flags and use your group results to help answer the discussion questions.

Group Discussion

A) For the United Arab Emirates flag, answer the following:
1. How many different ways did your group color the flag? How many matched?

2. How many different ways **could** the flag have been colored?

3. What is the probability that the flag you colored is the real flag of U.A.E.?

4. Suppose you remembered the correct color and position of one of the stripes on the real flag. Does this change the probability of drawing the correct flag? How? Does it matter if you remembered a vertical stripe or a horizontal one?

5. Describe the process your group used in determining each of the "counts" used in the previous questions. Did your group come up with a shortcut?

6. When you compared your flag to others in the group, either your flag matched with someone else's or it didn't. What is the probability that no one in your group colored the flag the same way?

B) For the flag of Zimbabwe answer questions 1 - 5 above. For question 4, consider the triangle to be a vertical stripe.

C) For the flag of Solvemeland, answer questions 1 - 5 above.
6. Suppose you remember that the green stripe is one of the three top ones. How does that change the probability of drawing the correct flag?
7. What if you remember the green flag is one of the two bottom ones?

Section 57: Counting Specific Orders

While working the previous activity, your group may have discovered a faster method for counting when the possibilities become extremely hard to list. One such method is called the multiple counting principle. When there are several decisions to make and each decision has multiple options, then the number of possible outcomes is the result of multiplying together the number of options for each decision.

Example 1: Mr. Boudreaux wants to carpet his office. He needs to choose a carpet company, a style of carpet and a color for the carpet. He has a choice of 4 carpet companies in his town, 7 styles of carpet and 20 colors. In how many different ways could Mr. Boudreaux carpet his office?

The first decision, the company, has 4 options; the second decision, the style, has 7 options; and the third decision, the color has 20 options. So, the total number of different possibilities can be found by multiplying $7 \cdot 4 \cdot 20$; which results in 560 possible ways to carpet the office.

Example 2: Suppose Mr. Boudreaux has 4 paintings that he wants to hang in his newly carpeted office. All four paintings fit on one wall if they are placed in a horizontal row. In how many ways can the paintings be placed on the wall to create a different wall scene?

The decisions that need to be made involve the placement of each painting on the wall. There are four placements but each placement does not have four options for paintings. Once a painting is positioned on the wall, it is eliminated from the possibilities for the other positions. No painting can hang in two places at the same time! Thus, if you think of the positioning of the paintings as going from left to right and Mr. Boudreaux starts hanging paintings on the left, then he has 4 choices in the first position, 3 choices in the second, 2 choices in the third and 1 choice in the final position. Using the multiple counting principle, he can create $4 \cdot 3 \cdot 2 \cdot 1$ or 24 different wall scenes.

When the options for a set of decisions come from only one group of possibilities (like the paintings), then a special case of the multiple counting principle, called a **permutation**, is created. Permutations count the number of ways a given set of objects can be arranged in different orders. Sometimes all of the objects are used in the arrangements but not always. However, in order to be called a permutation, there must be no possibility of replications in the choices.

Example 3: What if Mr. Boudreaux has 12 paintings to choose from but still could only hang four at a time?

Let the blanks below represent the positions of the four paintings. Then, the numbers in the blanks will represent the number of choices of paintings Mr. Boudreaux has for that particular position if he starts hanging the pictures from left to right.

$\underline{12} \cdot \underline{11} \cdot \underline{10} \cdot \underline{9}$

If we multiply these choices together, the result will be the possible number of arrangements Mr. Boudreaux can create. The product, 11880, is the permutation of the twelve objects using four at a time. Mathematically, the permutation of n objects r at a time is symbolized $P_{n,r}$ or ${}_nP_r$. (Note: There is a formula for calculating permutations. For most middle school problems, the formula is more difficult to work with than the actual problem.)

Looking back at the activity involving the flags, could this be considered a permutation problem or is it just multiple counting?

Homework: Section 57

1. In how many ways can 15 people line up to buy tickets for a concert?

2. Each person in Alex's secret club has a special code. The codes consist of four letters followed by two digits. The first letter must be A and the last digit must be even. No letter or digit can be used twice.
   a) How many different codes can be created?
   b) How many different codes can be created if repetition of letters or digits is allowed?

3. Wash day at the Price household is on Saturday. There is normally one load of whites, one load of darks, one load of pales and one load of delicates. The order in which the loads are washed varies from Saturday to Saturday to prevent boredom. In how many different ways can the order of the loads be changed?

4. Each year, fifty-one women vie for a national beauty title. In the end, only the top five places receive prizes - increasing in value from fifth place to first place. In how many different ways could the five prizes be awarded among the fifty-one women?

5. Vincent has 4 comic books, 6 coloring books, and 2 puzzle books. He likes to bring one of each type of book with him when he goes to the doctor's office. How many different sets of three books can he take?

6. Eight horses are running in this afternoon's race. What is the probability of choosing the correct order of finish of the top three?

7. The odds in favor of an event can be defined as the ratio of the number of ways the event can happen to the number of ways the event cannot happen. For example, the odds in favor of rolling a four on a single roll of a fair die are 1 to 5 because there is only one way to roll a four and five ways to not roll a four. The probability of rolling a four is 1/6. What is the relationship between the odds in favor and the probability of occurrence?

8. The groups in a particular math class consist of a leader, a recorder, an interpreter and a reporter. All groups are required to identify the person assigned to each job on any materials turned in to the teacher. Yesterday, one group forgot to make the identifications on their paper, so the teacher randomly assigned the jobs. What is the probability that the teacher's assignment of jobs matched the group's true assignment?

9. If the odds in favor of picking a blue marble out of a hat are 3 to 5, then what is the probability of picking a blue marble out of the hat?

10. The display rack at Blue's Archery holds three bows. The shop has 25 different bows. How many different displays can be made using three bows at a time from the 25 bows?

11. Fill in the blanks to complete the sequence and explain how you obtained you answers:
    1, 1, 2, 3, 5, ____, ____, ____, ____

Section 58: Counting Different Groups

Example 1: Smalltown High School recently held tryouts for the quiz bowl team. Coach Calhoun had to choose a team of four from the twenty students who tried out. How many different teams could the coach create?

If we approach this problem in the same manner as in the last section, we will get $20 \cdot 19 \cdot 18 \cdot 17$ or 116280 teams. However, this approach includes some repetitions that should not be counted. The order in which the students are chosen is not significant to the question. A team consisting of Mary, Joe, Sue and Todd is the same as a team consisting of Joe, Todd, Mary and Sue. We no longer want to count all the arrangements; we want a subset of those arrangements. Such a subset is called a **combination**. So, how can we eliminate the repetitions due to arrangement?

To start, how many ways can one team of four be arranged? Since there are 4 positions to be filled by four people, we can say $4 \cdot 3 \cdot 2 \cdot 1$ or 24 ways. These 24 arrangements are repeated for each of the possible teams to produce the total arrangements of 116280. So, if we divide the total arrangements by the number of arrangements for one team, we will obtain the different teams:

(Total arrangements) = (number of arrangements for 1 team)(number of teams)

$$116280 = 24 \cdot (\text{number of teams})$$

$$\frac{116280}{24} = \text{number of teams}$$

$$4845 = \text{number of teams}$$

Another way to look at it is:

$$\frac{20 \cdot 19 \cdot 18 \cdot 17}{4 \cdot 3 \cdot 2 \cdot 1} = 4845 \text{ teams}$$

Symbolically, the combination of n objects r at a time is given as $C_{n,r}$ or ${}_nC_r$. In Coach Calhoun's case, it would be $C_{20,4}$.

Example 2: A team of six astronauts is to be chosen from a group of 40 astronauts. How many different teams can be chosen?

Since the order in which the astronauts are chosen is not important, we need to eliminate all arrangements of the same team. Thus, we want a combination of the 40 people using 6 at a time or:

$$C_{40,6} = \frac{40 \cdot 39 \cdot 38 \cdot 37 \cdot 36 \cdot 35}{6 \cdot 5 \cdot 4 \cdot 3 \cdot 2 \cdot 1} \qquad \frac{\text{all possible arrangements}}{\text{all arrangements of 1 team}}$$

$$= 3838380 \text{ teams}$$

At this point, it may be helpful to introduce a mathematical symbol that appears on most scientific calculators that will aid in the calculation of these products. That is the ! or factorial symbol. x! is a shorthand way to write the product of x times each integer between x and 0. For example, $4! = 4 \cdot 3 \cdot 2 \cdot 1$ and $6! = 6 \cdot 5 \cdot 4 \cdot 3 \cdot 2 \cdot 1$. Also, by definition, $1! = 1$ and $0! = 1$. See if you can rewrite any of the previous homework problems using this notation.

The big question now is how to tell the difference between problems where you must use a particular method - permutations or combinations. You must discover the answer on your own by working more problems!

Homework 58

1. Kelly has ten decorative bottles. She wants to give two of the bottles to her friend Katie. How many different ways could Kelly choose the two bottles to give to Katie?

2. Keith's math class needs to send a committee of four to a special meeting. If there are 28 students in the class, how many different committees could be formed?

3. Dave's Diner orders sodas by the case. Dave always checks a sample of three from the case for underfilling and cracks. How many samples of size three can Dave select from a case of 24 sodas?

4. How many five card hands can be dealt from a standard deck of cards?

5. What is the probability of being dealt a 13-card bridge hand that contains:
   a) all red cards
   b) 3 clubs, 2 hearts, 7 diamonds and 1 spade

6. The faculty of a particular school is composed of 17 males and 28 females. A team of five is chosen from the faculty to meet with the PTA. How many different teams can be formed which consist of:
   a) 2 males and 3 females
   b) 3 males and 2 females
   c) at least one male
   d) at most two females

7. A bag contains 5 apples, 4 oranges and 3 bananas. Cassie grabs three pieces of fruit from the bag. What is the probability that she grabs:
   a) 2 apples and one banana
   b) three bananas
   c) an orange, an apple and a banana
   d) at least one apple

8. Fill in the blanks to complete the sequence:

   a) 2 , 6 , 8 , 14 , 22 , ____ , ____

   b) ____ , ____ , 1 , 6 , 6 , 36 , 216

   c) 4 , 7 , 5 , 8 , 6 , 9 , ____ , ____

9. In order to win one of the lottery games in Louisiana, a player must match 5 numbers chosen from the numbers 1 to 44.
   a) What is the probability of winning for a player who chooses one set of numbers?
   b) What if the player chooses 20 different sets of numbers?

10. A dart is thrown at the target drawn. What is the probability of landing in the shaded region? Assume that the dart will hit the target. Radius of the circle is 3.5 inches and the inner square is centered in the circle.

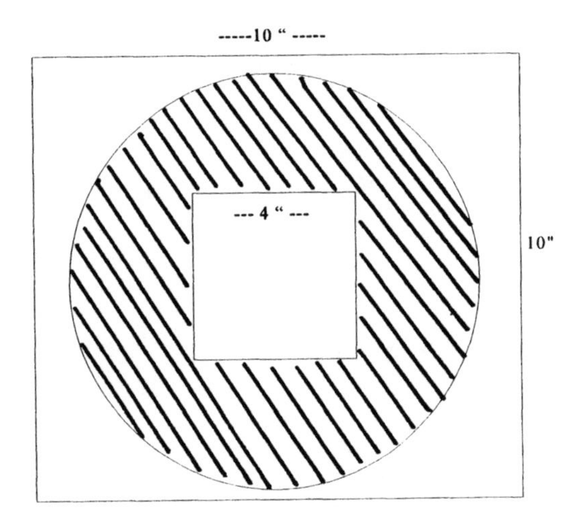

Section 59: Geometric Probability

Probability questions which involve geometric shapes do not usually produce answers which arise from counting. Most of the time, the probability relates to the area of the shape or some part of the shape. Problem 10 in Section 58 is such a problem. The probability of being in the shaded region is the ratio of the area of the shaded region to the area of the large square.

In general, the probability of being in a particular region of a geometric shape is the ratio of the area of the region to the area of the entire shape.

Activity 1

Each group has a paper target used for archery practice. Do not write or mark on the target. Measuring instruments are available for your use.

Assume that a random shot is made and it hits the target - ignoring the outer white edge of the target. Scores are recorded as shown on the next page. Maximum score is 5 - minimum is 1 if we assume the target face is hit.

Using the large target, answer the following:

a) What is the probability of landing in the center white section? (Scoring a 5)

b) What is the probability of landing in the "X-ring"? (Scoring a 5 in the inner white circle)

c) What is the probability of scoring 2 or 3?

d) If we want the probability of scoring a 5 to double, what would be the new radius of the center white section?

Using a small target on the back - assume one shot is made at one particular small target.

a) What is the percent decrease in area from the large target to one of the small targets?

b) Do the chances of scoring 5 increase or decrease when you switch to a small target on a particular shot? Why?

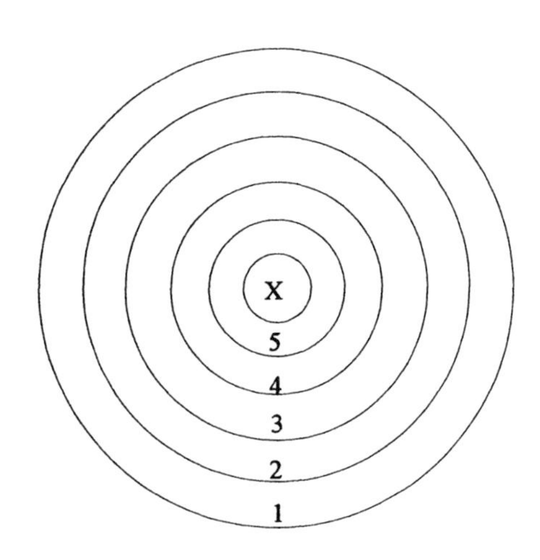

Each ring scores the points shown. The inner "X-ring" is also worth 5 points.

On the smaller targets, the rings worth 1,2, and 3 points are eliminated. You only score 5 or 4.

Homework: Section 59

1. An archery target consists of a gold circle with radius of 5 inches surrounded by a red ring whose diameter is 20 inches. Next, there are three progressively larger rings - blue, black, and white. The diameter of each is 10 inches larger than the previous one. John is a beginner so his aim is poor and, if his arrow hits the target, its location is completely at random. John shoots and then yells, "I hit it." What is the probability that
   a) his arrow landed in the gold circle?
   b) his arrow landed in the white area?
   c) his arrow landed in the red or in the blue area?

2. The game of baseball darts is played on a square board marked like a baseball diamond. Each side of the diamond is 15 inches long. The bases are 3.5 inch squares in each corner. There is a home-run circle in the center of diameter 2 inches. "Hits" are made by landing on the bases or the home-run circle. Landing anywhere else in the diamond is considered an "out". Mona is blindfolded and throws a dart which lands on the board in a random manner.
   a) What is the probability that she is out?
   b) If she already has two outs, what is the probability that her next throw will result in another out?

3. Eddie's Pool Hall features some very unique square dart boards, as shown below. Find the probability of a random shot, which hits the board, landing in a shaded region of the board.
   a) Circle has radius 1 inch. Quarter circles have straight edges of 3 inches. Square is 20 inches on a side.

   b) Circle has radius 1 inch. Inner square has sides of 5 inches. Width of one strip is 1 inch. Length of one strip is 7.5 inches. Square is 20 inches on a side.

c) Outer square is 20 inches on a side. Inner square (not shaded) is 10 inches on a side. Shaded squares touch the other ones at the midpoints of each side.

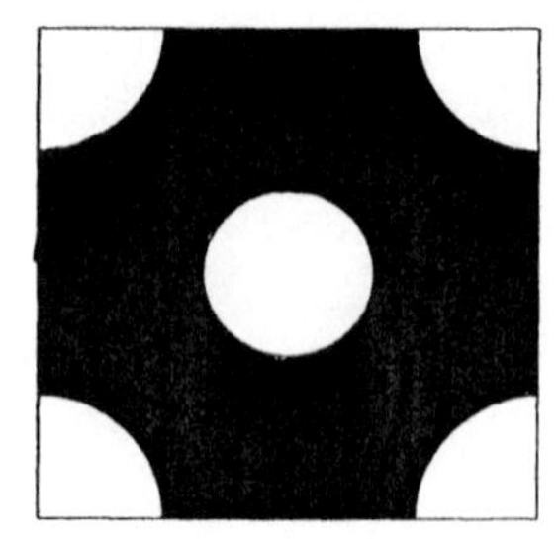
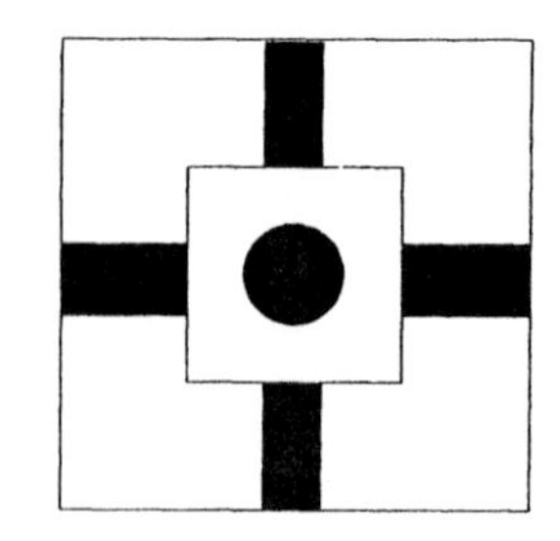

4. 0, 1, 4, 9, ...
   a) If the pattern continues, what could the next two numbers be? Explain how you found them.
   b) Without finding all the ones in between, explain how to find the tenth number in the pattern.

5. 2, 6, 18, 54, ...
   a) If the pattern continues, what could the next two numbers be? Explain how you found them.
   b) Without finding all the ones in between, explain how to find the eighth number in the pattern.

6. 216, 125, 64, 27, ...
   a) If the pattern continues, what could the next two numbers be? Explain how you found them.
   b) Without finding all the ones in between, explain how to find the sixteenth number in the pattern.

7. In Lou's diner there are some rectangular tables that seat three people on each of the long sides and one person on each end.
   a) How many people can be seated if five of these tables are placed end-to-end?
   b) How many people can be seated at a table created from t rectangular tables?
   c) Can you come up with a rule to help Lou figure out how many rectangular tables he needs to seat n people?

Section 60: The Normal Distribution

Suppose we wanted to find the area of the shaded region shown below:

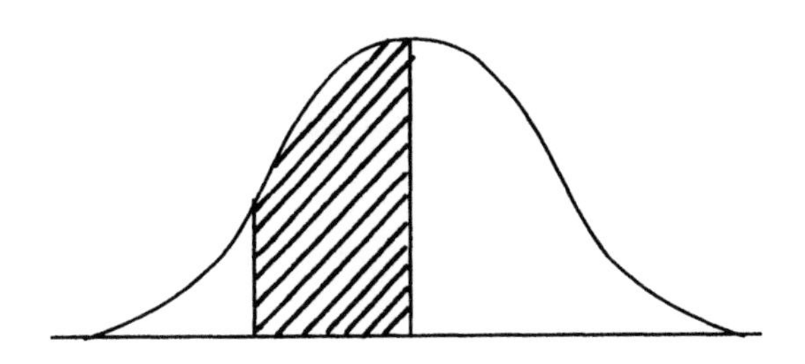

The shape created by the shading is not one of our standard geometric shapes. In order to find the area, we need to use methods taught in calculus. However, the bell-shape may seem quite familiar. There are numerous sets of data which, when graphed, produce the bell shape. This shape is a common characteristic of what are called **Normal probability distributions**. Because this shape occurs so often, statisticians have produced a table of probability values which relate to one particular Normal distribution - where the mean is 0 and the standard deviation is 1. Now, of course there are very few cases where a set of data will produce a bell-shape graph and have a mean of 0 and standard deviation of 1. But, every set of data that fits a Normal curve will have several common characteristics.

Each Normal probability distribution is defined by its own mean and standard deviation - that's the unique part. The common part is that for any given mean and standard deviation, distributions which follow a Normal curve will have about 68% of its data within one standard deviation of the mean. Approximately 95% of the data will be within two standard deviations of the mean and about 99.5% will be within three standard deviations. Knowing that the data comes from a Normal distribution gives extra information about your sample. The probabilities of being within those deviations are the same as the previously mentioned percentages.

The Normal distribution which has a mean of 0 and a standard deviation of 1 is called the standard Normal distribution. In order to use the tables created for Normal distributions, the data must be converted into standard Normal data. However, it is not necessary to convert the entire data set - we need only to convert the scores we may be interested in. The score we are converting to is called a **z-score**. For any Normal distribution, each data value can be thought of as being away from the mean by a certain number of standard deviations. These deviations are called z-scores. A z-score of 2 means that the value assigned to that z-score is 2 standard deviations more than the mean value. A z-score of -1.8 means that the value assigned is 1.8 standard deviations less than the mean. We used the concept of z-score without identifying it in problem 8, section 1.5. To compute a z-score for a given piece of data, x, subtract the population mean, μ, from x and then divide by the population standard deviation, σ.

$$z = \frac{x - \mu}{\sigma}$$

Our Normal distribution table on pages 281-282 will give us the probability of z being less than a value. For example, suppose we want the probability that z is less than 1.62. From the table, we see z = 1.62 relates to a value of .9474. The entire area under the curve is equal to 1 square unit. The area from $-\infty$ to 1.62 is .9474 square units. From the relationship between area and probability it follows that the probability that z is less than 1.62 is .9474.

Example 1: Find the area under the Normal curve that relates to each of the following situations. In all cases it is helpful to draw a picture of the curve with the z's located on the number line.

a) $z < -1.43$
b) $z > 2.61$
c) $1.5 < z < 2.97$

Answers:

a) The position of z is shown on the curve. Since we are interested in the area to the left of z (shaded) and the table gives us that area, we can read directly from the table the value of z = -1.43. The table value is .0764. Therefore, the area is .0764 square units and the probability is .0764.

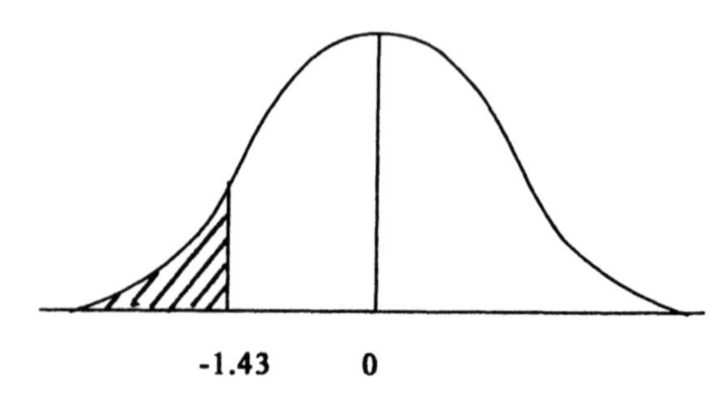

b) The position of z is shown on the curve. This time the region that we are interested in is to the right of z. Our table only gives us information about the left side of z. However, since the entire region represents 1 square unit, we can think of this as a complementary situation. The area to the right can be found by subtracting the area to the left from 1 square unit. From the table, the area to the left is .9955 square units. So, the area to the right is 1 square unit minus .9955 square units or .0045 square units.

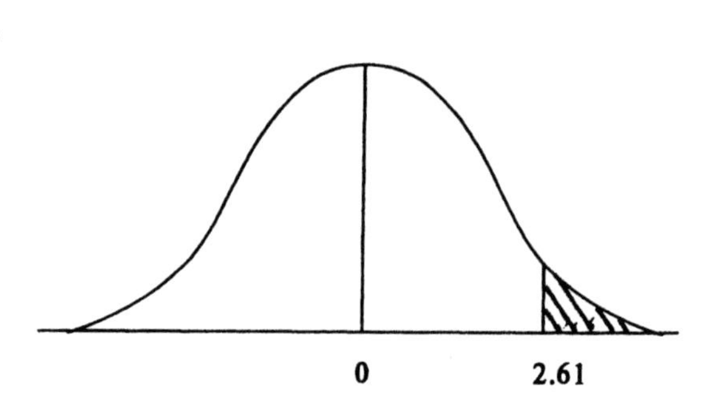

c) In this case, there are two z-scores that need to be addressed. The region we are interested in is between the two scores. Remember that the table only gives regions to the left and those regions go all the way to $-\infty$. How can we get the region we want? What if we subtract the area to the left of 1.5 from the area to the left of 2.97. This difference should then be the region of overlap which is want we want. So, from the tables we get, z = 1.5 relates to .9332 square units and z = 2.97 relates to .9985 square units. Therefore, our area of interest is .9985 - .9332 or .0653 square units.

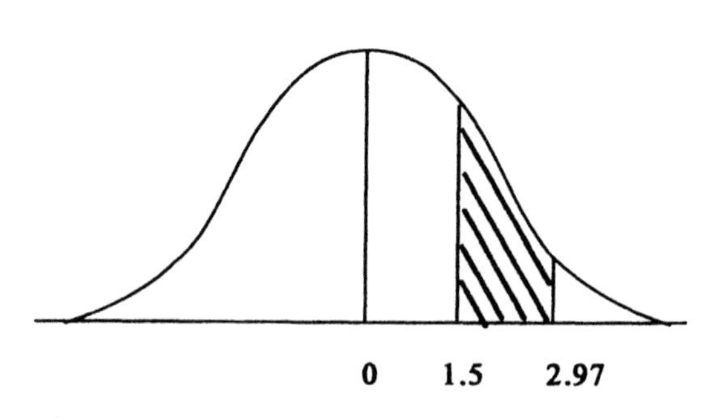

Example 2: Lyn came home from school upset that she was the shortest person in her class. Lyn is 59 inches tall. The average height of the class is 64 inches with a standard deviation of 2.5 inches. Heights are generally considered to follow a Normal distribution. What percent of the class would you expect to be shorter than Lyn?

A pictorial representation of the information given in the problem is usually a good place to start. Since the heights are considered Normal, draw a generic Normal curve with the average height as the mean and Lyn's height in relation to that mean.

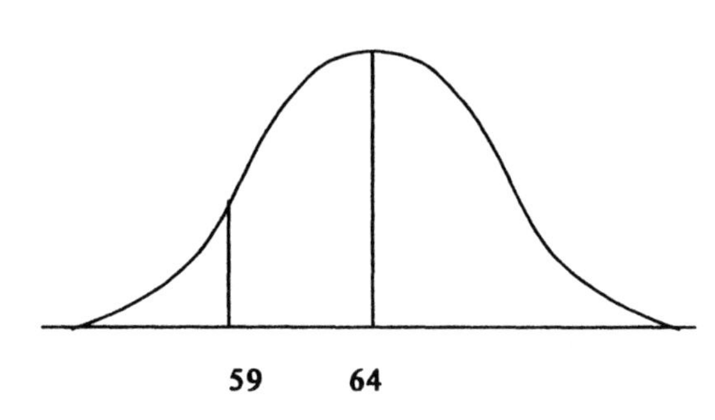

Then shade the region of the curve that relates to the question asked. In this case, the question concerns those heights that are <u>less than</u> Lyn's height, so shade the region to the left of 59 inches.

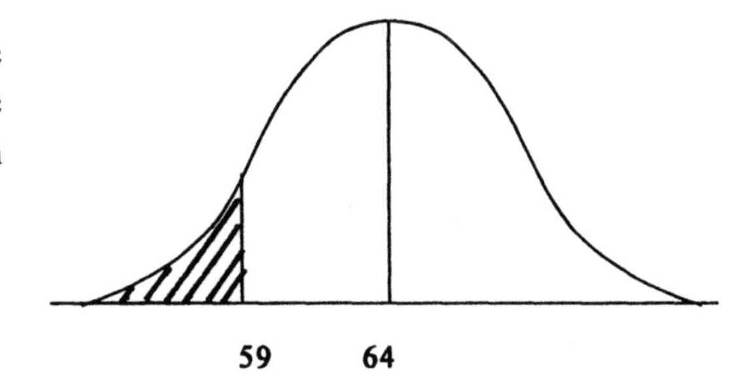

Remembering that our Normal distribution table only works for cases where the mean is 0 and the standard deviation is 1, we need to convert the height of 59 inches to its related z-score. In other words, how many standard deviations away from the mean is 59 inches? 59 is 5 inches below the mean of 64. Since this set of data has a standard deviation of 2.5 inches, that makes 59 inches exactly 2 deviations below the mean. Thus, z = -2.00. By formula:

$$z = \frac{59 - 64}{2.5} = -\frac{5}{2.5} = -2.00$$

Now our picture can be redrawn using the z-scores as the horizontal axis.

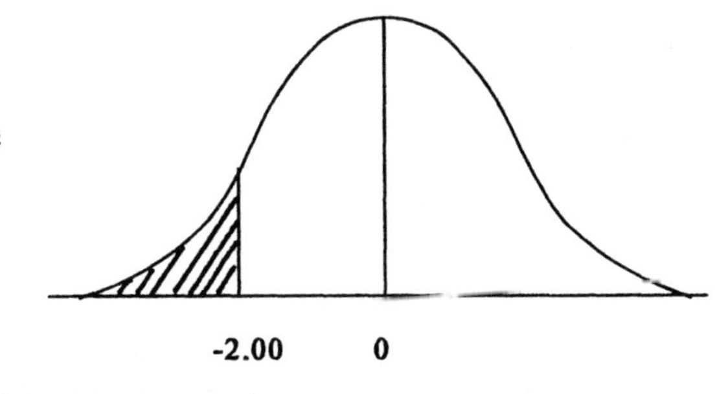

Then, from the table we see that the area of the region to the left of -2.00 is .0228 square units. This means that the probability of someone being shorter than Lyn is .0228 or 2.28% of the class is expected to be shorter than Lyn. But remember, as in all probability problems, what is expected does not always happen. (What upset Lyn was that she was the TEACHER!)

Homework: Section 60

1. Using the standard Normal distribution table, find the following probabilities:
   a) $P(z < 1.54)$
   b) $P(-.32 < z < .98)$
   c) $P(z > -1.06)$

2. Using the standard Normal distribution table, find the values of c which produce the desired probabilities:
   a) $P(z < c) = .0087$
   b) $P(z > c) = .7642$

3. Brand X cereal boxes contain an average of 13 oz. of cereal with a standard deviation of 0.25 oz. What is the probability of a box of Brand X cereal having less than 12.5 oz. of cereal? (Assume a Normal distribution for the number of oz.)

4. The length of time required for one dose of a particular headache cure to provide relief follows a Normal curve with mean time of 10.5 minutes and standard deviation of 2.4 minutes. Find the probability that the relief time after one dose is:
   a) longer than 12 minutes
   b) between 9 and 11 minutes
   c) less than 5 minutes

5. Professor Richardson likes to grade "on a curve". He wants 10% of the class to receive A's, 20% to receive B's, 40% to receive C's, 20% to receive D's, and 10% to receive F's. Assuming that he is talking about a Normal curve, what cutoff scores would constitute the various grades if the class average is 72 with a standard deviation of 7?

6. Brian, the manager of the local cinema, noticed that the amount of money spent by patrons on refreshments seemed to follow a Normal distribution. The average amount spent was $5.25 with a standard deviation of $1.75. If Brian's assumptions are correct,
   a) what percentage of his patrons spend between $7 and $10 on refreshments?
   b) what price range represents the amount spent by the middle 50% of his patrons?

7. Using a standard Normal distribution table, how can we find $P(z = 2.6)$? Does this value exist? Could we approximate it?

8. A particular IQ test produces test scores that follow a Normal distribution with a mean score of 100 and a standard deviation of 15. In order to join the Brainiak Club, a person must have a score of at least 150. In a city of 50,000 people, how many would you expect to be members of the Brainiak Club?

Cumulative Standard Normal Probabilities

$P(Z \le z)$

z	.09	.08	.07	.06	.05	.04	.03	.02	.01	.00
-3.4	.0002	.0003	.0003	.0003	.0003	.0003	.0003	.0003	.0003	.0003
-3.3	.0003	.0004	.0004	.0004	.0004	.0004	.0004	.0005	.0005	.0005
-3.2	.0005	.0005	.0005	.0006	.0006	.0006	.0006	.0006	.0007	.0007
-3.1	.0007	.0007	.0008	.0008	.0008	.0008	.0009	.0009	.0009	.0010
-3.0	.0010	.0010	.0011	.0011	.0011	.0012	.0012	.0013	.0013	.0013
-2.9	.0014	.0014	.0015	.0015	.0016	.0016	.0017	.0018	.0018	.0019
-2.8	.0019	.0020	.0021	.0021	.0022	.0023	.0023	.0024	.0025	.0026
-2.7	.0026	.0027	.0028	.0029	.0030	.0031	.0032	.0033	.0034	.0035
-2.6	.0036	.0037	.0038	.0039	.0040	.0041	.0043	.0044	.0045	.0047
-2.5	.0048	.0049	.0051	.0052	.0054	.0055	.0057	.0059	.0060	.0062
-2.4	.0064	.0066	.0068	.0069	.0071	.0073	.0075	.0078	.0080	.0082
-2.3	.0084	.0087	.0089	.0091	.0094	.0096	.0099	.0102	.0104	.0107
-2.2	.0110	.0113	.0116	.0119	.0122	.0125	.0129	.0132	.0136	.0139
-2.1	.0143	.0146	.0150	.0154	.0158	.0162	.0166	.0170	.0174	.0179
-2.0	.0183	.0188	.0192	.0197	.0202	.0207	.0212	.0217	.0222	.0228
-1.9	.0233	.0239	.0244	.0250	.0256	.0262	.0268	.0274	.0281	.0287
-1.8	.0294	.0301	.0307	.0314	.0322	.0329	.0336	.0344	.0351	.0359
-1.7	.0367	.0375	.0384	.0392	.0401	.0409	.0418	.0427	.0436	.0446
-1.6	.0455	.0465	.0475	.0485	.0495	.0505	.0516	.0526	.0537	.0548
-1.5	.0559	.0571	.0582	.0594	.0606	.0618	.0630	.0643	.0655	.0668
-1.4	.0681	.0694	.0708	.0721	.0735	.0749	.0764	.0778	.0793	.0808
-1.3	.0823	.0838	.0853	.0869	.0885	.0901	.0918	.0934	.0951	.0968
-1.2	.0985	.1003	.1020	.1038	.1056	.1075	.1093	.1112	.1131	.1151
-1.1	.1170	.1190	.1210	.1230	.1251	.1271	.1292	.1314	.1335	.1357
-1.0	.1379	.1401	.1423	.1446	.1469	.1492	.1515	.1539	.1562	.1587
-0.9	.1611	.1635	.1660	.1685	.1711	.1736	.1762	.1788	.1814	.1841
-0.8	.1867	.1894	.1922	.1949	.1977	.2005	.2033	.2061	.2090	.2119
-0.7	.2148	.2177	.2206	.2236	.2266	.2296	.2327	.2358	.2389	.2420
-0.6	.2451	.2483	.2514	.2546	.2578	.2611	.2643	.2676	.2709	.2743
-0.5	.2776	.2810	.2843	.2877	.2912	.2946	.2981	.3015	.3050	.3085
-0.4	.3121	.3156	.3192	.3228	.3264	.3300	.3336	.3372	.3409	.3446
-0.3	.3483	.3520	.3557	.3594	.3632	.3669	.3707	.3745	.3783	.3821
-0.2	.3859	.3897	.3936	.3974	.4013	.4052	.4090	.4129	.4168	.4207
-0.1	.4247	.4286	.4325	.4364	.4404	.4443	.4483	.4522	.4562	.4602
-0.0	.4641	.4681	.4721	.4761	.4801	.4840	.4880	.4920	.4960	.5000

Cumulative Standard Normal Probabilities

$P(Z \leq z)$

z	.00	.01	.02	.03	.04	.05	.06	.07	.08	.09
0.0	.5000	.5040	.5080	.5120	.5160	.5199	.5239	.5279	.5319	.5359
0.1	.5398	.5438	.5478	.5517	.5557	.5596	.5636	.5675	.5714	.5753
0.2	.5793	.5832	.5871	.5910	.5948	.5987	.6026	.6064	.6103	.6141
0.3	.6179	.6217	.6255	.6293	.6331	.6368	.6406	.6443	.6480	.6517
0.4	.6554	.6591	.6628	.6664	.6700	.6736	.6772	.6808	.6844	.6879
0.5	.6915	.6950	.6985	.7019	.7054	.7088	.7123	.7157	.7190	.7224
0.6	.7257	.7291	.7324	.7357	.7389	.7422	.7454	.7486	.7517	.7549
0.7	.7580	.7611	.7642	.7673	.7704	.7734	.7764	.7794	.7823	.7852
0.8	.7881	.7910	.7939	.7967	.7995	.8023	.8051	.8078	.8106	.8133
0.9	.8159	.8186	.8212	.8238	.8264	.8289	.8315	.8340	.8365	.8389
1.0	.8413	.8438	.8461	.8485	.8508	.8531	.8554	.8577	.8599	.8621
1.1	.8643	.8665	.8686	.8708	.8729	.8749	.8770	.8790	.8810	.8830
1.2	.8849	.8869	.8888	.8907	.8925	.8944	.8962	.8980	.8997	.9015
1.3	.9032	.9049	.9066	.9082	.9099	.9115	.9131	.9147	.9162	.9177
1.4	.9192	.9207	.9222	.9236	.9251	.9265	.9279	.9292	.9306	.9319
1.5	.9332	.9345	.9357	.9370	.9382	.9394	.9406	.9418	.9429	.9441
1.6	.9452	.9463	.9474	.9484	.9495	.9505	.9515	.9525	.9535	.9545
1.7	.9554	.9564	.9573	.9582	.9591	.9599	.9608	.9616	.9625	.9633
1.8	.9641	.9649	.9656	.9664	.9671	.9678	.9686	.9693	.9699	.9706
1.9	.9713	.9719	.9726	.9732	.9738	.9744	.9750	.9756	.9761	.9767
2.0	.9772	.9778	.9783	.9788	.9793	.9798	.9803	.9808	.9812	.9817
2.1	.9821	.9826	.9830	.9834	.9838	.9842	.9846	.9850	.9854	.9857
2.2	.9861	.9864	.9868	.9871	.9875	.9878	.9881	.9884	.9887	.9890
2.3	.9893	.9896	.9898	.9901	.9904	.9906	.9909	.9911	.9913	.9916
2.4	.9918	.9920	.9922	.9925	.9927	.9929	.9931	.9932	.9934	.9936
2.5	.9938	.9940	.9941	.9943	.9945	.9946	.9948	.9949	.9951	.9952
2.6	.9953	.9955	.9956	.9957	.9959	.9960	.9961	.9962	.9963	.9964
2.7	.9965	.9966	.9967	.9968	.9969	.9970	.9971	.9972	.9973	.9974
2.8	.9974	.9975	.9976	.9977	.9977	.9978	.9979	.9979	.9980	.9981
2.9	.9981	.9982	.9982	.9983	.9984	.9984	.9985	.9985	.9986	.9986
3.0	.9987	.9987	.9987	.9988	.9988	.9989	.9989	.9989	.9990	.9990
3.1	.9990	.9991	.9991	.9991	.9992	.9992	.9992	.9992	.9993	.9993
3.2	.9993	.9993	.9994	.9994	.9994	.9994	.9994	.9995	.9995	.9995
3.3	.9995	.9995	.9995	.9996	.9996	.9996	.9996	.9996	.9996	.9997
3.4	.9997	.9997	.9997	.9997	.9997	.9997	.9997	.9997	.9997	.9998

Chapter VII Review

1. A pair of standard dice are rolled. What is the probability that the sum rolled is :
   a) less than five
   b) six and the second die is a three
   c) six or the second die is a three
   d) six given the second die is a three

2. A die is made that has two faces marked with 2's, three faces marked with 3's, and one face marked with a 5. If this die is thrown once, find the probability of:
   a) getting a 2
   b) not getting a 2
   c) getting an odd number

3. If there are 10 chips in a box - 4 red, 3 blue, 2 white and 1 black - and two chips are drawn without replacement, what is the probability that:
   a) the chips are the same color
   b) exactly one chip is red
   c) at least one chip is red
   d) neither is red
   e) the second is red given the first is blue

4. If the odds in favor of buying a blue car are 3 to 4, then what is the probability of buying a blue car?

5. Three-fifths of a particular class are brunettes. Two-thirds of the brunettes are female, while one fourth of the rest of the class are female. Find the probability that a person in the class chosen at random is:
   a) female
   b) brunette but not female
   c) brunette or female

6. What is the probability of correctly guessing the order in which five songs will be played at a concert?

7. In how many ways can the questions on a 10-item test be arranged in different orders?

8 A survey was conducted at the local mall concerning spending habits. One question dealt with purchases of compact discs over the last year. The results are in the table.

	No. of CDs purchased			
Age (yrs)	0	1 - 5	6 - 10	over 10
under 21	4	9	15	18
21 - 40	9	12	13	3
over 40	20	7	5	1

A person is chosen at random from those surveyed. Find the probability that the person:

a) is under 21 yrs.
b) purchased 6 - 10 CDs.
c) purchased 6 - 10 CDs given that he or she is under 21 yrs.
d) is over 40 yrs. and purchased over 10 CDs.
e) purchased over 10 CDs or is over 40 yrs.
f) is over 40 yrs. given he or she purchased over 10 CDs.

9. Dura-ready batteries work on the average for 28.5 hours with a standard deviation of 6.8 hours. Assuming a normal distribution, what is the probability that a random Dura-ready battery lasts:
   a) less than 24 hours.
   b) at least 30 hours.
   c) between 20 and 35 hours.

10. The Math Department needs to select a team of four students to participate in a mathematics contest. Possible team members include 5 graduate students, 3 undergraduate math majors, 4 math education majors and Bubba.
   a) How many different teams can be formed?
   b) How many different teams contain exactly 2 graduate students, 1 math major and 1 math education major?
   c) What is the probability that Bubba is on the team?
   d) What is the probability that there is at least one graduate student on the team?

11. Margaret built a circular garden in her backyard that has a diameter of 10 ft. In the center of the garden is a circular bird bath that has a diameter of 3 ft. If a squirrel on a tree limb directly over the garden drops an acorn at random and it lands in the garden, what is the probability it does not hit the bird bath?

12. What is the probability of being dealt a five-card poker hand that contains:
   a) exactly 3 aces?
   b) 4 aces and a face card?
   c) no face cards?

13. A bag contains six slips of paper on which one of the letters A,T,S,R,E or D is written. One slip is chosen at random, the letter recorded, and then the slip is returned to the bag. This is repeated four times. What is the probability that the letters written, in order, spell "READ"?

14. Becky rented six movies over the weekend. In how many different ways can she watch the movies? (Assume she only watches a movie once.)

15. Marvin's sock drawer contains 4 blue socks and 6 black socks. Marvin chooses two socks without looking at them. What is the probability that they are both blue?

16. Timing mechanisms for a particular watch are shipped to the factory 12 to a box. The company which makes the mechanisms knows that there are two defective ones in each box of 12 shipped to the factory. However, they also know that the factory foreman will only check two of the mechanisms from the box. If both work, he keeps the shipment; otherwise, it goes back to the company. What is

the probability that the factory keeps the shipment?

17. There are twelve dancers on the CHS dance team. In how many different ways can they line up to take a team picture?

18. Use the Venn Diagram shown to compute the following:
    a) P(A)
    b) P(B)
    c) P(A and B)
    d) P(A or B)
    e) P(not A and not B)

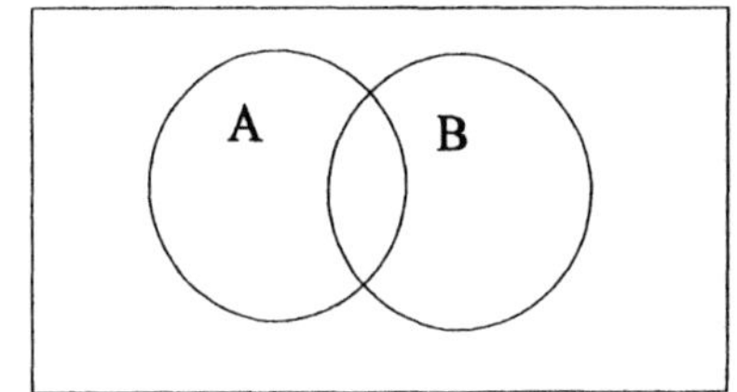

19. A survey of 40 first graders showed that 10 liked only chocolate ice cream, 8 liked only vanilla ice cream and 2 liked both. What is the probability that a student chosen from this class:
    a) likes chocolate ice cream?
    b) likes vanilla ice cream?
    c) likes vanilla and chocolate ice cream?
    d) likes vanilla or chocolate ice cream?
    e) doesn't like chocolate and doesn't like vanilla ice cream?

20. Jessica is planning her wardrobe for a weekend trip and has some color-coordinated separates. She has packed a beige skirt and matching blouse, a pair of blue pants with matching blouse, and white blouse.
    a) How many outfits can she create with the clothes she has packed?
    b) Jessica figures that by wearing a jacket or not wearing a jacket with a given top and bottom, she will have two different outfits. So she decides to add a beige jacket (that must be worn with a blouse) to her suitcase. How many outfits can she create now?
    c) At the last minute, Jessica also grabs a red sweater that matches everything in her suitcase. (The sweater must be worn with a blouse and cannot be worn with the jacket.) How many outfits can she make now?
    d) One day on her trip, Jessica decides to wear the skirt and the jacket. How many outfits can she choose from?
    e) One day on her trip, Jessica decides to wear the pants, how many outfits can Jessica create?

21. 19, 14, 9, 4, ...
    a) If the pattern continues, what could the next two numbers be? Explain how you found them.
    b) Without finding all the ones in between, explain how to find the twentieth number in the pattern.

22. $10, \frac{10}{3}, \frac{10}{9}, \ldots$
    a) If the pattern continues, what could the next two numbers be? Explain how you found them.
    b) Without finding all the ones in between, explain how to find the ninth number in the pattern.

23. Fill in the blanks to complete the sequence and explain how you obtained you answers:
    a) 10, 14, 20, 28, ____, ____, ____, ____ b) 20, 21, 19, 22, 18, ____, ____, ____, ____

Chapter VIII

Section 61: Natural Numbers and Whole Numbers

In your previous mathematics courses you learned about various sets of numbers, such as the integers and the rational numbers. In addition, you studied operations that act on these sets. The operations of addition and multiplication have some special properties, such as commutativity, which you have known use since elementary school. In this chapter we will examine familiar number systems under the four arithmetic operations and "rediscover" their properties. In order to really understand and appreciate these properties, we will also consider some non-familiar sets and operations.

The first numbers that a child learns are the counting numbers: 1, 2, 3, ... This collection of numbers is also called "the set of natural numbers", or N.

$N = \{1, 2, 3, 4, \ldots\}$.

If a number is in N we say that it is an element of N, the set of natural numbers, and we use the symbol $\in$ to mean "is an element of". For example, $1004 \in N$ but $-23 \notin N$.

After the natural numbers children learn about zero. If we include this number along with all of the natural numbers, we get the set of whole numbers, W. So $W = \{0, 1, 2, 3, 4, \ldots\}$. It is clear that $0 \in W$ but $0 \notin N$.

In the primary grades students learn the how to add, subtract, multiply and divide elements of these sets. Each of these operations is applied to two elements of the set and the result is another number - not necessarily in the set.

Example 1: 0, 4, 9, and 12 are all elements of the set of whole numbers.

a) $0 + 4 = 4$ and $4 \in W$.

b) $9 + 12 = 21$ and $21 \in W$.

c) $9 - 4 = 5$ and $5 \in W$.

d) $9 - 12 = -3$ but $-3 \notin W$.

e) $4 \times 12 = 48$ and $48 \in W$.

f) $9 \times 0 = 0$ and $0 \in W$.

g) $12 \div 4 = 3$ and $3 \in W$.

h) $9 \div 4 = 2.25$ but $2.25 \notin W$.

Notice that when we subtract or divide two whole numbers, there is no guarantee that the result will be another whole number. However, our experience tells us that when we add or multiply two whole numbers, the answer is another whole number. Whenever we perform an operation on a set of numbers it is important to know whether the result will be in the set. When we are guaranteed that it is, we say that the set is **closed under the operation**. The closure property depends on the set and on the operation so it is essential that both the operation and the set are named. We say that the set of whole numbers is closed under addition, or addition is closed in the set of whole numbers. However, the set of whole numbers is not closed under subtraction or division.

It is often easier to prove that a statement is false than it is to prove that it is true. Consider the statement, "All mathematics teachers at this school have brown hair." To prove that this is a true statement we would have to verify that each math teacher does indeed have brown hair. Only then can we say with certainty that the statement is true. If, however, we find only one teacher with hair that is not brown, we know immediately that the statement is false. This teacher is called a "counterexample".

If we wish to prove that a set is not closed under an operation, we only need to find two elements in the set whose answer is not in the set. This counterexample is our proof that the closure property does not hold.

Example 2:

a) Is the set of whole numbers closed under division?
No. $9 \in W$ and $4 \in W$ but $9 \div 4 = 2.25$ and $2.25 \notin W$.

b) Is the set of odd natural numbers closed under addition?
No. 7 and 11 are both odd natural numbers but $7 + 11 = 18$ and 18 is not an odd number.

What happens if we try to find a counterexample to a statement and we are unsuccessful? Does that mean that the statement is true? Not necessarily. We can only suspect that it is. In this case we have three options:

1) Accept the statement as true. This is rarely done. If so, we call the statement an "axiom".

We cannot prove that N is closed under addition, but we accept this property as an axiom. Likewise, we also accept as facts that the set of whole numbers is closed under addition and that both N and W are closed under multiplication.

2) Use mathematics to prove that the statement is true.

Example 3: Is the set of even natural numbers closed under addition?

We suspect that the sum of any two even natural numbers is an even natural number. Let us prove it. (Recall: An even number is a multiple of 2. So an even natural number is any number that can be written as 2 times another natural number.)

Proof: Let m and n be two even natural numbers. Then $m = 2k$ and $n = 2j$ for some natural numbers k and j. Now $m + n = 2k + 2j = 2(k + j)$. The natural numbers are closed under addition so $k + j \in N$. We have written $m + n$ as the product of 2 and a natural number. Therefore, $m + n$ is an even natural number.

So we have proven that the set of even natural numbers is closed under addition.

3) State that although we cannot prove that the statement is either true or false, we think that it is true.

If we do not have a counterexample and we cannot prove that a statement is true, it is very important that we do not claim that it true. We have only failed to prove that it is false. We will see an example of this later in this section.

Students quickly learn that $3 + 4 = 4 + 3$. This property is called the **commutative property of addition**. It means that the order in which the numbers are added is not important. We accept as axioms that addition and multiplication are commutative in both N and W.

Example 4: Define a new operation, *, on W by $b * c = 2b + c$. Is the operation * commutative in W?

To show that * is commutative, we must pick two arbitrary whole numbers b and c and show that $b * c$ is always the same as $c * b$.

Let us try an example to see what happens: Pick $b = 5$ and $c = 8$. Then

$$\begin{aligned} 5 * 8 &= 2(5) + 8 = 10 + 8 \\ &= 10 + 8 \\ &= 18 \end{aligned} \qquad \text{and} \qquad \begin{aligned} 8 * 5 &= 2(8) + 5 = 16 + 5 = 21. \\ &= 16 + 5. \\ &= 21 \end{aligned}$$

But $18 \neq 21$; so $5 * 8 \neq 8 * 5$. Therefore, * is not commutative in W. (Actually, * would fail to be commutative in any set except the one element set, $\{0\}$.)

Operations such as addition and multiplication are often called "binary" operations because each operation is performed on two numbers at a time. If we wish to perform an operation on more than two numbers, we must do it in stages – two numbers at a time; and often we must identify which two numbers must be used first. This is done with parentheses. For example, $(2 + 10) + 5$ means add two and ten first than add five to the result: $(2 + 10) + 5 = 12 + 5 = 17$. While $2 + (10 + 5)$ means add ten and five first and then add this result to two: $2 + (10 + 5) = 2 + 15 = 17$. For addition we are not surprised that changing the grouping does not change the result. This feature is called the **associative property of addition**. Likewise, we learned long ago that $5(4 \cdot 7) = (5 \cdot 4)7$, that is, in a multiplication problem the grouping is unimportant. This is because multiplication in the set of whole numbers is associative. We accept as axioms that the operation of addition is associative in N and W and that multiplication is associative in N and W.

Example 5: Define # on N by $d \# f = d + 3f$. Is the operation # associative in N?

Again we begin with a specific example to understand exactly what the operation does. Maybe this will be a counterexample. Since we are considering the associative property, we need to pick three numbers and test to see if changing the grouping changes the answer. Let's pick 5, 7, and 2. We need to calculate $(5 \# 7) \# 2$ and $5 \# (7 \# 2)$ and compare the answers.

$$\begin{aligned} &(5 \# 7) \# 2 \\ &= (5 + 3 \times 7) \# 2 \\ &= (5 + 21) \# 2 \\ &= 26 \# 2 \\ &= 26 + 3 \times 2 \\ &= 26 + 6 \\ &= 32 \end{aligned} \qquad \begin{aligned} &5 \# (7 \# 2) \\ &= 5 \# (7 + 3 \times 2) \\ &= 5 \# (7 + 6) \\ &= 5 \# 13 \\ &= 5 + 3 \times 13 \\ &= 5 + 39 \\ &= 44 \end{aligned}$$

Clearly, the result is not the same. We know that # is not associative in N since $5, 2, 7 \in N$ but $(5 \# 7) \# 2 \neq 5 \# (7 \# 2)$ since $32 \neq 44$.

Example 6: We can define an operation on a finite set by using a table. The table below defines the operation □ on the set S = {a, b, c}

□	a	b	c
a	a	b	c
b	b	c	a
c	c	a	b

a) What is b □ c?

b) What is c □ b?

c) Is □ commutative in S?

d) Is □ associative in S?

a) To find b □ c we must locate the row containing b and the column containing c and find the square where they intersect. So b □ c = a.

b) To find c □ b we must locate the row containing c and the column containing b and find the square where they intersect. So c □ b = a.

c) We don't know yet. All we know from parts a) and b) is that for the specific elements b and c, the order is unimportant. To be sure that □ is communicative we must test all possible pairs. (There are six to check, since we need not verify that a □ a = a □ a, etc.) What are they?

So you can prove that □ is commutative.
Can you tell by looking at the table if the operation is commutative?

d) It would take much more work to verify that □ is associative. You would have to verify 27 cases including (b □ b) □ b = b □ (b □ b) which we will do.

(b □ b) □ b
= c □ b
= a

b □ (b □ b)
= b □ c
= a

Again one example does not prove that □ is associative. We would have to prove all 27 cases before we can say that it is. Let us try a couple of more examples:

Is (a □ c) □ b = a □ (c □ b)?

Is (b □ a) □ c = b □ (a □ c)?

Since we have not found a counterexample and we have not proved that □ is associative, the best we can say is that we think it might be associative.

Homework: Section 61

1. Answer each of the following and **prove your answers**:
   a) Is subtraction commutative in N?
   b) Is division commutative in N?
   c) Is division associative in W?
   d) Is subtraction associative in W?

2. Define # on W by b # c = b + 3c. (Justify your answers.)
   a) Is W closed under #?
   b) Is # commutative in W?
   c) Is # associative in W?
   d) What is 5 # 0? What is 16 # 0? What is w # 0, where w is any whole number?
   e) What is 0 # 6? What is 0 # 9? What is 0 # 0? What is 0 # w where w is any whole number?

3. Let $a @ b = \frac{(a + b)}{2}$ (Justify your answers.)
   a) Is @ closed in W?
   b) Is @ closed in the set of even whole numbers?
   c) Is @ closed in the set of odd whole numbers?
   d) Is @ commutative?
   e) Is @ associative?

4. Define * on N by s * t = 10s - 2t. (Justify your answers.)
   a) Is N closed under *?
   b) Is * commutative?
   c) Is * associative?

5. Let ⊙ be an operation on the set of natural numbers by $a \odot b = a^b$. (Justify your answers.)
   a) Is N closed under ⊙?
   b) Is ⊙ commutative?
   c) Is ⊙ associative?
   d) What is 4 ⊙ 1? What is 91 ⊙ 1? What is n ⊙ 1, where n is any natural number?
   e) What is 1 ⊙ 5? What is 1 ⊙ 10? What is 1 ⊙ n, where n is any natural number?

6. Let T = {u, v, w} and # be defined on T: (Justify your answers.)

#	u	v	w
u	u	v	w
v	v	u	w
w	w	w	w

a) Is # closed in T?

b) Is # commutative?

c) Is # associative?

d) What is u # u? What is v # u? What is w # u?

e) What is u # u? What is u # v? What is u # w?

Section 62: The Set of Integers

In the previous section, we that the set of whole numbers is formed by including zero with the natural numbers. The arithmetic operations on W are easily defined by using the operations that were already known for N. We only need to worry about operations involving the new element.

For addition, $w + 0 = w$, $0 + w = w$, and $0 + 0 = 0$.

Multiplication involving the new element is simple: $0 \cdot w = 0$ and $w \cdot 0 = 0$ for any whole number w. This property is often called the "First Multiplicative Property of Zero"; the product of zero and any number is zero."

Zero has a second property with regard to multiplication. The first one is that the answer to a multiplication problem will always be zero if one of the numbers is zero. The second assures us that the <u>only</u> way that the answer of a multiplication problem can be zero is if one of the numbers is zero. The Second Multiplicative Property of Zero can be stated this way: "If the product of two or more numbers is zero, then one of the numbers must be zero.

One is also very special number with regard to multiplication: the product of one and a number is simply that number. One the **identity element for multiplication**, or we say that one is the multiplicative identity. It is very important that the operation be named. Each operation on a set may or may not have a corresponding identity.

Example 1: a) Is there an identity for addition in W?
b) Is there an identity for addition in N?

a) We know that the sum of any whole number and zero is the number, that is, $w + 0 = w$ and $0 + w = w$ for any whole number w.
Moreover, 0 is an element of W so 0 is the identity for addition in W.

b) For any natural number n, $n + 0 = n$ and $0 + n = n$. However, 0 is <u>not</u> an element of N so addition has no identity in N.

Notice that in order for an element of a set to be an identity for some operation on the set, two criteria must be met. First the identity must be in the set and second, the identity must "work from either side".

Example 2: Consider Problem 2 of the previous homework where # is defined on W by b # c = b + 3c. Is there an identity for # in W?

You found that 5 # 0 = 5, 16 # 0 = 16, and w # 0 = w for any whole number w. Since $0 \in W$ and w # 0 = w for any whole number w, we call zero a "right identity" for # in W. In this case, when zero operates on a number from the right the number is unchanged.

You also discovered that 0 # 6 = 18 $\neq$ 6. In fact, for any whole number w, 0 # w = 3w. When zero operates on a number from the left, the result is not always the number. We say that zero is not a left identity. Even though $0 \in W$ and it is a right identity, zero is not an identity for # in W because it fails to leave the number unchanged when it operates on either side.

The previous example shows that it is possible for a set to have a "right identity" or a "left identity" for a given operation, but the word "identity" is reserved for a two-sided identity.

Since we will be creating new sets from old ones, we need to agree on some terminology. Let us look at an example first.

Example 3:

Let A = {2, 3, 6, 7, 18} and B = {2, 4, 6, 7, 9}.

a) What is the collection of numbers that these sets have in common?
b) What is the set formed by taking elements that appear in either set?
c) Which elements of A are not elements of B?
d) What is the collection of numbers that are in B but not in A?

a) Both sets contain the numbers 2, 6, and 7. So the desired set is {2, 6, 7}.

b) The numbers 2, 3, 6, 7, 18, 4, and 9 are elements of one set or the other. The new set is {2, 3, 4, 6, 7, 9, 18}. Notice that when listing the elements of a set the order is not important although we often list numbers in ascending order. Also, since a set is a collection of items, we do not list an element more than once.

c) The elements of A are 2, 3, 6, 7, and 18. Since 2, 6, and 7 are also in B, the new set cannot contain these elements. The answer is {3, 18}.

d) We start with the elements of B: 2, 4, 6, 7, and 9. We then eliminate the numbers that are also elements of A, namely, 2, 6, and 7. The resulting set is {4, 9}.

Each of the sets described in Example 3 above have a special name:

a) The set created by including all the elements that two sets, S and T, have in common is called the **intersection** of the sets and is denoted as $S \cap T$. In the example above $A \cap B$ = {2, 6, 7}.

If two or more sets have no elements in common we say that their intersection is the empty set. The symbol for the empty set is $\varnothing$.

b) The set that is composed of elements that appear in either set S or in set T is called the **union** of S and T. $S \cup T$ is the notation for this set. In the example, $A \cup B = \{2, 3, 4, 6, 7, 9, 18\}$.

c) Whenever we discuss elements that are not in a given set S, we are considering the **complement** of the set. When the universal set is known, the complement of S is denoted by an overlined S, $\overline{S}$, or with S'. However, it is possible to find the complement of a set S with respect to any set T. T - S is the complement of S with respect to T and consists of elements that are in T but not in S. In the Example 3, $A - B = \{3, 18\}$.

d) The complement of A with respect to B is $B - A = \{4, 9\}$.

Example 4:

Let $R = \{a, b, c, d\}$ $S = \{a, b, e, f, g\}$ $T = \{c, e, g, h, i\}$

a) $R \cap S = \{a, b\}$ b) $R \cup S = \{a, b, c, d, e, f, g\}$

c) $R - S = \{c, d\}$ d) $S - R = \{e, f, g\}$

d) $R \cap T = \{c\}$ e) $S \cup T = \{a, b, c, e, f, g, h, i\}$

f) $T - S = \{c, h, i\}$ g) $R - T = \{a, b, d\}$

h) $(R \cap S) \cup T = \{a, b\} \cup \{c, e, g, h, i\} = \{a, b, c, e, g, h, i\}$

i) $(R \cap S) \cap T = \{a, b\} \cap \{c, e, g, h, i\} = \varnothing$

Until now we defined a set by listing the elements of the set. However, we often define a set by describing a general element of the set. This is called set-builder notation. For example, if $w \in W$, $-w$ is the opposite of w. Let S be the set that consists of all the opposites of the whole numbers. So $S = \{0, -1, -2, -3, ...\}$. We can also define S as $\{x \mid x \text{ is an opposite of a whole number}\}$ or $S = \{x \mid x = -w \text{ for some } w \in W\}$.

What is $W \cup S$? It is $\{..., -4, -3, -2, -1, 0, 1, 2, 3, ...\}$. This new set is called the **set of integers**. We will now examine this important set more closely.

The set of integers consists of all the whole numbers and their opposites. We will use the symbol Z for the set of integers. Other books may use a different symbol such as I. So

$$Z = \{..., -3, -2, -1, 0, 1, 2, 3, ...\}.$$

We have created the set of integers by using the set of whole numbers. We will now define operations on the integers by using the results of operations on whole numbers.

Each of the operations requires two numbers. Let $y, z \in Z$. There are three possibilities:

i) One (or both) of the numbers is zero.
ii) The numbers have the same sign, that is, they are both positive or they are both negative.
iii) The numbers have opposite signs; one is positive and the other is negative.

The magnitude or absolute value of any integer is a whole number. So operations involving the magnitudes of integers are operations on the set of whole numbers. The results of these operations are known.

Addition:

i) For any $z \in Z$, $z + 0 = 0 + z = z$.

ii) If two integers have the same sign, their sum is obtained by adding their magnitudes and using their common sign, that is, the sum of two positive integers is positive and the sum of two negative integers is negative.

iii) If the numbers have opposite signs, the magnitude of their sum is found by subtracting the smaller magnitude from the larger magnitude. The sign of their sum is the same as the sign of the integer with the larger magnitude.

Subtraction: To find the difference between two integers, add the opposite of the second to the first: $y - z = y + (-z)$

This definition of subtraction allows us to easily change any difference to a sum.

Example 5: Find each of the following:

a) $15 + (-23)$ b) $-12 + 34$ c) $-16 - 5$ d) $21 - (-6)$

a) We subtract magnitudes: $23 - 15 = 8$. Since -23 has the larger magnitude, the sum is negative. Thus $15 + (-23) = -8$.

b) We subtract magnitudes: $34 - 12 = 22$. Since 34 has the larger magnitude, the sum is positive. $-12 + 34 = 22$.

c) By the definition of subtraction, $-16 - 5$ is the same as $-16 + (-5)$. We add magnitudes: $16 + 5 = 21$. Since both the integers are negative, the sum is negative. $-16 - 5 = -16 + (-5) = -21$

d) $21 - (-6) = 21 + (6) = 27$

In a similar manner, we define multiplication and division on the integers using their magnitudes. The magnitude of the answer if found by multiplying or dividing. We need only learn how to find the sign of the result.

i) The product of zero and any integer is zero.

ii) The product of a positive and a negative integer is negative.

iii) The product of two negative integers is positive.

Recall that if z is an integer, the z may be zero, positive, or negative. So the opposite of z, -z may be zero, negative, or positive. The following properties summarize multiplication of opposites in Z:

Law of Signs I: $(-x)(y) = x(-y) = -(xy)$

"If the opposite of a number is multiplied by another number, the result is the same as multiplying the two numbers first and then taking the opposite."

Law of Signs II: $(-x)(-y) = xy$

"The product of the opposites of two numbers is the same as the product of the two numbers."

Homework: Section 62

1. a) Is there a multiplicative identity in W?

   b) Is there a multiplicative identity in Z?

2. Does Z have an identity for division?

3. Is there an identity for subtraction in Z Justify your answers.

4. Explain how the fact that one is the multiplicative identity for any real number is used to write a fraction in another form: $\frac{3}{4} = \frac{15}{20}$

5. Let $s, t \in Z$. What, if anything, can you conclude about s and t in each of the following:

   a) $3 \cdot 0 = s$ b) $2 \cdot s \cdot 5 = 0$ c) $s \cdot t = -12$

   d) $s + t = 0$ e) $t \cdot s = 0$ f) $s + 6 = 0$

   g) $7 \cdot (-t) = -28$ h) $-(t \cdot s) = 21$

6. Let P = {5, 7, 9, 11, 13} Q = {6, 8, 10, 12} R = {7, 8, 9, 10, 11, 12}

   a) $P \cup Q$ b) $R \cap Q$ c) $P \cup R$ d) $P \cap Q$ e) $R - P$ f) $Q - R$

7. Define the operation $\odot$ by $a \odot b = a^b$.

   a) Does N have an identity under $\odot$?

   b) Is Z closed under $\odot$?

   c) Is W closed under $\odot$?

8. The table below defines the operation □ on the set S = {a, b, c}

□	a	b	c
a	a	b	c
b	b	c	a
c	c	a	b

   Is there an identity for □ in S?

9. Identify each of the following sets:

   a) $N \cup \{0\} =$ ____ b) $W \cap N =$ ____ c) $W \cup N =$ ____

10. Let $a @ b = \frac{(a + b)}{2}$ and let E be the set of even integers. Is there an identity for @ in E?

11. Define ◊ on Z by $w \lozenge v = -3wv$.

    a) Is Z closed under ◊?

    b) Is ◊ commutative?

    c) Is ◊ associative?

    d) Is there an identity for ◊ in Z?

12. Define * on Z by $s * t = 6s - 6t$.

    a) Is Z closed under *?

    b) Is * commutative?

    c) Is * associative?

    d) Is there an identity for * in Z?

Section 63: Inverses

We call 2 and -2 opposites because they are the same distance from zero on the number line but they are on opposites sides of zero. Another way to verify that 2 and -2 are opposites is to note that $2 + (-2) = 0$. This means that one number "undoes" the other by addition. Because of this, we also call -2 the "additive inverse" of 2. Likewise 2 is the additive inverse of -2. Two integers are **additive inverses** if their sum is zero.

In general what does it mean to say that an element has an inverse in a set? Let * be an operation on a set S and let $s \in S$. An inverse of s must "undo" s when we perform the operation * on the two elements. By "undo", we mean that the result is the identity of the operation in the set S. 4 "undoes" -4 by addition since $-4 + 4 = 0$ and 0 is the additive identity. In the set of real numbers $\frac{2}{3}$ "undoes" $\frac{3}{2}$ under multiplication since $\frac{3}{2} \times \frac{2}{3} = 1$ and 1 is the multiplicative identity. In addition, the inverse of a number must "undo" the number from either side. If the operation is not commutative, we must verify that the inverse is both a left inverse and a right inverse. Finally, the inverse itself must be an element of the set S.

We summarize these below:

Let * be an operation on a set S and let e be the identity for * in S.

Let s be an element of S. We say that t is the inverse of s if the following hold:

1) $t \in S$
2) $s * t = e$ and $t * s = e$ (e is the identity for *.)

Example 1: Define # on Z by $a \# b = a + b - 2$.

a) Does 7 have an inverse in Z?

b) Does every integer have an inverse under #?

a) First, we must find the identity for # in Z. The identity is 2. We know this because $2 \in Z$ and, for any $z \in Z$,

$$z \# 2 = z + 2 - 2 = z \quad \text{and} \quad 2 \# z = 2 + z - 2 = z$$

Next we need to find a number, v, that satisfies the conditions: $7 \# v = 2$ and $v \# 7 = 2$.

$7 \# v = 7 + v - 2 = 2$. If we solve this, we see that $v = -7 + 4 = -3$. So -3 is the right inverse of 7.

$v \# 7 = v + 7 - 2 = 2$. If we solve this, we see that $v = -7 + 4 = -3$. So -3 is the left inverse of 7.

So -3 is the inverse of 7 for the operation # in Z.

b) Next we need to find a number that satisfies the condition: z # ▒ = 2 and ▒ # z = 2.
z # ▒ = z + ▒ - 2 = 2. If we solve this, we see that ▒ = -z + 4.

▒ # z = ▒ + z - 2 = 2. If we solve this, we see that ▒ = -z + 4.

This gives us a "candidate" for the inverse of z. (When looking for a candidate it is often helpful to try a few specific numbers and look for a pattern.) We now <u>prove</u> that for any $z \in Z$, $-z + 4$ is its inverse under #.

1) Since -z is an integer whenever z is an integer and the integers are closed under addition, $-z + 4 \in Z$.

2)

$$z \# (-z + 4) = z + (-z + 4) - 2 = z + (-z) + 4 - 2 = 2$$

$$(-z + 4) \# z = (-z + 4) + z - 2 = 2.$$

Thus, every element of z has an inverse under # in Z and that inverse is $-z + 4$.

Homework: Section 63

1. a) Which elements of N have multiplicative inverses?

   b) Which elements of Z have multiplicative inverses?

2. The table below defines the operation □ on the set S = {a, b, c}

□	a	b	c
a	a	b	c
b	b	c	a
c	c	a	b

In the previous section you discovered that a is the identity for □ in S. Does each element of S have an inverse under □?

3. Let Z^- be the set of negative integers and Z^+ be the set of positive integers. Label each of the following as true or false and explain your answer.

_____________ a) $Z^+ \cup Z^- = Z$

_____________ b) $Z \cap Z^+ = Z^+$

_____________ c) $Z - Z^- = N$

_____________ d) $Z^+ \cup W = Z^+$

4. Let T = {u, v, w} and # be defined on T:

#	u	v	w
u	u	v	w
v	v	u	w
w	w	w	w

a) Is there an identity for # in T?
b) If x represents an arbitrary element of T, what is x # w?
c) If $r \in T$, what is w # r?
d) Does every element in T have an inverse?

5. Define ◇ on Z by $w \diamond v = w + v - 5$.
   a) Is Z closed under ◇?
   b) Is ◇ commutative?
   c) Is ◇ associative?
   d) Is there an identity for ◇ in Z?
   e) Does every element of Z have an inverse under ◇?

6. Define @ on Z by $c @ d = c - d + 3?$.
   a) Is Z closed under @?
   b) Is @ commutative?
   c) Is @ associative?
   d) Is there an identity for @ in Z?
   e) Does every element of Z have an inverse under @?

Section 64: Rational Numbers

In this section, we continue our discussion of familiar sets and operations. As we have discovered, properties of an operation depend on both the operation and the set upon which the operation acts. So we treat the operation and the set together as a system. To avoid confusion we often denote a system as (S, *) where S is the set and * is the operation. Some of the systems we have studied include (N, +), (N, ·), (W, +), (W, ·), (Z, +), and (Z, ·). Each of these systems is closed. The grouping of the elements does not affect the answer so the operation is associative in each of these systems. Each system is commutative since the order of the elements being operated upon is unimportant.

Let us compare (N, +), (W, +), and (Z, +). (N, +) does not have an identity. Zero is the identity for (W, +) and zero is the only element of (W, +) with an inverse. Zero is the identity for (Z, +) and every element of (Z, +) has an inverse. Once an additive system has inverses, subtraction is closed. (We subtract a number by adding its inverse.) Can we accomplish the same objective for multiplication?

One is the identity in (N, ·), in (W, ·), and in (Z, ·). In (N, ·) and in (W, ·), one is the only element with an inverse. Both 1 and -1 have an inverse in (Z, ·). We will now try to create a set based on the integers in which every element has a multiplicative inverse.

A rational number is any number <u>that can be written</u> as the quotient of an integer and a nonzero integer. Some examples of rational numbers are

$$-\frac{3}{4};\ \frac{9}{-7};\ -5 = \frac{-5}{1};\ 33\% = \frac{33}{100};\ .\overline{45} = \frac{5}{11}.$$

As seen in this example, any integer is a rational number. In fact, any terminating or repeating decimal is a rational number. We will use the letter Q to denote the set of all rational numbers. (While Q is a common symbol, other books may use other symbols.) We cannot list the elements of the set of rational numbers in the same manner that we did for N, W, and Z. It is easier to use set-builder notation:

$$Q = \{q \mid q = \frac{a}{b} \text{ where a and b are integers and } b \neq 0\}.$$

Any rational number may be written in many different ways.

For example, $-\frac{7}{3} = \frac{-7}{3} = \frac{7}{-3} = \frac{14}{-6} = \frac{-56}{24}$.

We need a way to determine if two different numerals represent the same rational number: $\frac{a}{b} = \frac{c}{d}$ if and only if ad = bc and we say that $\frac{a}{b}$ and $\frac{c}{d}$ are equivalent.

Now we are ready to define addition and multiplication on Q. Recall that addition can only be performed on "like things". So we can only define addition of two rational numbers that have been written using the same denominator. (If the rational numbers have different denominators, write them using a common denominator first.)

Let $\frac{a}{b}$ and $\frac{c}{b}$ be two elements of Q, we define $\frac{a}{b} + \frac{c}{b}$ to be $\frac{a+c}{b}$. Note that both the numerators are integers and the denominator is not zero; we have defined addition in Q using addition in Z.

Example 1: Add: $\frac{2}{15} + \left(-\frac{4}{25}\right)$

First write each number with a denominator of 75. $\frac{2}{15} = \frac{10}{75}$ and $-\frac{4}{25} = \frac{-12}{75}$.

ow add $\frac{10}{75} + \left(\frac{-12}{75}\right) = \frac{10 + (-12)}{75} = \frac{-2}{75} = -\frac{2}{75}$.

Let us investigate the properties of addition in the set of rational numbers. It appears from the definition that the set of rational numbers closed under addition. To prove it we must choose two <u>arbitrary</u> elements of Q, say r and q. Since we can write each of these numbers as the quotient of integers with a common denominator, we simply choose $r = \frac{x}{y}$ and $q = \frac{z}{y}$ where x, z ∈ Z and y is a nonzero integer.

Then $r + q = \frac{x}{y} + \frac{z}{y} = \frac{x+z}{y}$ by definition of addition.

Now x and z are elements of Z and Z is closed under addition so x + z is an element of Z. Moreover, y is a nonzero integer. Hence $r + q = \frac{x+z}{y} \in Q$. Therefore, the set of rational numbers is closed under addition.

Is addition associative in the set of rational numbers? It is helpful to try a specific example first.

Example 2: Is $\left(\frac{3}{4} + \frac{8}{4}\right) + \frac{10}{4} = \frac{3}{4} + \left(\frac{8}{4} + \frac{10}{4}\right)$?

$$\left(\frac{3}{4} + \frac{8}{4}\right) + \frac{10}{4} = \frac{(3 + 8)}{4} + \frac{10}{4} = \frac{11}{4} + \frac{10}{4} = \frac{11 + 10}{4} = \frac{21}{4}$$

$$\frac{3}{4} + \left(\frac{8}{4} + \frac{10}{4}\right) = \frac{3}{4} + \frac{(8 + 10)}{4} = \frac{3}{4} + \frac{18}{4} = \frac{3 + 18}{4} = \frac{21}{4}$$

In this problem we used the definition of addition in Q, associativity of addition in Z, and the definition of addition in Z. It appears that we should be able to prove that addition is associative in the set of rational numbers. Let p, q, r $\in$ Q. p, q, and r can be written using the same nonzero denominator: $p = \frac{x}{w}$; $q = \frac{y}{w}$; and $r = \frac{z}{w}$.

$$(p + q) + r = \left(\frac{x}{w} + \frac{y}{w}\right) + \frac{z}{w}$$

$= \frac{(x + y)}{w} + \frac{z}{w}$ by definition of addition in Q

$= \frac{(x + y) + z}{w}$ by definition of addition in Q

$= \frac{x + (y + z)}{w}$ since addition is associative in Z

$= \frac{x}{w} + \frac{(y + z)}{w}$ by definition of addition in Q

$= \frac{x}{w} + \left(\frac{y}{w} + \frac{z}{w}\right)$ by definition of addition in Q

$= p + (q + r)$

Therefore, addition is associative in the set of rational numbers.

The proof that addition is commutative in Q is similar and is left as an exercise.

We now prove that zero is the identity for (Q, +). Let $r = \frac{c}{d} \in Q$.

$$r + 0 = \frac{c}{d} + \frac{0}{d}$$

$$= \frac{c + 0}{d} \quad \text{by definition of addition is Q}$$

$$= \frac{c}{d} \quad \text{since 0 is the additive identity in Z}$$

Since addition is commutative, we have $0 + r = r + 0 = r$ for any rational number r in Q. Therefore, 0 is the identity for (Q, +).

It is left as an exercise for the student to prove that every rational number has an additive inverse.

It is now necessary to define multiplication in the set of rational numbers. Let $\frac{a}{b}$ and $\frac{c}{d}$ be two elements of Q, we define $\frac{a}{b} \cdot \frac{c}{d}$ to be $\frac{a \cdot c}{b \cdot d}$. Here we have used multiplication in Z to define multiplication in Q.

It is not difficult to prove that the set of rational numbers is closed under multiplication and that multiplication is associative in Q. Let us prove that multiplication in Q is commutative. Let $r = \frac{x}{y}$ and $q = \frac{z}{w}$ where $w, x, y, z \in Z$ and $y \neq 0$ and $w \neq 0$.

$$r \cdot q = \frac{x}{y} \cdot \frac{z}{w} = \frac{x \cdot z}{y \cdot w} \quad \text{by definition of multiplication in Q}$$

$$= \frac{z \cdot x}{w \cdot y} \quad \text{since multiplication is commutative in Z}$$

$$= \frac{z}{w} \cdot \frac{x}{y} \quad \text{by definition of multiplication in Q}$$

$$= q \cdot r$$

The proof that one is the multiplicative identity for the set of rational numbers is similar to the proof that zero is the identity in (Q, +). (If m is a nonzero integer, then $\frac{m}{m} = \frac{1}{1} = 1$ by definition of equivalent rational numbers since $m \cdot 1 = m \cdot 1$.) Moreover, it is easy to see that zero has the same multiplicative properties in Q as it does in Z and the Laws of Signs also apply in (Q, ·).

We now turn our attention to the existence of inverses in (Q, ·) For any rational number r, we must find an element of Q such that $r \cdot \square = 1$ and $\square \cdot r = 1$. Since multiplication is commutative in Q we need only prove one of these, and the other one follows. It is immediately apparent that we will have some difficulty finding an inverse for zero since $0 \cdot \square = 0$ for every rational number we might try. Thus $0 \cdot \square$ can <u>never</u> be equal to one. Because of this, zero does not have an multiplicative inverse in Q.

We concentrate on finding an inverse for each nonzero rational number. Let r be a nonzero rational number. Then $r = \frac{x}{y}$ where both x and y are nonzero integers. The candidate for the inverse of r is

$p = \frac{y}{x}$. Since x, y $\in$ Q and x $\neq$ 0, p $\in$ Q. Since y $\neq$ 0, p $\neq$ 0.

$$r \cdot p = \frac{x}{y} \cdot \frac{y}{x} = \frac{x \cdot y}{y \cdot x} \quad \text{by definition of multiplication in Q}$$

$$= \frac{y \cdot x}{y \cdot x} \quad \text{since multiplication is commutative in Z}$$

$$= 1$$

Thus r and p are multiplicative inverses in Q.

Once the existence of additive inverses is proven, it is possible to define an operation that "undoes" addition, namely subtraction. The difference of two elements is the sum of the first and the additive inverse of the second: a - b = a + (-b). In a similar manner, the existence of multiplicative inverses for nonzero rational numbers allows us to define and operation that "undoes" multiplication, namely division. The quotient of two rational numbers is the product of the first and the multiplication inverse of the second. If $s \neq 0,\ \ r \div s = r \cdot \frac{1}{s}$.

Homework: Section 64

1. Fill in the blank with the appropriate set:

   a) $Z \cup Q =$ _______ b) $Z - Z^{-} =$ _______

   c) $Q \cap N =$ _______ d) $Z \cap Q =$ _______

   e) $(Q \cap Z) \cup W =$ _______ f) $Z^{+} - N =$ _______

2. Prove that addition is commutative in the set of rational numbers.

3. Prove that every element of Q has an additive inverse.

4. Prove that (Q, ·) is closed.

5. Prove that multiplication is associative in the set of rational numbers.

6. Prove that one is the multiplicative identity in Q.

7. Let $T = Z \cup \{ \frac{1}{z} \mid z \text{ is a nonzero integer}\}$. Every element of T is a rational number.

   a) Does every nonzero element of T have a multiplicative inverse?
   b) Is T closed under rational number addition?
   c) Is T closed under rational number multiplication?
   d) What is $2 \cdot \frac{1}{3}$? Is this product an element of T? Does this product have a multiplicative inverse in T?
   e) A set with additive inverses, (Z, +), is created by forming the union of W with the set containing all of the additive inverses of W. T is created by forming the union of the integers with the set containing all of the multiplicative inverses of nonzero integers. Why is T not the best choice for a set in which every nonzero element has a multiplicative inverse?

8. Let $Q^* = Q - \{0\}$ and $\cdot$ be the usual multiplication in Q.
   a) Is $(Q^*, \cdot)$ closed?
   b) Is there a multiplicative identity in $(Q^*, \cdot)$?
   c) Does every element in $(Q^*, \cdot)$ have an inverse?

9. Define the operation * on Q by $r * s = \frac{r}{2} \cdot \frac{s}{2}$ where $\cdot$ is the usual multiplication of rational numbers.
   a) Is (Q, *) closed?
   b) Is (Q, *) commutative?
   c) Is (Q, *) associative?
   d) Does (Q, *) have an identity?
   e) Does every nonzero element of (Q, *) have an inverse?

10. Let T = {u, v, w} and # be define on T:

#	u	v	w
u	w	u	v
v	u	v	w
w	v	w	u

a) Is # closed in T? b) Is # commutative? c) Is # associative?
d) Is there an identity for # in T? e) Which elements of (T, #) have an inverse? Name the inverse.

11. Let @ be defined on the set M = {r, s, t, v} by the table below:

@	r	s	t	v
r	r	r	r	r
s	r	v	t	s
t	r	t	v	t
v	r	s	t	v

a) Is @ closed in M?
b) Is @ commutative?
c) Is @ associative?
d) Is there an identity for @ in M?
e) Which elements of (M, @) have an inverse? Name the inverse.

Answers to Selected Homework Exercises

Chapter I

Section 1: 1. a) 2434 b) 18,831 c) 225,229 2. a) VV ⟨⟨⟨ VV ⟨⟨ V b) VV ⟨⟨⟨ VV ⟨⟨ V c) VV ⟨⟨⟨ VV ⟨⟨ V

3. a) 16,238 b) 175,081 4. a) •••/— ⊕ ••/—/— b) • •• = ⊕

Section 2: 1. a) 0, 1 b) 0, 1, 2, 3, 4, 5, 6 c) 0, 1, 2, 3, 4, ..., 8, 9, T, E, ☆
2. a) 1132 b) 103 c) 2697 d) 1644 e) 69 f) 618
3. a) 101111_{two}, 142_{five}, 52_{nine}, 32_{fifteen}
b) 1011111_{two}, 340_{five}, 115_{nine}, 65_{fifteen}
c) 10100011_{two}, 1123_{five}, 201_{nine}, $T♧_{\text{fifteen}}$

Section 3: 1. 2 x 2 x 13, 1 x 4 x 13, 1 x 1 x 52, 1 x 2 x 26 2. No, because the only factors of p are 1 and p. 3. 2, 76, 1, 38 4. 1, 2, 4, 5, 7, 10, 14, 20, 28, 35, 70, 140 5. 21, 0, 84 6. 148 7. 1, 2, 3,4,6,12; 12 is greatest 8. 90; 90, 180, 270, 360, 450, etc. 9. 1 10. 0 11. 79; 127 12. 140 = 2 x 2 x 5 x 7; 324 = 2 x 2 x 3 x 3 x 3 x 3; 441 = 3 x 3 x 7 x 7 13. 11 people

Section 4: 1. 137 2. 0, 490, 980, 1470, 1960, etc. 3. 28 4. $114♧_{\text{fourteen}}$ 5. 9 6. 11; 781; 2959 7. (a), (c) 9. 17 10. 7 11. 2•5•7 12. $2^2 \cdot 3^4 \cdot 11$ and $2^5 \cdot 3^5 \cdot 19^2$ 13. a) none b) one c) three 15. 89

Section 6: 1. 4416_{twelve} 2. 727 3. 181 4. 0, 14, 28, 42, 56, 70, ... 5. 1, 3, 5, 15 6. 1, 5, 7, 35 7. 109 9. 4:36 p.m. 10. 50

Section 7: 1. a) $\frac{3}{8}$ b) $\frac{11}{4}$ or $\frac{11}{20}$ c) $\frac{5}{12}$ 3. Parts are not equal 4. 133 5. a) $\frac{69}{8}$ b) $\frac{69}{72}$
6. 4500 8. True 9. Column J 10. Graph (c)

Section 8: 1. $\frac{6}{15}, \frac{10}{15}, \frac{4}{15}, \frac{10}{15}, \frac{5}{15}$ 2. $\frac{4}{7}; \frac{55}{66}; \frac{1}{3}$ 3. 68 miles per hour 5. 20456_{seven}
6. 1844 apples 8. 8 months 9. 10 gals 10. 30 sec.

Section 9: 3. $\frac{35}{15}, \frac{17}{7}, 2\frac{18}{36}, 2\frac{4}{5}, \frac{17}{6}$ 4. $\frac{39}{24}; \frac{16}{24}; \frac{108}{24}$; impossible; $\frac{36}{24}; \frac{456}{24}; \frac{53}{24}$
5. $\frac{13}{8}; \frac{2}{3}; \frac{9}{2}; \frac{9}{20}; \frac{3}{2}; \frac{17}{6}; \frac{9}{5}$ 6. 63 yards 7. 118,000 8. 5440 9. $30☆T_{\text{thirteen}}$
10. 3836 14. Ed; Bob

Section 10: 1. $\frac{8}{20}$ or $\frac{2}{5}$ 2. $\frac{250}{420}$ or $\frac{25}{42}$ 3. $\frac{6}{36}$ or $\frac{1}{6}$ 4. 1; 0 5. a) $\frac{3}{8}$ b) $\frac{2}{8}$ or $\frac{1}{4}$
c) $\frac{4}{24}$ or $\frac{1}{6}$ 6. 2000 times 7. $\frac{1{,}400{,}000}{2{,}000{,}000}$ or $\frac{7}{10}$ 8. a) 9 cups b) 4 cups c) yes d) 2 to 3
9. 18 10. $\frac{1}{5}$ 11. 45 mi-per-hr

Section 11: 1. 65 badges 2. $200☆_{\text{thirteen}}$ 4. 13 to 3 5. 1, 2, 3, 5, 6, 7, 10, 14, 15, 21,30, 35, 42, 70, 105, 210 6. a) $\frac{90}{360}$ or $\frac{1}{4}$ b) 120 to 150 or 4 to 5 c) $\frac{120}{360}$ or $\frac{1}{3}$
30,240 8. 15 9. a) $\frac{26}{52}$ or $\frac{1}{2}$ b) $\frac{13}{52}$ or $\frac{1}{4}$ c) $\frac{4}{52}$ or $\frac{1}{13}$

d) $\frac{2}{52}$ or $\frac{1}{26}$ e) $\frac{8}{52}$ or $\frac{2}{13}$ f) $\frac{12}{52}$ or $\frac{3}{13}$ 10. a) > b) < c) =

11. a) $\frac{5}{9}$; $\frac{5}{4}$; $\frac{4}{9}$ b) 3 groups of 51 each; 9 groups of 17 each; 17 groups of 9 each; 51 groups of 3 each c) 17 12. 2 x 2 x 21 or 2 x 6 x 7 or 2 x 3 x 14 or 4 x 3 x 7 13. all true 14 . a) 40 to 30 or 4 to 3 b) 11 to 5 c) 1 to 1 15. 1330 miles

Section 12: 1. $\frac{5}{2}$ pairs per min; 150 pairs per hour; 1200 pairs per day; 6000 pairs per week; constant rate 2. $\frac{7}{12}$ beats per second; 35 beats per minute; isolated rate 3. about 20 minutes 4. 225 irons 5. 3522, 1761, 1174 6. 20 7. about \$90 or \$100 8. 0, 58, 116, 174, etc; no 9. $17\frac{1}{2}$ minutes 10. 27 mg 11. graph on left: 3 to 4; 18% graph on right: 3 to 4; $37\frac{1}{2}\%$ 12. $\frac{2}{5}$ cm per pound

Section 13: 2. a) $7\frac{93}{100}$ b) $\frac{2051}{10000}$ c) $4\frac{7}{1000}$ d) $\frac{69381}{100000}$ 3. a) < b) > c) >

4. a) 3, 5 b) 2 c) 5, 7 d) 2, 3 e) 2, 5, 11 5. a) $\frac{21875}{100000}$ b) $\frac{24}{100}$ c) impossible

d) impossible e) $\frac{24}{1000}$ 6. $\frac{4}{7}, \frac{8}{14}, \frac{12}{21}, \frac{16}{28}, \ldots$ 7. eight 8. 0, 48, 96, 144, etc.

9. 2, 3, 5, 7 10. 1404 11. 14 12. 504; 1008, 1512, 2016, 2520, etc.
13. a) 1.096 b) .875 c) 4.275 d) 3.4
14. a) 5^3 b) 16 c) 25 d) 5^7 e) 2^8
15. a) 1000 b) 3870 c) .076 d) 1 kw operating for 1 hour
e) 2.4; 14.4¢ f) 2.4¢

Section 14: 1. a) .06 b) 135% c) 27.6% d) 8.44% e) 730% f) 175% g) 87.5% h) 8.8% i) 625% 2. a) .48 b) .125 c) 2.8 d) .157 e) .003 3. a) = b) < c) < d) > e) < f) < 5. 500 6. 40△ 7. 3, 6, 9, 12, 18, 24, 36, 48, 72, 144 8. .28 9. 1299 things 10. .000000048 11. 60 sq cm 12. 40% 13. 8840 14. 50 15. a) $\frac{7}{12}$ laps per min.

b) 35 laps per hour c) 36 minutes d) $\frac{7}{6}$ laps 16. $33\frac{1}{3}\%$ 18. ≈ 132 cups 19. 45%

20. \$6549.25 21. \$620 22. 4% increase 23. 16% loss 24. 1950 people 25. 9 to 4

26. a) 65% b) $43\frac{3}{4}\%$ c) 50%

Section 15: 1. a) $-3\frac{2}{7}$ b) - 45% c) 65 d) -1248 e) 0 2. 5 and -5 3. - 2.7

4. 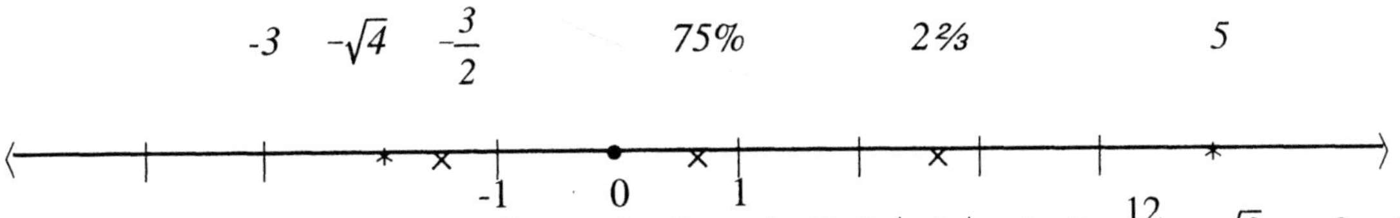

5. a) < b) < c) < d) > 6. 3 or -3 7. 0, | -3 |, −(−4), $\frac{12}{3}$, $\sqrt{9}$ 9. 0; No smallest integer 10. 35 spaces 11. −14 or −15 ; $\frac{3}{8}$, or $\frac{5}{12}$, or $\frac{11}{24}$, etc. 12. 15%
13. $\frac{4}{36}$ or $\frac{1}{9}$ 14. b) Temperature decreases 2° for every 1300 ft increase in altitude; rate of decrease is $\frac{2}{1300}$ degrees-per-ft. c) 84° ; ≈ 64.8° d) 8000 ft ; 15,800 ft 15. 66 inches

Chapter I Review: 1. 112211_{three} 2. 60% 3. $\frac{49}{35}$; $\frac{14}{35}$; $\frac{21}{35}$; impossible; $\frac{20}{35}$ 4. 1121 things

I. 5. all true except (b), (g), (i); (k) 7 to 4) 6. 2 x 2 x 39 or 2 x 6 x 13 or 2 x 3 x 26 or 3 x 4 x 13 7. any four of: 1, 2, 3, 4, 6, 8, 12, 16, 24, 32, 48, 96 8. No 9. Yes; 314 r 6 or 314.5 10. Eight 11. 0, 14, 28, 42, 56, etc. 12. 18 13. 2876 14. 1 15. 11 x 431 17. 0, 46, 92 18. 1061 19. Eight eggs 20. 70 21. $2^2 \cdot 3 \cdot 7$ 22. No 23. 2 and 5 24. Even; $0 = 2 \cdot 0$
26. Ten machines 27. $\frac{2}{5}$ hour; $\frac{1}{60}$ day; $\frac{1}{420}$ week 28. $\frac{1}{3}$

IV. 1. $\frac{21}{250}$ 2. 30 pounds 3. ≈21.8 miles per gallon 4. 64 5. 26% 6. 3 hrs
7. 2 hrs 40 min 8. $490 9. $17.55 10. 6,025,000 11. $9,600 12. ≈ 65 miles-per-hr 13. $48,750 14. 2⅔ cups of flour; ≈33 cookies 15. $4,800 16.$124 billion 17. 15% increase 18. 304 19. $38.70 20. 19 to 11 21. 15% 22. yes 23. 44.1 ft 24. a) $\frac{7}{12}$

b) $125 million for welfare; $500 million for services; $250 million for highways

c) $4\frac{1}{6}$% d) ≈ $104,167,000

Chapter II

Section 16: 1. $3\frac{1}{2}$ 2. 24 spaces 3. 5.3 5. a) 2.4375 b) .139 c) 1.5 6. −14, |−8|, −(−3), 0 7. 1, 2, 4, 13, 26, 52 8. 83 and 167 9. 840 10. 60%; 340%; .6%; 225%; 82.53%; 110% 11. a,b, and f are true 12. a) −8 b) 17 or −17 c) 3 d) 9 e) −7 f) −23 g) 5 h) 4 or -4 14. 35 degrees 15. $66

16.
```
                    j+m          m+k          j+k
<-------------------•--*-*----•----•----**----------•----*-------------->
                    m   -j    0    j    -m          k
```

Section 17: 1. 58 2. 33 spaces 3. a) $(-11)+(-6)$ b) $5+3$ c) $-7+2$ d) $8+(-19)$ e) $2\frac{1}{4}+(-5)\frac{1}{6}$ f) 9.8 + 13.4 4. 0 5. a) positive b) positive c) negative d) negative e) positive f) positive g) negative 7. a) −24 b) −16 c) 2 8. 2 ½ hours 9. $2 \times 3 \times 5^3$ 10. 113 12. 1, 2, 3, 6, 9, 18, 27, 54 13. a) $j > 26$ b) $-c$ c) $m+(-8)$ d) $2\frac{1}{7}+k$ e) y+ 7 14. a) $\frac{1734}{1000}$ b) $\frac{2835}{10000}$ c) $\frac{79}{17}$ d) $\frac{8}{1000}$

15. a) $\frac{33}{24}$ b) impossible c) $\frac{14}{24}$ d) impossible

e) $\frac{48}{24}$ f) $\frac{0}{24}$

16. a) $\frac{30}{59}$ b) $\frac{11}{59}$ c) greater than 26 d) $\frac{17}{59}$ e) 0 f) 1

17. 6

18.
```
     c+d                                  b-a              b-c  b+a
<------*--------•--------------------•--------•-------*---•--------•--------***-------->
                d               c    0        a           b        -d
```

Section 18: 2. a) $\frac{41}{17}$ b) $\frac{174}{1000}$ c) $\frac{4097}{1000}$ d) $\frac{3}{10000}$ 3. a) .7 b) .24 c) 2.65 d) .0075 4. 37 ½ % 5. $\frac{28}{100}$ or $\frac{7}{25}$ 6. a) $\frac{2}{5}$ b) $\frac{4}{15}$ c) $3\frac{1}{2}$ 8. 2½ hours 10. 304 11. 4849 things 12. 1200 seconds or 20 minutes 13. $32\text{ T }00_{\text{twenty}}$ 14. 8 bags of 30 peaches 15. a) 7 or −7 b) 12 or −12 c) .9 or −.9 d) $\frac{5}{8}$ or $-\frac{5}{8}$ e) no such number f) $\sqrt{14}$ or $-\sqrt{14}$

Section 19: 1. 7×3; $7 \cdot 3$; (7)(3) 3. 295 6. sum 7. hundred thousands 8. 250%

9. fortieths 10. $\frac{7}{26}$ 11. 65 to 35 or 13 to 7 12. difference 13. thousandths

14. product 15. a) Multiplication is associative b) Distributive property c) Addition is commutative Challenge a) 134_{seven} b) 213_{eight} c) 1201_{five}

Section 20: 2. All True except b), f), i), and j) 3. 1 to 2 4. 200112_{three} 5. any three of these: 1, 2, 3, 4, 6, 8, 12, 16, 24, 48 6. a) 0 b) $\sqrt{20}$ or $-\sqrt{20}$ c) any number d) 0 e) $\frac{5}{7}$ or $-\frac{5}{7}$

7. 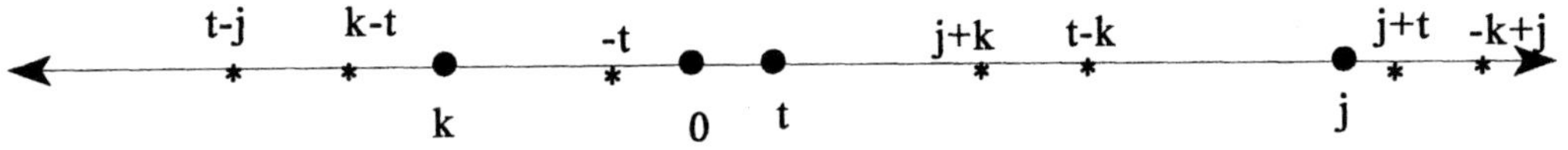

8. .0525; .625; .0225; .0000028 9. 7 to 5 10. $1E00_{twelve}$ 11. 3537_{eight} 12. 32042_{five}
13. −8 14. thirty

Section 21: 2. $\frac{-25}{8}$; $\frac{17}{100}$; $\frac{93}{1000}$; $\frac{4147}{100}$; $\frac{102}{100}$; $\frac{5}{1000} = \frac{1}{200}$ 3. 8; 1; 4; 0; 14; 2

4. 2.25; .095; .875; 4.2; $11.\overline{6}$ 5. 18.2%; $37\frac{1}{2}$%; $\frac{3}{4}$% or .75%; 125%; $172.\overline{2}$%

6. a) 8600 b) .1 c) 4.68 d) 10 e) 7,582,000
7. a) = b) > c) > d) > e) < f) <

8. $\frac{27}{36}$ or $\frac{3}{4}$ 9. Sue; $\frac{2}{15}$ mph greater 10. 45 oz 11. 1, 2, 11, 22

12. $-4\frac{1}{6}$ 13. \$36 14. $\frac{7}{24}$ 15. $\frac{20}{36}$ or $\frac{5}{9}$ 16. 18 to 7

17. $\frac{16}{52}$ or $\frac{4}{13}$ 18. 390 19. $\frac{133}{15}$ spaces 20. $\frac{1}{12}$

21. $\frac{36}{28}$; $\frac{11}{28}$; $\frac{35}{28}$; impossible; $\frac{20}{28}$ 22. about 36%

Section 22: 2. 130%; $91\frac{2}{3}$% or $91.\overline{6}$ % or about 92%; 8.47% or about 8.5%; 560%

3. a) > b) > c) < d) > 4. 1.7 5. 32% 6. 65% 7. 797

8. $\frac{1}{52}$ b) $\frac{13}{52}$ or $\frac{1}{4}$ c) $\frac{6}{52}$ or $\frac{3}{26}$ d) $\frac{16}{52}$ or $\frac{4}{13}$

9. .6% 10. 1, 2, 3, 4, 6, 7, 8, 9, 12, 14, 18, 21, 24, 28, 36, 42, 56, 63, 72, 84, 126, 168, 252, 504
11. a) 1451_{six} b) $33TE_{twelve}$ c) 21513_{seven} d) 304_{five}

12. a) $\frac{28}{18}$ b) $\frac{4}{18}$ c) impossible d) $\frac{0}{18}$ 13. Any three of: 0, 392, 784, 1176, etc.

14. a) 52 b) 6 c) −5 d) −2 e) −1 f) 3 g) 26 15. 15% 16. (a), (b)

17.

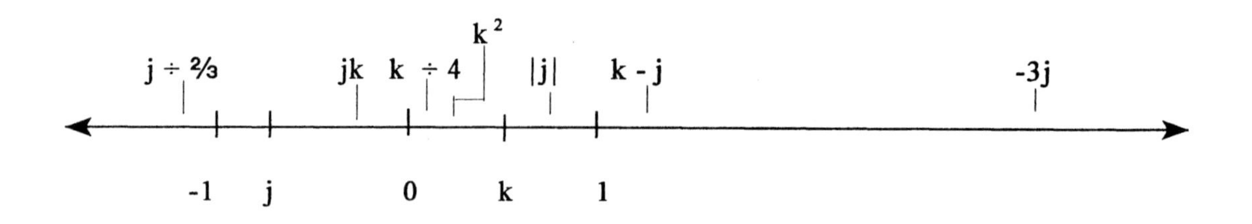

18. a) 480 b) .8 c)4 hours d) 12

Cumulative Review: 6. 1, 3, 5, 15 13. a) $\frac{4}{7}$ b) 0 c) no d) 3 to 4 33. 4567

34. 12 35. 5¼ cup 36. 450 37. $135.50 38. ≈½ tablet 39. 1 hr 10 min.
40. $6.30 41. 41 days 42. $96.01 43. 15% 44. 256% 45. $648 46. 27.2 million
47. -600 48. 25 boys 49. 90 houses 50. 56 51. a) $7.25 b) 57 min 52. 5 ft 3½ in
53. 35% 54. $2.70 55. 28 56. 81 57. a) 23°; 1.6° b) 5600 ft c) summer

58. 2nd hour 59. a) 2 hrs b) 29.2% c) $\frac{1}{6}$ d) 14.3% e) $\frac{9}{24}$ or $\frac{3}{8}$

f) 6 hrs; 7 hrs g) 3 hrs 36 min h) 5 hrs 60. a) $125 billion b) $235 billion
c) $50 billion d) $20 billion e) $95 billion; ≈ 70%

Chapter III

Section 23: 1. a) 1, 4, 7 b) 40% c) 72 d) 87 e) 192 out of 200 f) -200 g) $\frac{7}{25}$
2. 13 people 3. a) 0 b) $\frac{3}{15}$

Section 24: 1. a) area b) length c) area d) capacity e) weight f) capacity g) length
h) volume 2. a) length b) weight c) capacity d) length e) area f) capacity
g) length h) weight i) length j) capacity k) weight l) area m) volume n) length
o) capacity p) capacity q) capacity 3. a) unknown; area; volume 6. $90 7. 17 to 18

Section 25: 1. a) 2 or 3 inches b) ≈6 or 7 inches c) about an inch d) ≈50 or 60 miles
e) about 6 ft 2. a) ≈ 6 or 7 cm b) ≈ 350 km c) ≈ 1 meter d) 6 or 7 cm e) ≈ 3 m
3. a) 122 - 123 mm b) 18 cm 4. a) 700,000 b) 27 c) $\frac{2}{3}$ d) 0.0216; 216 e) 17
5. 36 6. 1/3 in 7. No 8. 63.36 9 . 100 meters; 300 in 10. Russian

Section 26: 1. ≈ 560 yds 3. ≈ 1300 revs 5. 4 spools 7. 87.6
8. a) 45 b) 8.57 c) $5\frac{2}{3}$ d) 16,300 e) 0.008 f) 13.5 g) 56 h) 78 i) 4
j) 30 or $32\frac{1}{2}$ k) 400 l) 30 m) 15 - 16 n) $\frac{15}{52}$ o) c, b, a, e, d p) $52 q) 9

Section 27: 3. a) 27 b) $5\frac{5}{16}$ c) 6300 d) 32 e) 13 f) 10 g) 7000 h) 3.8; 0.038
i) 30 - 33 j) 56 4. a) weight b) length c) not a measurement d) weight e) not a measurement f) area g) weight h) not a measurement i) weight j) volume 5. ½

Section 28: I. 1. 9 sq cm 2. 21 sq cm 3. 20 sq cm 4. 8 sq cm 5. 44 sq cm 6. 30 sq cm II. 1. a) 75% b) ≈ 80% c) 50% 2. about 6 to 19 3. a) 504 b) 96,000 4. ≈ 1 acre 5. 10 or 11 sq ft 6. 72 cm

Section 29: 1. a) 3170 b) 263 c) 5 d) 16 e) 64 f) 10 g) 78,900 h) 6.6 i) 27 j) 4; 32 k) 240 or 250 l) 10 ½ m) 60 n) 1.7; 0.017 2. a) length b) capacity c) area d) weight/mass e) capacity f) weight g) capacity h) weight/mass i) volume j) area k) capacity l) capacity m) weight/mass n) volume o) length p) area 3. a) 60 cm b) 3.4 kg c) 28 sq in d) 20 liters e) 400 ml f) 3 fl oz 4. ≈ 9500 gal 5. ≈ 57

Chapter III Review:

I. 1. capacity 2. weight 3. capacity 4. area 5. weight 6. capacity/weight 7. capacity 8. area 9. weight 10. weight 11. capacity 12. weight 13. capacity 14. weight/length 15. capacity

II. 1. feet/meter 2. ton/kilogram 3. mile/kilometer 4. lb/kg 5. yd or ft/meter 6. inch/cm 7. inch/mm 8. light year or mile/km 9. sq ft or sq yd/sq m 10. cu ft or cu yd/cu meter 11. sq mi/sq km 12. cu in/cu cm 13. sq in /sq cm 14. gallon/liter 15. acre/sq. meter 16. cu ft/cu m 17. gallon/liter 18. sq mi/sq km 19. fl oz/ml 20. fl oz/ ml 21. sq in/sq cm 22. cu in /cu cm or cu m

IV. 1. No 2. 65 yd 3. Joe 4. 64 liters 5. $28.60 6. 58 acres 7. $4307 8. 27 cu yds 9. $700 10. 28°C - shorts 5°C - heavy coat 20°C - room temp. 20°F - heavy coat 45°F - jacket 79°F - room temp. 11. 348 (not 351) 12. a) about 14 cm b) 15.6 cm or 156 mm c) 14.8 cm or 148 mm 13. $6.47 14. American Beer 15. a) 21 - 22 sq cm or 3½ sq in b) 18 sq cm or 2½ sq in

V. A. 1. 23 ft 2. 60 mm 3. 0.075 kg 4. 3¼ lb 5. 14,784 yd 6. 304.2 cm 7. 126 lb 8. 374,000 cm 9. 6⅓ ft 10. 43,000 mg 11. 0.00028 g 12. 102 oz 13. 0.4 mi 14. 21,300 mg 15. 0.08 ton 16. 8.73 m 17. 2¾ yd 18. 104 cm 19. 0.00092 km 20. 224 oz 21. 126 sq ft 22. 0.3780 sq m 23. 14.8 ml 24. 26 qt 25. 2⅓ Tbs 26. 6.5 pt 27. 12,400 ml 28. 4 cu yd 29. 6,690,816 sq ft 30. 0.007296 cu m 31. 336 sq in 32. 1270 sq mm 33. 380 ml 34. 5¼ qt 35. 3.5 sq ft 36. 0.625 cu ft 37. 4.970 liters 38. 432 sq in 39. 31,000 sq cm 40. 26 fl oz

V. B. 1. 40 cm 2. 0.95 kg 3. 616 km 4. 3.1 oz 5. 16 m 6. 21,000 mg 7. ≈21,000 lb 8. 230 in 9. 9.8 oz 10. 16,000 yd 11. 30 m 12. 241 g 13. 660 cm 14. 9.6 in 15. 34 mm 16. 31 oz 17. 75 cm 18. 4 km 19. 5600 m 20. 350 g 21. 240 ml 22. 660,000 sq ft 23. 250 cu ft 24. 50 l 25. 380 cu cm 26. 300 sq cm 27. 21,000 sq m 28. 720 ml 29. 540 ml 30. 350 acre 31. 20°C 32. 35°C 33. 177°C 34. -20°C 35. 88°F 36. 14°F 37. 144°F 38. -7°C

Chapter IV

Section 31: 2. a) ray b) segment c) line 3. a) segment b) triangle c) line d) number e) ray 4. MV + VQ = MQ 5. plane; triangle 8. a) infinite b) one c) none d) infinite e) one 13. 6 units or 16 units 15. a) segment b) circle c) semi-sphere 16. a) true b) true c) true d) false e) true f) false

Section 32: 1. a) KQ, TK, AT, etc. b) $\triangle$CTK, $\triangle$KTM, $\triangle$CTL, etc. c) K, L, T, Q, or A d) one e) AT, TE, LT or TC f) $\triangle$ATE and $\triangle$ETC or $\triangle$ETC and $\triangle$MTK g) point M h) AK i) QT j) $\overline{AT}$, $\overline{CT}$, $\overline{LT}$ k) 95° 2. a) plane b) segment c) angle d) line e) circle (for plane) or sphere f) point g) adjacent angles h) segment i) angle j) ray k) line 3. a) 10.9 cm; 8.5 cm; 5.8 cm; 3.5 cm; 4.0 cm b) 18.5°; 24°; 36°; 90°; 59.5° c) 3.5 cm 5. 16 cm 6. b) one

Section 33: 1. no 2. a) 60° b) 60° c) 90° d) no 3. a) $m_1 \perp m_3$ b) $m_1 = m_3$ or $m_1 \parallel m_3$ c) $m_1 = m_3$ or $m_1 \parallel m_3$ 4. a) 90° b) $\overline{CA}$ c) 140° 5. x = 20° 6. x = 30°; y = 115° 7. x = 101° 8. 50 cm

Section 34: 1. a, g are convex polygons 2. square 3. 1080°; 4500° 4. a) 2340° b) 156° 5. a) 170° b) (n-2)180/n 6. 0, 2, 5, 9, 14, etc. 7. a) no b) yes

Section 35 - 36: 3. a) $\triangle$AKQ ≅ $\triangle$ATQ b) VPRL ≅ NPRA c) PVMAQ ≅ PCKTQ 6. x = 105° 7. x = 70°

Section 37: 3. a) 72° b) 180° c) 36° d) 180° 4. a) CVSQP ≅ TABKN b) $\triangle$ACT ≅ $\triangle$GDO c) MCKPT ≅ ABNVQ or MCKPT ≅ AQVNB d) HTAM ≅ EVOL

Section 38: 6. a) $\angle$K = $\angle$R = $\angle$V = $\angle$T = 90°; RV = KT, KR = TV; KR ∥ TV, KT ∥ RV; TR = KV b) SR = ST, $\angle$R = $\angle$T = 30° c) AT = RK, TQ = RM, AQ = KM; $\angle$A = $\angle$K, $\angle$T = $\angle$R, $\angle$Q = $\angle$M d) VA = VK e) PT = TQ = QA = AC = CP; $\angle$P = $\angle$T = $\angle$Q = $\angle$A = $\angle$C = 108° f) AB = BC = CD = DA; $\angle$A = $\angle$C, $\angle$B = $\angle$D; AB ∥ CD, BC ∥ AD; AC ⊥ BD; $\angle$A + $\angle$D = 180° 7. x = 6

Section 39: 1. $\triangle$ABC ~ $\triangle$DEC; x = 8.75; y = 10.5 2. 7.5 3. $\triangle$ABE ~ $\triangle$CBD; $\angle$C; $18\frac{2}{3}$ 4. $41\frac{1}{4}$ft 5. $\triangle$STQ; 2.4 6. a) m$\angle$ABF b) m $\angle$AFB c) $\triangle$BAF 7. ≈200 m 8. ≈240 ft 9. a) 16 sq units b) 12 units; 144 sq units c) No. Area of Square B is 9 times area of Square A. 10. no 11. a) yes b) 32 bags 12. All pairs of matching angles are equal, and all pairs of matching sides have same ratio. 13. Matching sides can have different lengths. 14. yes 15. no 16. Matching sides must have same length. 17. a) Sides increased by 24%; area increased by ≈54% b) 65%

Chapter IV Review:

I. 1. pentagon; perpendicular lines; triangle; rhombus; trapezoid; n-gon 2. none of them 3. 180°; 72°; 90°; 360°; 120°; 180° 5. x = 50° y = 60° 6. yes; $\triangle$KTA ≅ $\triangle$CQM 7. no; yes 8. 39°; $\overline{QV}$ 9. a) one b) one c) one d) one e) infinite number 10. equal; perpendicular 11. ≈10 cm 12. no 14. $\triangle$KQA~$\triangle$MAT; KM = 21 units 15. x = 110° 16. 12 sides; 81 cm 17. 28 muks; 42 muks 18. yes 25. AKQCVT; $\triangle$TQA ~ $\triangle$MCA

II. 4. 2.4 cm 10. ≈1025 ft

Chapter V

Section 40: 2. 7.2 units 3. 429 sq units 4. 108 sq units 5. 84 sq units 6. 15 sq units

Section 41: 1. 40π sq units ≈ 126 sq units 2. 9π sq units ≈ 28 sq units 3. a) ≈13 sq cm b) ≈36 sq cm 4. 96/π ≈31 sq cm

Section 42:
I. 1. 24 mi 2. 40 ft 3. 54 sq units 4. 4.8 units 5. x = 4 units
6. $25 + 80\sqrt{3}$ sq units ≈ 164 sq units 7. 180 sq units 8. $12\sqrt{3}$ sq cm ≈ 21 sq cm
9. 72 sq units 10. 336 sq ft 11. 128 sq in 12. $25\sqrt{3}$ sq in ≈ 43 sq in
Section 43: 1. ≈38 sq cm 2. ≈230 sq cm 3. 74 sq cm (slant height: $\sqrt{53}$) 4. three
5. 24 units 6. 8 sq units 7. $18 + 6\sqrt{6} - 6\sqrt{3} \approx 22.30$ units 8. $\frac{16}{49}$

Section 44: 1. ≈140 sq cm 2. $\sqrt{189} \approx 13.7$ units high; 280 sq units
I. 3. 900π sq cm ≈ 2800 sq cm 4. 450π lb ≈ 1400 lb 5. $100 6. 13 cans 7. ≈39 sq in
8. ≈47 sq in 9. 1300 gal 10. floor: 48,000 sq ft; windows: 13,500 sq ft 12. ≈38 sq in
13. paper: ≈68 sq cm; height: ≈4.3 cm
Section 45: 1. $21 2. 4800 gal 3. $47,000 4. ≈170 cu cm 5. 6^+ cups
6. 48 kg 7. ≈2.4 qt 8. 480 sq cm 9. 48 gal 10. canvas: 7200 sq ft; floor: 4300 sq ft
12. 30 sq units 13. 72 sq ft 14. 20 sq ft 15. 400 g 16. 6.72 cm
17. ≈41 rev 18. paper: 120 sq cm; 5.0 fl oz 19. 5 gal of paint; 6600 gal of water 20. 58 lb
21. 160 sq cm; 97 cu cm

Chapter VI
Section 46: 1. a) 8 b) 7 2. 4 3. 3 4. 37
Section 47: 3. In order for Katie's average on four tests to be 93, she must have scored 130 on the unknown test - which is not possible. 4. a) 12, 14 b) 20 5. a) 32, 64 b) 1024 6. a) 9, 11 b) 19 7. a) 2 b) 6 c) 3
Section 48: 1. a) 299 pts and 300 pts b) mean is 292.4 pts. and median is 297 pts. 3. 100 4. the mode
5. mean is $27285.71; median is $18000; mode is $16000 6. 0 7. 71.5 8. a) 12 b) 24
Section 49: 1. a) 125.48 cm c) 291.1 pts. 2. x and y are opposites 3. 24 4. a) 110 b) 100 c) 174
6. a) 21 b) 81 7. a) 81, 243 b) 6561 8. a) 18, 20 b) 30 9. a) 14
Section 50: 1. about 7.6 2. the mean is that equal value and the standard deviation is 0 3. the mean is decreased by 10 but the standard deviation remains the same 4. the range is 8 pairs of shoes, the standard deviation is about 3.2 pairs of shoes 5. $\bar{x} = 3.7$ for all three machines; $s_1 = 2.0$, $s_2 = .8$, $s_3 = 2.3$. Therefore, Rosco should choose machine B. 6. x = 69.2, s = 15.4 7. a) 36.08 b) 29.12 c) 25.64 to 39.56 8. the History test 9. a) -5, -9 b) -45 10. a) 96, 192 b) 1536
Section 51: 1. a) 50th b) 22.5th c) 90th d) first quartile score is 4; third quartile score is 10. 3. a) B b) F 4. 1/10 5. 12 6. 21 7. 10
Chapter VI Review: 1. 32 2. Brown: 32.4%; Yellow: 25.4%; Orange: 16.9%; Green: 14.1%; Red: 8.5%; Blue: 2.8% 3. $\bar{x} \approx 7.64$ hrs., s ≈ 5.58 hrs. 4. a) 6.89 g b) 5 and 10 c) 5.5 g 5. 77 inches
6. 70, 73, 79 7. Modal size 8. Carol was actually 12 points away from an A. 9. a) 3, 6 b) 78

Chapter VII
Section 53: 1. a) 7/17 b) 1/17 c) 9/17 d) 10/17 e) 0 2. a) 1/2 b) 5/18 c) 5/36 d) 1/36 3. a) 1/2 b) 1/4 c) 1/26 d) 1/13 e) 3/13 4. a) 1/8 b) 3/8 c) 1/2 d) 1/4 5. a) 1/12 b) 1/3 c) 1/4 d) 1/2
Section 55: 1. a) 1/17 b) 4/663 c) 1/663 d) 1/2652 e) 1/1326 2. a) 1/3 b) 3/10 c) 8/15 d) 1/45 e) 44/45 3. a) 1/17 b) 1/17 c) 1/13 d) 1/26 4. 360 orders 5. 1/1000 6. a) 11/40 b) 49/80
7. 4/19; No, but it must be a positive multiple of 19. 8. a) 1/3 b) 1/6 c) 2/5 9. 1000 codes
10. 10 ways

Section 57: 1. 1307674368000 ways 2. a) 621000 codes b) 878800 3. 24 ways 4. 281887200 ways 5. 48 sets 6. 1/336 8. 1/24 9. 3/8 10. 13800 displays 11. 8, 13, 21, 34

Section 58: 1. 45 ways 2. 20475 committees 3. 2024 samples 4. 2598960 hands 5. a) .000016378 b) .000783679 6. a) 445536 teams b) 257040 teams c) 1123479 teams d) 329868 teams 7. a) 3/22 b) 1/220 c) 3/11 d) 37/44 8. a) 36, 58 b) 1/6, 6 c) 7, 10 9. a) 1/1086008 b) 5/271502 10. .225

Section 59: 1. a) .04 b) .36 c) .32 2. a) .77 b) .77 3. a) .92 b) .08 c) .38 4. a) 16, 25 b) 81 5. a) 162, 486 b) 1312 6. a) 8, 1 b) -729 7. a) 32

Section 60: 1. a) .9382 b) .4620 c) .8554 2. a) -2.38 b) -0.72 3. .0228 4. a) .2643 b) .3189 c) .0110 5. Lowest A: 80.96; lowest B: 75.64; lowest C: 68.36; lowest D: 63.04. 6. a) 15.53% b) $4.08 to $6.42 8. 20 people

Chapter VII Review: 1. a) 1/6 b) 1/36 c) 10/36 = 5/18 d) 1/6 2. a) 1/3 b) 2/3 c) 2/3 3. a) 2/9 b) 8/15 c) 2/3 d) 1/3 e) 4/9 4. 3/7 5. a) ½ b) 1/5 c) 7/10 6. 1/120 7. 10! = 3628800 ways 8. a) 46/116 = 23/58 b) 33/116 c) 15/46 d) 1/116 e) 54/116 = 27/58 f) 1/22 9. a) .2546 b) .4129 c) .7259 10. a) 715 teams b) 120 teams c) 4/13 or .31 d) 129/143 or .90 11. .91 12. a) 94/54145 or .0017 b) 1/216580 or .0000046 c) 27417/108290 or .25 13. 1/1296 14. 720 ways 15. 2/15 16. 15/22 17. 479001600 ways 18. a) .3 b) .25 c) .05 d) .5 e) .5 19. a) .3 b) .25 c) .05 d) .5 e) .5 20. a) 6 b) 12 c) 18 d) 3 e) 9 21. a) -1, -6 b) -64 22. a) $\frac{10}{27}, \frac{10}{81}$ b) $\frac{10}{3^8} = \frac{10}{6561}$ 23. a) 38, 50, 64 b) 23, 17, 24

Chapter VIII

Section 61: 1. a) no b) no c) no d) no 2. a) yes b) no c) no d) 5; 16; w e) 18; 27; 0; 3w 3. a) no b) no c) no d) yes e) no 4. a) no b) no c) no 5. a) yes b) no c) no d) 4; 91; n e) 1; 1; 1 6. a) yes b) yes c) it appears to be d) u; v; w e) u; v; w

Section 62: 1. a) yes, 1 b) yes, 1 2. no, only right identity 3. no, only right identity 5. a) s = 0 b) s = 0 c) s = 1; t = -12 or s = -1; t = 12 or s = 2; t = -6 or s = -2; t = 6 or s = 3; t = -4 or s = -3; t = 4 or repeat choices with s and t switched d) s and t are opposites e) t = 0 or s = 0 f) s = -6 g) t = 4 h) s = 1; t = -21 or s = -1; t = 21 or s = 3; t = -7 or s = -3; t = 7 or repeat choices with s and t switched 6. a) {5, 6, 7, 8, 9, 10, 11, 12, 13} b) R ∩ Q = {8, 10., 12} c) {5, 7, 8, 9, 10, 11, 12, 13} d) the empty set e) {8, 10, 12} f) {6} 7. a) no b) no c) no d) yes, a 8. yes, a 9. a) W b) N c) W 10. no 11. a) yes b) yes c) yes d) no 12. a) yes b) no c) no d) no

Section 63: 1. a) 1 b) 1, -1 2. yes 3. a) false b) true c) false d) false 4. a) yes, u b) w c) w d) no 5. a) yes b) yes c) is appears to be (It can be proven.) d) yes, 5 e) yes, the inverse of z is 10 - z 6. a) yes b) no c) no d) no, right identity only e) no

Section 64: 1. a) Q b) W c) N d) Z e) Z f) ∅ 7. a) yes b) no c) no d) 2/3; no; no 8. a) yes b) yes c) yes 9. a) yes b) yes c) yes d) yes, 4 e) yes, 8 - z is the inverse of z 10. a) yes b) yes c) it appears to be d) yes, v e) u and w are inverses and v is its own inverse 11. a) yes b) yes c) no d) yes, v e) each of these elements is its own inverse: t, s, v (r does not have an inverse.)